GENETIC POLYMORPHISMS AND SUSCEPTIBILITY TO DISEASE

GENETIC POLYMORPHISMS AND SUSCEPTIBILITY TO DISEASE

Edited by

M.S. Miller
*Wake Forest University School
of Medicine, NC, USA*

M.T. Cronin
ACLARA Biosciences Inc., CA, USA

London and New York

First published 2000 by Taylor & Francis
11 New Fetter Lane, London EC4P 4EE

Simultaneously published in the USA and Canada
by Taylor & Francis Inc.,
29 West 35th Street, New York, NY 10001

Taylor & Francis is an imprint of the Taylor & Francis Group

© 2000 Taylor & Francis

Typeset in Sabon by
Florence Production Ltd, Stoodleigh, Devon
Printed and bound in Great Britain by
St Edmundsbury Press, Bury St Edmunds, Suffolk

Every effort has been made to ensure that the advice and information
in this book are true and accurate at the time of going to press.
However, neither the publisher nor the authors can accept any legal
responsibility or liability for any errors or omissions that might be
made. In the case of drug administration, any medical procedure or
the use of technical equipment mentioned within this book, you
are strongly advised to consult the manufacturer's guidelines.

British Library Cataloguing in Publication Data
A catalogue record for this book is available from the British Library.

Library of Congress Cataloging in Publication Data
A catalog record for this book has been requested.

ISBN 0–7484–0822–3

To the people we love, who make our lives fun and interesting: our spouses, Helen Miller and Jim Cronin, and our children, Matthew and Adam Miller, and Daniel, Thomas, and Peter Cronin. Thanks for your love, your support, and being there even when the grants didn't get funded or the patents didn't go through.

Mark and Maureen

CONTENTS

CONTRIBUTORS

Ann B. Begovich, Ph.D., Research Leader, Department of Human Genetics, Roche Molecular Systems, Alameda, CA, USA

Gerard T. Berry, M.D., Professor of Pediatrics, Division of Biochemical Development and Molecular Diseases, The Children's Hospital of Philadelphia, University of Pennsylvania School of Medicine, Philadelphia, PA, USA

Keith W. Crawford, Ph.D., Assistant Professor, Department of Pharmacology, College of Medicine, Howard University, Washington, D.C., USA

Maureen T. Cronin, Ph.D., Director, Genomics Applications, ACLARA Biosciences Inc., Mountain View, CA, USA

Paolo Fortina, M.D., Associate Professor of Pediatrics, Division of Hematology, The Children's Hospital of Philadelphia, University of Pennsylvania School of Medicine, Philadelphia, PA, USA

Suzanne A. W. Fuqua, Ph.D., Professor of Medicine, Breast Center, Baylor College of Medicine, Houston, TX, USA

Torsten A. Hopp, Ph.D., Postdoctoral Fellow, Breast Center, Baylor College of Medicine, Houston, TX, USA

Mark O. Lively, Ph.D., Professor of Biochemistry, Molecular Genetics Program, Wake Forest University School of Medicine, Winston-Salem, NC, USA

William B. Mattes, Ph.D., Senior Scientist, Investigative/Genetic Toxicology, Pharmacia & Upjohn, Kalamazoo, MI, USA

Mark S. Miller, Ph.D., Associate Professor, Department of Cancer Biology, Comprehensive Cancer Center, Wake Forest University School of Medicine, Winston-Salem, NC, USA

Susan M. Mockus, Ph.D., Postdoctoral Fellow, Department of Pharmacology, University of Washington, Seattle, Washington, USA

Jorge R. Oksenberg, Ph.D., Assistant Professor, Department of Neurology, School of Medicine, University of California at San Francisco, CA, USA

CONTRIBUTORS

R. Mark Payne, M.D., Associate Professor, Department of Pediatrics, Division of Cardiology, Wake Forest University School of Medicine, Winston-Salem, NC, USA

Peter G. Shields, M.D., Section Chief, Molecular Epidemiology Section, Laboratory of Human Carcinogenesis, Division of Basic Sciences, National Cancer Institute, Bethesda, MD, USA

Kent E. Vrana, Ph.D., Associate Professor, Department of Physiology & Pharmacology, Wake Forest University School of Medicine, Winston-Salem, NC, USA

Peter J. Wedlund, Ph.D., Associate Professor, College of Pharmacy, University of Kentucky, Lexington, KY, USA

PREFACE

The idea for this book started with a Continuing Education Course, *Techniques for Determining Genetic Polymorphisms*, that Peter Wedlund and Mark Miller co-chaired at the 36th Annual Meeting of the Society of Toxicology in 1997. Presenters included Maureen Cronin, Bill Mattes, and Fred Farin. From questions and comments that were raised during the session, it became apparent that, although there was a plethora of literature on genetic polymorphisms and their biological effects, the data were scattered throughout the literature. More importantly, people working on the same problem in different fields were often unaware of advances being made in other disciplines that related directly to their own work. A centralized reference that would discuss genetic polymorphisms in relation to disease susceptibility that covered a variety of fields was clearly lacking.

During the course of the meeting, Elaine Stott, who at the time worked for Taylor & Francis, approached Mark and asked about the possibility of generating a book from the Continuing Education Course. Mark was dubious at first, since this appeared to be a rather large undertaking for an assistant professor who had just moved to a new institution. However, Elaine was very persistent and Mark finally agreed to develop the concept if he could (1) find a willing co-editor, and (2) get at least 10 people to initially agree to write chapters. Maureen was a natural choice of a co-editor, given her strong background in genetic polymorphisms and some of the newer technologies that were being developed, and several potential authors expressed their enthusiasm and willingness to participate. Thus, a book was born! We submitted a book proposal and moved ahead with the project. Early on, Elaine left Taylor & Francis, but we were now in the capable hands of Dilys Alam, who helped us navigate the, for us, unchartered waters of book editing.

This book is intended to serve as a reference source for established researchers currently involved in genetic research. Advances in molecular biology over the last decade have allowed unprecedented progress in our understanding of the role of genetics in determining human susceptibility to various diseases. The ability to clone, sequence, and map the alleles of specific genes has allowed us to correlate gene sequence and structure with function. Thus, allelic differences in genetic sequence, or polymorphisms, have been identified that code for proteins with altered function, and these differences have been correlated with differences in susceptibility to disease in a number of major organ systems throughout the body. However, most studies of genetic polymorphisms and disease susceptibility are "disease-based", and scientists working in the area of say, cancer research, may be unaware of important correlations between a putative cancer susceptibility gene and a gene identified in cardiac or neurological disease. Another consequence of human genetic diversity research being principally disease-based is that most of the available information has been about polymorphisms resulting in gross phenotypic variants. Only now are the more subtle phenotypes that result from genetic polymorphisms and their collective interactions being appreciated. As more and more advances are made, and greater portions of the genome are sequenced, the risk of subspecialization increases. Thus, a major impetus in putting together this book was to provide an overview of the role of genetic polymorphisms in disease susceptibility in a wide variety of diseases and fields, integrating into one volume relevant research in oncology, neurology, immunology, and cardiology.

Thus, in one book, we hope to provide genetic researchers from different disciplines with a broad but thorough overview of how allelic gene differences influence disease susceptibility in the human population. We hope this will provide researchers with an invaluable reference source and will allow individuals from different areas of specialization to find out about similar types of gene–disease correlations in other fields. We have tried to keep the chapters relatively short, so that individuals reading overviews outside their field are not overwhelmed by an excess of information; rather, we have tried to provide a thorough synopsis of the current advances in the various disciplines in an easily digestible format.

We have both enjoyed working on this project, and greatly appreciate the help and diligence of our co-authors. We hope that our readers will find this book as interesting and useful as we have.

Mark Miller and Maureen Cronin

1

TECHNIQUES FOR IDENTIFYING GENETIC POLYMORPHISMS: AN HISTORICAL PERSPECTIVE

Peter J. Wedlund

1.1 The importance of genetic variability

The importance of genetic variability is readily apparent to anyone who has ever searched for a familiar face in a crowd. We depend on individual physical traits to identify people within the mass of humanity that surrounds us. Yet physical traits are only a small part of what we are. What we believe, how we feel, what we value, what we don't value, what motivates us, determines our persistence and defines our identity. These less tangible personal traits are affected not only by our genetic make-up, but also by our environment and how we respond to it.

Physical and personal traits underscore a basic dichotomy that also exists in the study of genetic variability and its role in disease. Certain genetic variations lead to outward manifestations of disease. Like physical traits (hair, eye and skin color, facial features, body build, and height), they are easily recognized by the physical signs they produce. However, it is the interaction between individual genetic composition and the environment that determines the most prevalent and important diseases affecting the broader range of humanity. The many diseases that result from gene–environment interactions are, like personal traits, far more complex to understand.

The relationship between genetic variation and disease is presented pictorially in Figure 1.1. If one considers that there are an estimated 60 000 to 100 000 functional genes in the human genome (Dykes, 1996), it is a daunting task just to understand the role and function of each gene product. The ability to address how variations in one or more gene products interact with specific environmental factors to affect disease risk seems an almost insurmountable obstacle. One might equate it with attempting to understand the processes in the human mind that define our responses, behaviors and mannerisms to various social situations. Although difficult to study, progress has been made, and the successes provide some guidance on how to proceed with future research in this area. Before describing the challenges of this future research, it is worthwhile to look at the past to understand what has been learned.

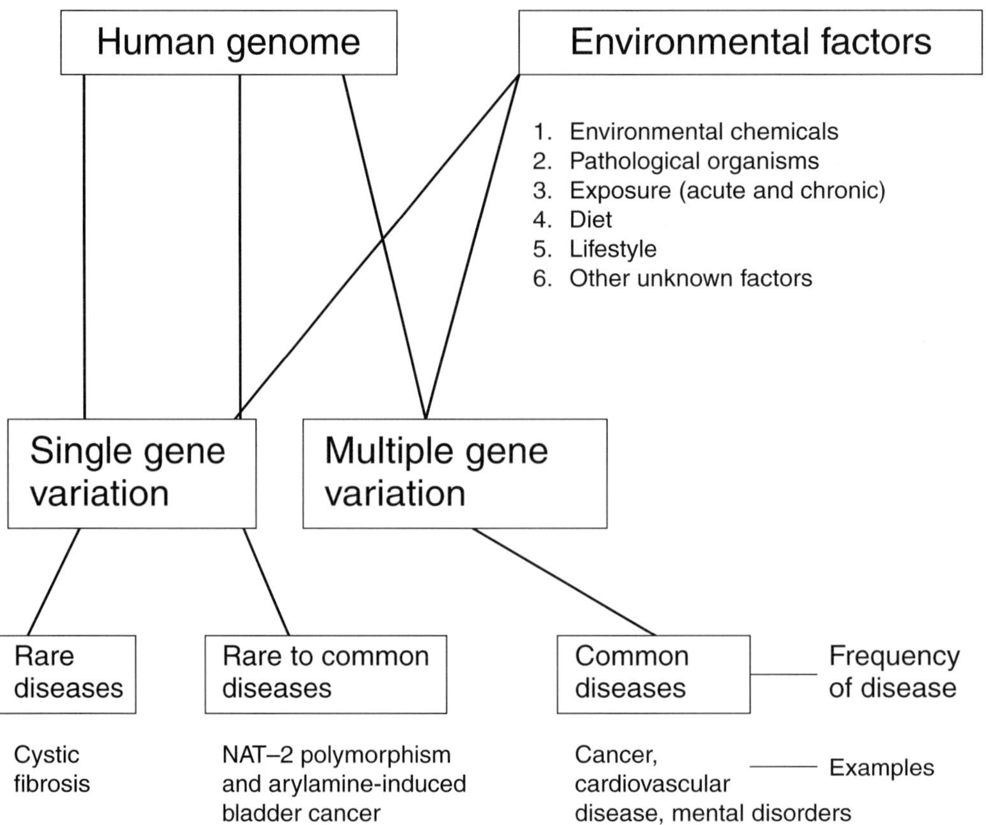

Figure 1.1 The relationship between variations in the human genome, environmental factors and disease.

1.2 Mendel and the laws of heredity

The foundation for modern-day genetics can trace its roots to Gregor Mendel's research on pea plants published in 1866 (Brown, 1990). It was here the basic laws of heredity were laid out and shown to follow a predictable course. The "unit factors" Mendel used to describe inheritance are now referred to as "genes", and the variations in the "unit factors" are now called "alleles". Mendel recognized that the physical traits (phenotypes) he recorded were the result of a pair of independent alleles. Furthermore, he understood that the relative proportion of phenotypes obtained when crossing pea plants followed a predictable ratio depending on the dominant or recessive nature of each allele. From these observations Mendel formulated his laws of heredity.

Mendel's work was really the first to demonstrate that it was possible to study heredity. Over the ensuing 134 years, tremendous progress has been made in our understanding of heredity, genetics, and the cause of genetic variation. During most of this time, research has focused on physical traits, and genetic variations that produced a measurable change in a specific trait. Traits that exist in more than one form are called "polymorphic" and are described as "polymorphisms".

1.3 Single gene variation diseases

Table 1.1 provides a list of selected genetic variations that have been directly associated with disease. Current estimates suggest that defects in 5 to 10% of the genes in the human genome are capable of directly causing disease in humans. This estimate may change as more information is gathered, but it is obvious that every gene in the human genome does not automatically produce some serious disorder if it is not functional. The human genome contains a remarkable degree of redundancy that prevents a catastrophe if certain genes exhibit diminished or absent expression. Among those single genes directly associated with disease, the majority are dominant and result in disease if a single copy of the gene is defective (Weatherall, 1991). More often than not, these genes code for an important structural protein or are part of a multi-protein complex. The other genes directly associated with disease are recessive and cause problems only if both gene copies are defective. These genes often have a role in metabolic or regulatory functions. Having too many copies of genes can also cause problems, as demonstrated by trisomy syndromes (extra copies of chromosomes; Down's syndrome, Edwards syndrome, Patau syndrome) and the importance associated with inactivating one of the X chromosomes in females (Gardner and Sutherland, 1989).

Two important points can be gleaned from Table 1.1. First, most diseases traceable to a single gene are rare. Each disease accounts for at most a few subjects per thousand or ten thousand individuals, although some are well recognized even by the general public (e.g. cystic fibrosis, Huntington's disease, phenylketonuria). Second, many diseases caused by a single-gene defect tend to follow a specific clinical course, with particular clinical features (physical traits). Many of these diseases are so recognizable based upon abnormal clinical values, that the mere mention of the disease often generates a visual picture of what to anticipate in affected patients. Perhaps more important is the limited ability to treat most single-gene diseases (e.g. Huntington's disease, cystic fibrosis, polycystic kidney disease), the expense of treating others (tens to hundreds of thousands of dollars per year) and the lifelong nature (Zimran, 1991; Hershfield, 1995) of these disorders. As a result, care for an individual affected by single-gene diseases is very costly on a per-patient basis and puts a major strain on health-care finances.

Diseases caused by single-gene variations have provided valuable information about cellular processes and about other disease mechanisms. For example, familial hypercholesterolemia led to the recognition that elevated cholesterol levels could cause coronary heart disease (Goldstein and Brown, 1987). Research on this single-gene disorder has resulted in a number of new drugs aimed at lowering cholesterol levels in the general population to decrease coronary heart disease risk. Similarly, the study of single-gene disorders that cause diabetes, kidney failure, aging, mental retardation, blindness, skeletal deformities and cancer have improved our understanding about the complexities of these processes. Thus, interest in single-gene disorders remains a very active research area, although in most instances their importance is limited to a relatively small and select population that is affected by them.

Table 1.1 Some monogenetic diseases, causes, gene and location, allele numbers, frequency of disease, trait type and clinical features

Disease	Cause	Gene, Location, Alleles	Frequency	Type	Clinical features
Cystic fibrosis	Sodium and chloride channel transporter defect	CFTR; 7q31.2 Hundreds of alleles	1/2000	Recessive	Thick mucus secretions, chest infections, blocked pancreatic duct
Tay–Sachs disease	Deficiency in hexosaminidase A enzyme	HEXA; 15q23–q24 More than 70 alleles	1–2/10 000	Recessive	Retardation, paralysis, dementia, blindness with possible early death
Retinoblastoma	Defect in a regulator gene that inhibits cell cycle progression and cell proliferation	RB1; 13q14.1–q14.2 More than 45 alleles	1/23 000	Dominant	Eye tumors and cancer in affected individuals
Adenosine deaminase deficiency	Deficiency in adenosine deaminase enzyme	ADA; 20q13.11 29 alleles	3–4/1000 and 1–2/1000 in Blacks	Recessive	Severe combined immunodeficiency, dysfunction of both T and B lymphocytes, prone to infections
Hypercholesterolemia	Defect in the low-density lipoprotein receptor	LDLR; 19p13.2–p13.1 More than 350 alleles	2/1000	Dominant	Elevated serum cholesterol levels, progression to coronary artery disease and death by middle age
Phenylketonuria	Defect in phenylalanine hydroxylase enzyme	PAH; 12q24.1 More than 105 alleles	1/8000 1/50 000 in Blacks	Recessive	Mental retardation, epilepsy
Huntington's disease	Huntington's gene, increase in CAG triplet causing progressive neuronal cell death	HD; 4p16.3 $(CAG)_n$ where n varies from 37 to 100 found in patients with Huntington's disease	4–7/100 000	Dominant	Involuntary jerky movements, rigidity, possible dementia, seizures and eventually death
Gaucher disease (pronounced Goshay)	Deficiency in acid-beta glucosidase	GBA; 1q21 More than 40 alleles	1/100 000	Recessive	Cerebral lipidosis, pain, fatigue, jaundice, bone damage, and anemia

1.4 Disease and single-gene–environment interactions

Some genes predispose subjects to disease, but only under selected environmental conditions. There are a few points worth noting about the data summarized in Table 1.2. First, most of the genetic variations shown in the table occur much more frequently than the genetic variations listed in Table 1.1. Consequently, more individuals are affected by gene–environment interactions than individuals who suffer from monogenetic disorders. It has been estimated that everyone may have more than five to ten genes that are lethal under the appropriate environmental conditions (Ledley, 1994). Second, the best known single-gene–environment interactions are those where the environmental agent is a drug or specific dietary component. Far less is known about environmental agents associated with disease where the environmental substance is a more ubiquitous chemical. The lack of information about environmental substances associated with genetic toxicity lies with the difficulty in identifying an environmental agent and relating its exposure directly to a specific disease process. Third, the data in Table 1.2 underscore that genetic variation alone may be insufficient to cause a disease manifestation. In the absence of exposure to an exogenous agent, a predisposed individual may live a normal healthy life. Indeed, genetic variation may in some instances be beneficial both to individuals and to the population, as suggested by the high incidence of thalassaemias and glucose-6-phosphate dehydrogenase deficiencies in areas where malaria is endemic (Beutler, 1990). It has also been suggested the cystic fibrosis gene may provide carriers with resistance to typhoid fever (Pier *et al.*, 1998). Finally, Table 1.2 data indicate that there can be many genetic variations (alleles) at a given gene locus to account for a specific gene–environment interaction. This implies that it might be necessary to characterize many different alleles of a gene to fully understand the gene–environment connection.

Recognition of the interaction between gene and environment is important, because it indicates that removal of an environmental exposure might eliminate the condition (e.g. stop smoking and decrease lung cancer risk and cardiovascular disease) (Kakvan and Greenberg, 1977; Jeremy *et al.*, 1995). Disease incidence can be radically decreased by simply avoiding exposure to the triggering agent (e.g. chronic beryllium disease, alcoholic cirrhosis, lead-induced porphyria, fava beans and hemolytic anemia) (Doss *et al.*, 1984; Beutler, 1990; Meyer, 1994). This is the basis for the Occupational Safety and Health Administration (OSHA) mandated Permissible Exposure Limits (PELs) for all man-made chemicals. Alternatively, by compensating for a genetic deficiency an adverse outcome might be prevented (e.g. placing lactase in milk products for lactose-intolerant individuals).

A comparison of disease prevalence in monozygotic and dizygotic twins has been used to estimate the relative importance of genetic variation relative to all other factors contributing to disease. Table 1.3, which shows the heritability associated with a number of common diseases, highlights the relative importance of both genetics and the environment in these conditions. Twin studies indicate that the appearance of some diseases may depend just as much upon exposure to the less tangible "other (environmental) factors" as upon a person's genetic make-up.

Natural and man-made chemicals pervade our lives so completely it is all but impossible to predict how any single chemical in such a milieu may affect disease

Table 1.2 Some monogenetic variations affected by environment, frequency and clinical outcomes

Gene variation	Location and alleles	Percent deficient	Environmental factor	Clinical response
Glucose-6-phosphate dehydrogenase (G6PDH)	Xq28; over 100 alleles	400 000 000 world wide with some deficiency	Fava beans	Hemolytic anemia
Delta-aminolevulinate dehydrogenase (ALAD)	9q34; 6 alleles	1%	Lead	Lead toxicity, lead neuropathy
Histocompatibility type B1 (HLA-DPB1)	6p21.3; 1 allele known	10% frequency of allele in Caucasians	Beryllium in workplace	Sensitivity to beryllium; development of chronic beryllium disease
Aldehyde dehydrogenase-2 (ALDH2)	12q24.2; 2 alleles Dominant trait	35–50% Asians	Alcohol	Facial flushing, tachycardia, fainting
Lactase deficiency	2q21; at least 2 alleles	6% Europeans; >60% Asians	Milk and milk products	Gas, diarrhea, gastro-intestinal upset
Flavin monooxygenase-3 (FMNO-3)	1q23–q25; 6 alleles	Unknown; rare	Triethylamine in fish, eggs, soy beans	Smell like dead fish, social ostracism leading to depression
Cytochrome P450-2D6 (CYP2D6)	22q13.1; 53 alleles	7–8% Caucasians 2% Blacks	Perhexiline, tricyclic drugs; cardiovascular medications	Perhexiline – liver toxicity; Tricyclic and cardiovascular drugs – excessive drug accumulation and toxicity
N-Acetyltransferase-2	8p23.1–p21.3; over 17 alleles	~50% Caucasians, Asians, Blacks	Isoniazid; procainamide	Hepatotoxicity from isoniazid and procainamide increased risk of overdose
Thiopurine methyltransferase	6p22.3; 8 alleles	0.3% Caucasians	Azathioprine; 6-mercaptopurine	Hematopoietic toxicity
Cytochrome P450-2C19 (CYP2C19)	10q24.2–q24.3; 5 alleles	3–4% Caucasians 13–20% Asians	Mephenytoin	Extreme sedation
Cytochrome P450-2C9 (CYP2C9)	10q24	0.25% Caucasians	Warfarin	Low rate of elimination and anticoagulation difficulties

Table 1.3 Diseases and the estimated role of genetics (heritability) on their appearance in the population

Disease category	Specific disease classification	Heritability estimate based on twin studies
Mental disorders	Schizophrenia	0.74
	Bipolar disorder	0.59
	Major depression	0.45
	Panic disorder	0.40
Cancer	Breast	0.56
	Endometrial	0.52
	Ovarian	0.40
Diabetes mellitus	Early onset	0.75
	Late onset	0.53
Coronary artery disease		0.65
Essential hypertension		0.62
Myocardial infarction	Females	0.60
	Males	0.26

risk within such a heterogeneous group as the human population. In a simple extract from a common vegetable, hundreds of chemicals can be detected, some of which are extremely toxic (Ames *et al.*, 1990). If this exposure is extrapolated to the numerous types of foods, spices, vegetables, food additives and other substances we breath, drink and swallow, humans are collectively exposed to millions of different chemicals in varying amounts on a daily basis. The effects of such daily sporadic exposure to an environmental milieu of chemicals that change and vary in amounts over days, months and years merely add to the difficulty in identifying which ones interact with genetic variations in the human population to affect the prevalence of disease. It is not surprising, therefore, that the Center for Disease Control (CDC) annually finds thousands of disease clusters in the United States for which there is a suspected, but no identifiable, environmental cause.

A less-daunting task than identifying environmental chemicals that are risk factors for disease is to determine whether a genetic variation is associated with a specific pathological process. Associating a gene variation with disease is a far less complex task than attempting to identify a specific environmental agent responsible for disease. Although such genetic associations do not prove that a specific genetic polymorphism is the cause of a disease, it may provide the basis for developing testable hypotheses and models that can address whether the potential linkage is real or not. It may also narrow the field of potential environmental substances that must be examined as risk factors for a disease or toxicity if the association is confirmed.

Understanding how genetic variation affects responses to chemicals also has potential therapeutic relevance. For example, there are marked differences in patient survival after the diagnosis and initiation of treatment of colorectal cancer. Many factors can affect patient survival, including the effectiveness of their cancer chemotherapy. 5-Fluorouracil (5-FU) has been a standard in the treatment of colorectal cancer for many years. 5-FU is incorporated into RNA and DNA and blocks thymidylate synthetase enzyme activity. 5-FU disrupts the balance in cell growth and ultimately leads to activation of the p53-mediated pathway of apoptosis in the cell. Since cancer cells grow more rapidly than healthy cells, they are more susceptible to 5-FU cell killing. However, many cancer cells carry mutations in their p53 genes that make the p53 apoptotic pathway ineffective. It has been reported that patients who express an inactive p53 gene at the time of their cancer diagnosis have a significantly shorter survival after the start of 5-FU therapy (Benhattar *et al.*, 1996). By employing drugs that disrupt cancer cell function by mechanisms other than activation of p53-mediated apoptosis, survival may be improved among these patients.

1.5 Disease and multiple-gene–environment interactions

With our improved understanding of cellular processes and disease, it has become obvious that a given disease can result from more than one mechanism. For example, cancer may arise by a number of independent means. The overstimulation of cell growth and division can result from mutations in genes (proto-oncogenes; *erb, myc, ras*) associated with normal cell growth (Savill, 1997). Alternatively, cancer can arise when tumor suppressor genes (*BRCA1; TP53; RB1; BCL2*) do not function properly to prevent normal cell growth from becoming abnormal (Knudson, 1985; P.L. Chen *et al.*, 1990; J. Chen *et al.*, 1998). Cancer also occurs when the enzymes associated with DNA repair are inoperative and fail to repair spontaneous or environmentally induced mutations in the DNA sequence (Cleaver, 1968; Kolodner, 1995; Shen *et al.*, 1998). Although the pathological process (e.g. cancer) is the same in each case, the initiating events and the gene(s) and environmental factors involved differ from patient to patient. Consequently, efforts to predict cancer by searching for variations in a single gene fail to be very predictive when applied to the general population.

Physiological processes that rely on a number of separate proteins may be susceptible to disruption from mutations in any of the genes that code for those proteins. Since many genetic variations only predispose a subject to disease under the right environmental conditions, many common diseases listed in Table 1.3 have a significant environmental component. This may also imply that genetic variations in metabolic and transport proteins that interact with environmental chemicals or specifically modulate target tissue exposure to them may be more logical choices to associate with various environmental exposure-based diseases.

1.6 Nomenclature

The explosion of genetic information and the sequencing of the human genome have resulted in the need for a more uniform approach to organizing and locating information in the human genome. The result has been an evolving nomenclature for describing genes and gene variations, gene structure, gene regulation and the types of mutations that affect gene expression. Highlights are presented below of some of the more salient points required to understand the material in subsequent chapters. However, for interested readers more detailed descriptions can be found in textbooks on molecular biology and biochemistry (Alberts *et al.*, 1989; Weatherall, 1991).

1.6.1 *Chromosomes*

The normal human carries 23 pairs of chromosomes in nearly every cell in his or her body. Mature red blood cells which have no cell nucleus or chromosomes are one exception and haploid reproductive cells are another. The autosomal chromosomes are labeled from one to 22 based on size, with chromosome one being the largest and chromosome 22 being the smallest. The X/X (female) or the X/Y (male) sex chromosomes make up the twenty-third pair. Each chromosome has unique light and dark bands produced by different staining methods, and a centromere (whose position helps to provide further identification of the chromosome). The centromere is the site of the spindle attachment during cell division, and it divides the chromosome into two arms of unequal length. By definition, the shortest arm of the chromosome is called "p" and the longest arm is called "q". Each arm is further subdivided into one to four regions depending on its size. For example, the q arm of chromosome one has four regions, while the short and long arms of chromosome 22 have only one region each. Each region of each arm is further subdivided by a series of bands. By convention, the numbering of the regions and the numbering of bands in each region increases as one moves along each arm from the centromere toward each end of the chromosome. Thus, the CFTR gene associated with cystic fibrosis is given the chromosomal location 7q31.2 (see Table 1.1). This indicates that the gene is located on the long arm (q) of chromosome seven in the third region away from the centromere of that arm, in the first band within region three (actually the first two-tenths of this first band). Genes located close to the ends of the p or q arms are given the location pter (for p terminal) or qter (for q terminal).

1.6.2 *Genes*

Genes are given an abbreviated name based on the function of the gene product (e.g. PAH for phenylalanine hydroxylase), based on nomenclature established by a recognized committee (e.g. *CYP* for Cytochrome P-450), or based on a morphological, biochemical or disease characteristic (e.g. HD for Huntington's disease) (Shows *et al.*, 1987). The nomenclature for human genes has been evolving toward a consensus that all gene names should reflect the function of the gene. Genes are currently given names that do not exceed five characters. Each gene name begins with the first letter of the gene name and the subsequent letters and/or Arabic numbers cannot duplicate any existing gene name. No superscripts, subscripts, spaces, dashes, Roman numerals or Greek letters are used in any gene name. Different genes of similar function are given Arabic numbers to distinguish them (e.g. *CYP1*; *CYP2*; *CYP3*; *CYP4*; etc.) or letter designations (e.g. lactate dehydrogenase *LDHA*; *LDHB*; *LDHC*). The final character in the gene name may be used to specify a characteristic of the gene (e.g. alpha and beta chains of hemoglobin are *HBA* and *HBB*, respectively). This standardization has grown from the Human Gene Mapping effort, which defined guidelines for human gene nomenclature and naming of allelic variants (Shows *et al.*, 1987).

1.6.3 *Alleles*

Allele designations are limited to four alphanumeric characters and are separated from the gene name by an asterisk. Alleles can be given Arabic numerals (e.g. *CYP2D6*1*; *CYP2D6*2*; *CYP2D6*3*; *CYP2D6*4*; etc.), letters (e.g. *AMY1*A*; *AMY1*B*), or a combination of letters and numbers. The growing recognition that many variations in the gene sequence may have little or no effect on gene expression or function has led to the increased use of the term allelic variant to describe modest variations in alleles that are of secondary importance to the primary allele sequence. In some instances, these variations are denoted by a letter following the number designation for the allele (e.g. *CYP2D6*4A*; *CYP2D6*4B*; *CYP2D6*4D*; *CYP2D6*4E*; *NAT2*5A*; *NAT2*5B*; *NAT2*5C*; etc.) (Vatsis *et al.*, 1995; Daly *et al.*, 1996).

1.7 Gene structure, mRNA transcription and mRNA transport

It is estimated that only about 10 000 genes are translated into proteins in any given cell, and of this number about 3000 genes are associated with low messenger RNA (mRNA) copy numbers in the cytoplasm. This implies that only a small fraction of the genes throughout the human genome has a significant functional role in any given cell. To understand this regulation and how genetic variation affects gene expression it is necessary to examine the structure of a typical gene as seen in Figure 1.2.

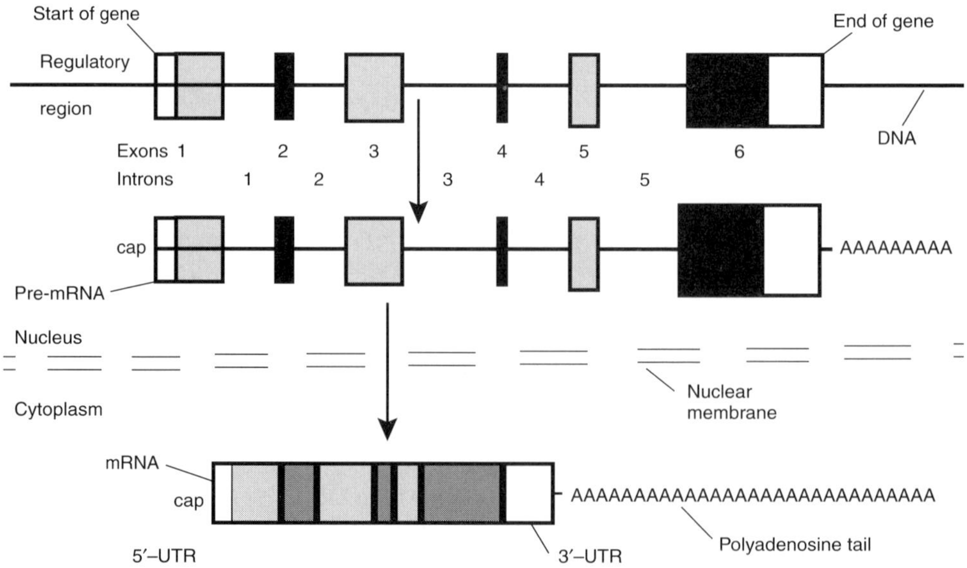

Figure 1.2 Diagram of a hypothetical human gene and the transcription and processing events that must occur before a functional mRNA is released into the cytoplasm. Abbreviations: mRNA, messenger RNA; 5′- and 3′-UTR, 5′ and 3′ untranslated regions.

Each gene is divided into several sections. Upstream of the start site for each gene is a regulatory region (5′ upstream gene regulatory area). A gene can vary in size from a few thousand nucleotides to over 1 000 000 nucleotides. Most genes are subdivided into segments called exons (in black boxes) and introns (thin line segment). The first and last exons contain so-called untranslated regions (in white boxes) that can extend anywhere from 50 to 100 nucleotides in length to several thousand nucleotides. Each gene ends about 20 to 25 base pairs downstream from a specific sequence that defines the termination site for gene transcription by the RNA polymerase enzyme.

The upstream regulatory region of the gene provides the initial point for attachment of RNA polymerase II (the enzyme that transcribes the human DNA into an RNA message, pre-mRNA). In order for the RNA polymerase to attach and begin transcribing the gene into pre-mRNA, a certain amount of protein scaffolding must be in place to allow the proper enzyme docking at the beginning of the gene. It is the transcriptional promoters, enhancers, inhibitors and other regulatory proteins that bind in this upstream region that determine whether genes are transcribed into a pre-mRNA in the cell and at what rate they are transcribed by RNA polymerase.

The proper docking of the RNA polymerase II enzyme is followed by the faithful transcription of the DNA sequence into a pre-mRNA molecule. This pre-mRNA molecule is further processed in the nucleus. First, a cap is placed on the first nucleotide base in the pre-mRNA molecule. After the RNA polymerase II terminates DNA transcription, a series of adenine nucleotides is added to the end of the message (poly A tail). Finally, all the introns in the pre-mRNA are removed before the mature mRNA is transported from the nucleus to the cytoplasm. A problem at any step in this process can prevent a functional mRNA from ever reaching the cytoplasm and the gene from being translated into protein.

1.8 Genetic variation impact on gene expression

The old adage that "anything that can go wrong will", is an appropriate one to describe the multitude of genetic variations that affect gene expression. Mutations in the 5′ upstream regulatory region of some genes have been described that affect RNA polymerase docking and attachment, preventing it from transcribing the gene (Bosma *et al.*, 1995). Mutations in the sequence for terminating the transcription of the gene may also affect whether a functional message is produced. Numerous mutations have been found in introns where they abut exons. These prevent proper intron removal from the pre-mRNA and the final mRNA export to the cytoplasm (Akli *et al.*, 1991; Beutler *et al.*, 1992). Insertion and deletion mutations of part or all of a gene can prevent gene expression (Gaedigk *et al.*, 1991; Hirschhorn *et al.*, 1996). However, the most prevalent mutations are single nucleotide changes in the DNA sequence. In short, any change in the DNA code that affects the ability of RNA polymerase to transcribe a gene, pre-mRNA processing, mRNA stability, or reading the mRNA code can affect gene expression. Some genes display a large number of different mutations within the population that affect final gene expression (beta-hemoglobin, glucose-6-phosphate dehydrogenase), while other genes have been observed to have only one or two major mutations (glutathione *S*-transferase mu-1).

1.9 Genetic variation: phenotype versus genotype

Until relatively recently, all genetic mutations and polymorphisms were identified based on some observable difference in a physical or chemical property in an organism. For example, differences in patient response to drugs, familial appearance of disease, differences in response to dietary substances, or the electrophoretic mobility of a protein or enzyme have all served as the stimulus for examining the molecular basis for these phenotypes. In each case, a variant phenotype was recognized within the population and subsequently investigated to better understand the underlying cause(s) of the variation. These studies relied on the recognition of phenotypic variation to stimulate investigation of the underlying responsible mechanism (Kalow, 1962; Evans, 1993).

It is now possible to detect sequence variation at the DNA level, and test genes directly for specific variations without ever evaluating a phenotype. This evolving capability has led many to embrace genetic testing as the primary method for evaluating variation in the population. Genetic testing has opened up the possibility of assessing how variations in any gene product may affect individual risk for disease even if there is no obvious clinical or accessible biochemical marker associated with the genetic variation. The ability to bypass a phenotype before proceeding to study genetic variation, however, does not eliminate the need to demonstrate association between genetic variation and gene function. A genetic test is a useful risk or disease marker only when an association exists between the genetic variation being tested and the gene's expression and *in vivo* function of its product. In the absence of this information, attempting to correlate a genotype with disease or toxicity is little more than a futile exercise. A number of factors may ultimately determine whether a phenotype or genotype is the most practical means of predicting disease risk in individual cases.

1.9.1 *Cost considerations*

The most appropriate method for defining inter-individual differences in a population depends on the goals and objectives for applying the test. Until genetic testing becomes more routine and less costly, practical considerations should be considered prior to selecting between a phenotype and genotype as the testing method of choice.

In the 1940s, a hereditary deficiency in the hydroxylation of phenylalanine was identified as one cause of mental retardation (Kaufman, 1976). This enzyme deficiency increases blood phenylalanine levels and results in increased urinary excretion of phenylpyruvic acid. The presence of urinary phenylpyruvic acid (phenotype) can be detected by adding ferric chloride to urine. In the presence of phenylpyruvic acid, the ferric chloride produces a green color. The total cost for testing urine by this method is less than one dollar per test. To date, over 40 variations in the phenylalanine hydroxylase gene have been found that are associated with a deficiency in phenylalanine hydroxylase. However, it is unlikely that a genetic test can be designed in the foreseeable future that can screen newborns as effectively and inexpensively as the simple phenotype test currently used by hospitals.

Deficiency in plasma cholinesterase activity was discovered in the 1950s after the short-acting muscle relaxant succinylcholine was reported to produce a prolonged apnea in certain patients. Patients can be phenotyped for this deficiency by a relatively inexpensive assessment of plasma cholinesterase inhibition by dibucaine (Kalow and Staron, 1957). For this test, a comparison is made between benzylcholine hydrolysis in diluted plasma before and after dibucaine administration by monitoring its hydrolysis with a UV spectrophotometer at 240 nm over 3 minutes. A percent inhibition of benzylcholine hydrolysis, called the dibucaine number, is determined. Subjects deficient in plasma cholinesterase have predictably low dibucaine numbers (less than 20), while the dibucaine number is greater than 70 for patients with normal plasma cholinesterase activity. A determination of this phenotype can be made for under five dollars per plasma sample. While at least 12 alleles of the plasma cholinesterase enzyme have been described, the dibucaine test remains a far more cost-effective means for detecting this polymorphism than genetic testing.

1.9.2 *Practical considerations*

A phenotype may not always be the preferred solution to assess genetic variation. The caffeine–halothane contracture test to assess susceptibility to malignant hyperthermia in response to general anesthetics is a relatively complicated and labor-intensive test to perform (Evans, 1993). It requires collecting a muscle biopsy from the quadriceps, and performing a test at 37°C under rather rigorous standardized conditions. The objective is to measure muscle tension in the presence of escalating amounts of halothane, while exposing the muscle to a low electrical stimulation to induce twitches. A less invasive, less complex genetic test would be preferred to this method for defining response to general anesthetics.

The CYP2D6 enzyme deficiency can be detected at a modest cost by administering 30 to 60 mg of dextromethorphan orally followed by an 8-hour urine collection. The dextromethorphan urinary metabolite is hydrolyzed with strong acid or beta-glucuronidase and, after a short extraction, examined by thin-layer chromatography. Total processing costs are no more than $30 to 40 per sample, compared with a genetic test that currently ranges in cost from $100 to 450 per test, depending on the laboratory doing the work (Guttendorf *et al.*, 1988; Chen *et al.*, 1996). However, phenotyping may be inaccurate in patients receiving other medications, so a *CYP2D6* genetic test may be preferred since it is not affected by concurrent drug administration. Thus, genotyping may be indicated when it insures predictive reliability of the results.

Direct assessment of genetic variation still faces some daunting obstacles before it supplants phenotyping as the sole predictor of variable population responses to medications. A phenotype measures the output of a gene product without characterizing the exact reason why a gene product may have an abnormal expression. For example, a phenotype for CYP2D6 activity requires only a single measure of *in vivo* enzyme function. However, for a *CYP2D6* genotype to provide an equally accurate prediction of that metabolic activity it is necessary to test each DNA sample for at least 13 variant gene alleles. Although this can be done for a small gene such as *CYP2D6*, it is more difficult for genes that span hundreds of thousands of nucleotides or where tens to hundreds of different alleles are known to be expressed. These are practical issues that must be considered before applying genetic testing.

1.10 Conclusions

During the last 134 years since Mendel published his original work on the laws of heredity much has been learned. DNA has been identified as the source of all genetic information in higher life forms, and the DNA code for relating the nucleic acid sequence to primary amino acid sequence in proteins has been deciphered. The basic structure of the gene has been characterized and complete DNA sequences for some organisms have been described. Today, vast stretches of human DNA genomic sequence are accessible from an office or a home computer. The ability to describe genetic variation no longer depends on finding observable physical traits, but extends directly to nuclear and mitochondrial DNA. Variations in human DNA provide insight into historical migration patterns of people around the world, define human races and the similarities and differences between humans and other life forms.

We now recognize the importance of genetic variation as an essential element of our species' survival, but imbedded in that genetic variation also resides individual risk for disease. Rare inherited diseases result from genetic variations in critical gene products. However, a far more frequent cause of disease is from gene–environment interactions. Normally benign genetic variations can lead to toxicity and disease under the proper environmental conditions. A major objective during the next century will be to determine how genetic variations interact with environmental factors to produce disease. Some of the tools and data required to advance in that direction are addressed in the following chapters of this book.

References

Alberts, B., Bray, D., Lewis, J., Raff, M., Roberts, K. and Watson, J.D., 1989, Macromolecules: structure, shape and information, in *Molecular Biology of the Cell*, 2nd edn, pp. 87–106, New York: Garland Publishing.

Akli, S., Chelly, J., Lacorte, J., Poenaru, L. and Kahn, A., 1991, Seven novel Tay–Sachs mutations detected by chemical mismatch cleavage of PCR-amplified cDNA fragments. *Genomics*, **11**, 124–134.

Ames, B.N., Profet, M. and Gold, L.S., 1990, Nature's chemicals and synthetic chemicals: comparative toxicology. *Proc. Natl Acad. Sci., USA*, **87**, 7782–7786.

Benhattar, J., Cerottini, J.P., Saraga, E., Metthez, G. and Givel, J.C., 1996, *Int. J. Cancer*, **69**, 190–192.

Beutler, E., 1990, The genetics of glucose 6-phosphate dehydrogenase deficiency. *Seminars in Hematology*, **27**, 137–164.

Beutler, E., Gelbart, T., Kuhl, W., Zimran, A. and West, C., 1992, Mutations in Jewish patients with Gaucher disease. *Blood*, **79**, 1662–1666.

Bosma, P.J., Chowdhury, J., Bakker, C., Gantla, S., de Boer, A., Oostra, B.A., Lindhout, D., Tytgat, G.N.J., Jansen, P.L.M., Elferink, R.P.J. and Chowdhury, N., 1995, The genetic basis of the reduced expression of bilirubin UDP-glucuronosyltransferase I in Gilbert's syndrome. *New Eng. J. Med.*, **333**, 1171–1175.

Brown, T.A., 1990, What Mendel discovered, in *Genetics: a Molecular Approach*, pp. 267–282, London: Chapman and Hall.

Chen, J., Silver, D.P., Walpita, D., Cantor, S.B., Gazdar, A.F., Tomlinson, G., Couch, F.J., Weber, B.L., Ashley, T., Livingston, D.M. and Scully, R., 1998, Stable interaction between products of the BRCA1 and BRCA2 tumor suppressor genes in mitotic cells. *Molec. Cell*, **2**, 317–328.

Chen, P.L., Chen, Y., Bookstein, R. and, Lee, W.H., 1990, Genetic mechanisms of tumor suppression by the human p53 gene. *Science*, **250**, 1576–1580.

Chen, S.Q., Chou, W.H., Blouin, R.A., Mao, Z.P., Humphries, L.L., Meek, C.Q., Martin, W.L., Hays, L.R., Neill, J.R. and Wedlund, P.J., 1996, The cytochrome P4502D6 (CYP2D6) enzyme polymorphism: screening costs and influence on clinical outcomes in psychiatry. *Clin. Pharmacol. Ther.*, **60**, 522–534.

Cleaver, J.E., 1968, Defective repair replication of DNA in xeroderma pigmentosum. *Nature*, **218**, 652–656.

Daly, A.K., Brockmoller, J., Broly, F., Eichelbaum, M., Evans, W.E., Gonzalez, F.J., Huang, J.D., Idle, J.R., Ingelman-Sundberg, M., Ishizaki, T., Jacqz-Aigrain, E., Meyer, U.A., Nebert, D.W., Steen, V.M., Wolf, C.R. and Zanger, U.M., 1996, Nomenclature for human CYP2D6 alleles. *Pharmacogenetics*, **6**, 193–201.

Doss, M., Laubenthal, F. and Stoeppler, M., 1984, Lead poisoning in inherited delta-aminolevullinic acid dehydratase deficiency. *Int. Arch. Occup. Environ. Health*, **54**, 55–63.

Dykes, C.W., 1996, Genes, disease and medicine. *Br. J. Clin. Pharmacol.*, **42**, 683–695.

Evans, D.A.P., 1993, Genetic factors in drug therapy, in *Clinical and Molecular Pharmacogenetics*, pp. 1–626, Cambridge: Cambridge University Press.

Gaedigk, A., Blum, M., Gaedigk, R., Eichelbaum, M. and Meyer, U.A., 1991, Deletion of the entire cytochrome P450 CYP2D6 gene as a cause of impaired drug metabolism in poor metabolizers of the debrisoquine/sparteine polymorphism. *Am. J. Hum. Genet.*, **48**, 943–950.

Gardner, R.J.M. and Sutherland, G.R., 1989, in *Chromosome Abnormalities and Genetic Counselling*, 2nd edn, pp. 95–100, Oxford: Oxford University Press.

Goldstein, J.L. and Brown, M.S., 1987, Regulation of low-density lipoprotein receptors: implications for pathogenesis and therapy of hypercholesterolemia and atherosclerosis. *Circulation*, **76**, 504–507.

Guttendorf, R.J., Wedlund, P.J., Blake, J. and Chang, S.L., 1988, Simplified phenotyping with dextromethorphan by thin layer chromatography: application to clinical laboratory screening for deficiencies in oxidative drug metabolism. *Ther. Drug Monitor.*, **10**, 490–498.

Hershfield, M.S., 1995, PEG-ADA: An alternative to haploidentical bone marrow transplantation and an adjunct to gene therapy for adenosine deaminase deficiency. *Hum. Mutat.*, **5**, 107–112.

Hirschhorn, R., Yang, D.R., Puck, J.M., Huie, M.L., Jiang, C.K. and Kurlandsky, L.E., 1996, Spontaneous *in vivo* reversion to normal of an inherited mutation in a patient with adenosine deaminase deficiency. *Nature Genet.*, **13**, 290–295.

Jeremy, J.Y., Mikhailidis, D.P. and Pittilo, R.M., 1995, Cigarette smoking and cardiovascular disease. *J. R. Soc. Health*, **115**, 289–295.

Kakvan, M. and Greenberg, S.D., 1977, Cigarette smoking and cancer of the lung: a review. *R. I. Med. J.*, **60**, 588–591.

Kalow, W., 1962, Pharmacogenetics, in *Heredity and the Response to Drugs*, pp. 1–205, Philadelphia: W.B. Saunders.

Kalow, W. and Staron, N., 1957, On distribution and inheritance of atypical forms of human serum cholinesterase, as indicated by dibucaine numbers. *Canad. J. Biochem. Physiol.*, **35**, 1305–1320.

Kaufman, S., 1976, Phenylketonuria: biochemical mechanisms. *Adv. Neurochem.*, **2**, 1–132.

Knudson, A.G., 1985, Hereditary cancer, oncogenes and anti-oncogenes. *Cancer Res.*, **45**, 1437–1443.

Kolodner, R.D., 1995, Mismatch repair mechanisms and relationship to cancer susceptibility. *Trends Biochem. Sci.*, **20**, 397–401.

Ledley, F.D., 1994, Therapeutic promise of molecular genetics. *J. Invest. Dermatol.*, **103**, 2S–5S.

Meyer, K.C., 1994, Occupational and environmental lung disease. Beryllium and lung disease. *Chest*, **106**, 942–946.

Pier, G.B., Grout, M., Zaidi, T., Meluleni, G., Mueschenborn, S.S., Banting, G., Ratcliff, R., Evans, M.J. and Colledge, WH., 1998, *Salmonella typhi* uses CFTR to enter intestinal epithelial cells. *Nature*, **393**, 79–82.

Savill, J., 1997, Role of molecular cell biology in understanding disease. *Br. J. Med.*, **314**, 203–206.

Shen, M.R., Jones, I.M. and Mohrenweiser, 1998, Nonconservative amino acid substitution variants exist at polymorphic frequency in DNA repair genes in healthy humans. *Cancer Res.*, **38**, 604–608.

Shows, T.B., McAlpine, P.J., Boucheix, C., Collins, F.S., Conneally, P.M., Frezal, J., Gershowitz, H., Goodfellow, P.M., Hall, J.G., Issitt, P., Jones, C.A., Knowles, B.B., Lewis, M., McKusick, V.A., Meisler, M., Morton, N.E., Rubinstein, P., Schanfield, M.S., Schmickel, R.D., Skolmick, M.H., Spence, M.A., Sutherland, G.R., Traver, M., Cong, N.V. and Willard, H.F., 1987, Guidelines for human gene nomenclature. *Cytogenet. Cell Genet.*, **46**, 11–28.

Vatsis, K.P., Weber, W.W., Bell, D.A., Dupret, J.M., Price-Evans, D.A., Grant, D.M., Hein, D.W., Lin, H.J., Meyer, U.A., Relling, M.V., Sim, E., Suzuki, T. and Yamazoe, Y., 1995, Nomenclature for N-acetyltransferase. *Pharmacogenetics*, **5**, 1–17.

Weatherall, D.J., 1991, The structure, organization and regulation of human genes, in *The New Genetics and Clinical Practice*, 3rd edn, pp. 1–368, Oxford: Oxford University Press.

Zimran, A., Hadas-Halpern, I., Abrahamov, A., 1991, Enzyme-replacement therapy for Gaucher's disease. *New England J. Med.*, **325**, 1810–1811.

2

PCR: METHODS AND LIMITATIONS

William B. Mattes and Mark S. Miller

2.1 Background

Just as the inventions of the steam engine, spinning machine and the cotton gin sparked the Industrial Revolution of the eighteenth and nineteenth centuries, so the inventions of molecular cloning, rapid DNA and protein sequence determination, and polymerase chain reaction fostered the current revolution in molecular genetics. The first two of these genetic tools allowed (1) the isolation of large quantities of specific gene fragments, and (2) determination of the specific DNA sequence of those fragments. Polymerase chain reaction (PCR) makes use of the DNA sequence to accomplish the same result as molecular cloning, but with a fraction of the effort. Because of its ability to amplify a target with such simplicity, the PCR is a cornerstone in current genetic analysis.

The PCR essentially uses a DNA polymerase to increase the amount of a target DNA exponentially, doubling every "cycle" of the process. As Saiki (1989) notes "reduced to its most basic terms, PCR merely involves combining a DNA sample with oligonucleotide primers, deoxynucleotide triphosphates, and the thermostable *Taq* DNA polymerase in a suitable buffer, then repetitively heating and cooling the mixture for several hours until the desired amount of amplification is achieved." However, all too common experiences with the PCR include problems such as mis-priming at non-target sites, primer dimer formation, failed denaturation, mutant PCR product, and contamination. Thus, Saiki (1989) goes on to observe that "In fact, the PCR is a relatively complicated and, as yet, incompletely understood biochemical brew where constantly changing kinetic interactions among the several components determine the quality of the products obtained." It is the goal of this chapter to enable the reader to understand that biochemical brew well enough to achieve the best possible PCR results.

2.2 Basic features

2.2.1 Requirements for DNA synthesis

The first word of "polymerase chain reaction" indicates a critical component of the PCR process, namely DNA polymerase. This enzyme can make a complementary

17

copy of a single strand of DNA in the presence of deoxynucleotide triphosphates (dNTPs) and a divalent cation (Mg^{2+} or Mn^{2+}). However, all known DNA polymerases require as a starting point for polymerization a primer region, i.e. a short region of double-stranded DNA, that is then extended in a 5′ to 3′ direction (see Figure 2.1). Thus, to copy a single strand of DNA a short oligonucleotide must be hybridized, or annealed, to its complementary region on the template strand. This requirement for a primer is key for the PCR.

2.2.2 *The exponential process – requirement for two primers*

The second critical feature of the PCR is its analogy to the chain reaction in nuclear fission. In both processes, each single reaction gives rise to two products, each of which is a reactant in subsequent reactions. This is accomplished in the PCR by the use of two oligonucleotide primers. If only one primer were used, in one polymerization cycle one copy of the template strand would be made. Following denaturation of this product and annealing of another primer to the template strand, another copy could be made. However, the key to the PCR's exponential process is that both strands of a double-stranded DNA molecule are copied because two oligonucleotide primers are used, each complementary to one of the DNA strands (Figure 2.2). Furthermore, their position is chosen such that DNA synthesis from one primer overlaps the position of the other primer (see Figure 2.3). Thus, after three cycles of denaturation, annealing, and elongation (Figure 2.4), a short-double stranded molecule is produced whose length and sequence is dictated by the positions of the primers on the original template. In each subsequent cycle, each of these short products will give rise to two identical copies, and the chain reaction is underway. Ultimately, it is this product that is measured in analyses of the reaction.

2.2.3 *Requirement for a thermostable DNA polymerase*

A third development in the PCR, as we know it today, was the application of the thermostable DNA polymerase from *Thermus aquaticus* (*Taq* DNA polymerase)

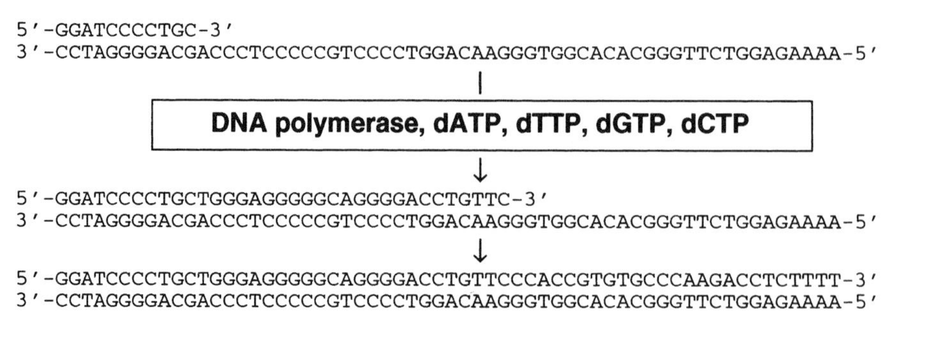

Figure 2.1 Basic elements of DNA synthesis.

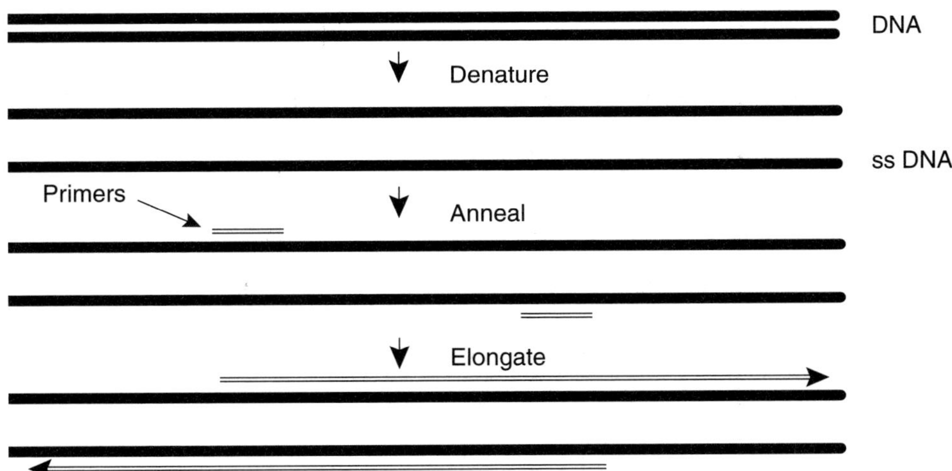

Figure 2.2 Cycle 1 of the PCR.

(Mullis, 1990). When the PCR was first discovered *E. coli* DNA polymerase was used. Because the polymerase from *E. coli* exhibits maximal enzyme activity at 37°C and is destroyed by high-temperature incubations, one had to add fresh enzyme after each denaturation cycle. With the discovery of *Taq* DNA polymerase and the concomitant introduction of instruments to carry out precise heating and cooling cycles, the PCR process became relatively automated and, once the reaction was started, could be carried out without user intervention.

2.3 Issues and pitfalls

The utility of the PCR, like that of any other analytical method, is limited by the ratio of signal to noise. The signal is, of course, the short DNA product (or "amplicon") as noted above. Noise can be a variety of undesirable products including "primer dimer" and non-target amplifications. Another perspective is to separate signal-to-noise problems into two components: specificity, or the ratio of the amount of desired product to the amount of "other" products, and yield or efficiency, the amount of desired PCR product produced in each cycle. It should be noted that these two are not synonymous, since conditions that result in the maximum efficiency of product formation may result in decreased specificity. In addition, a third consideration comes into play in the PCR, that of fidelity, defined as the sequence accuracy of the DNA product. Successful PCR experimental design pays close attention to these considerations.

The determinants of signal and noise in the PCR may be divided into those that are intrinsic to the PCR set-up and those that are intrinsic to the experimental performance. The former include primer design, the thermal profile, and reaction components, while the latter include sample quality and contamination. These will be discussed in detail in subsequent sections.

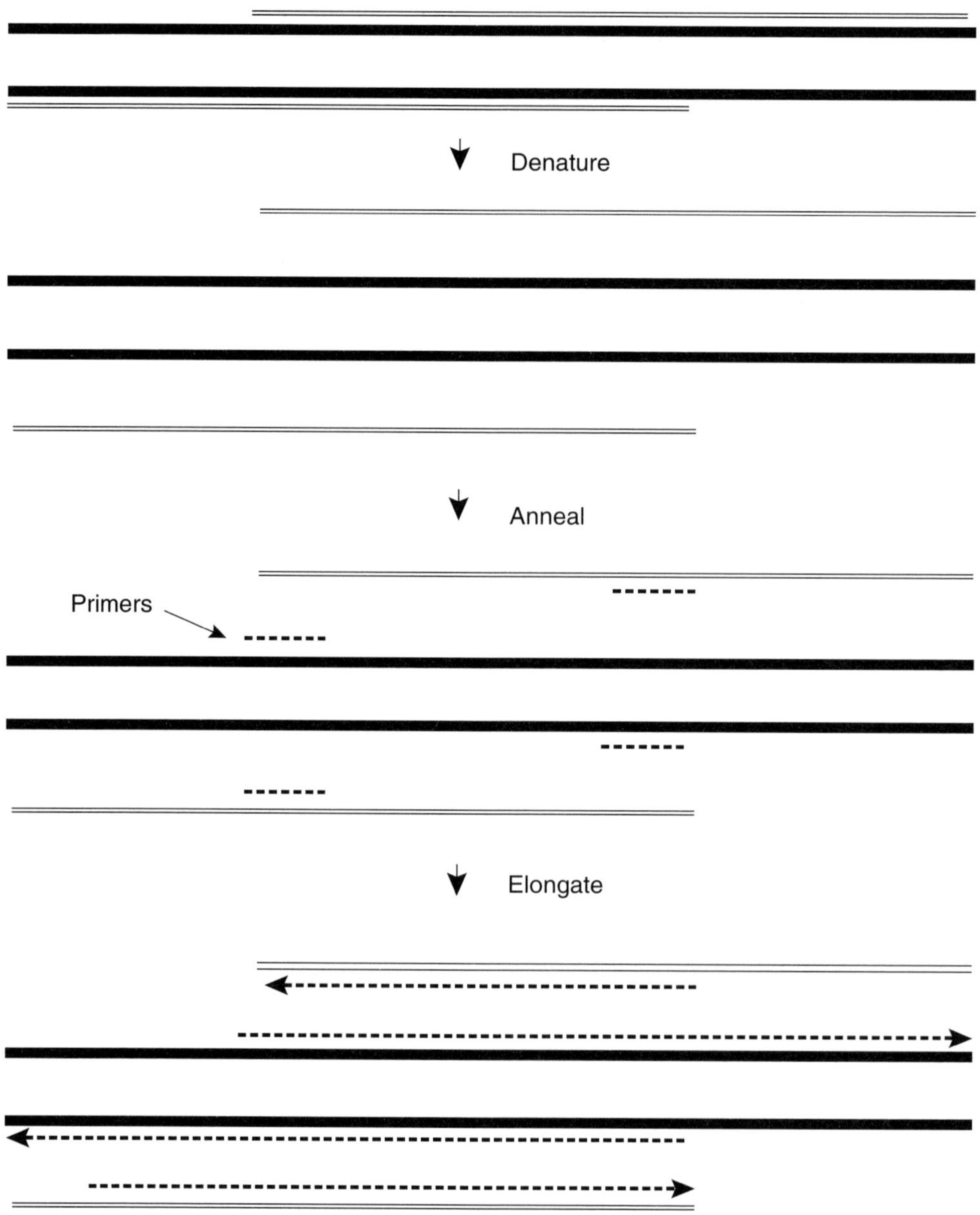

Figure 2.3 Cycle 2 of the PCR.

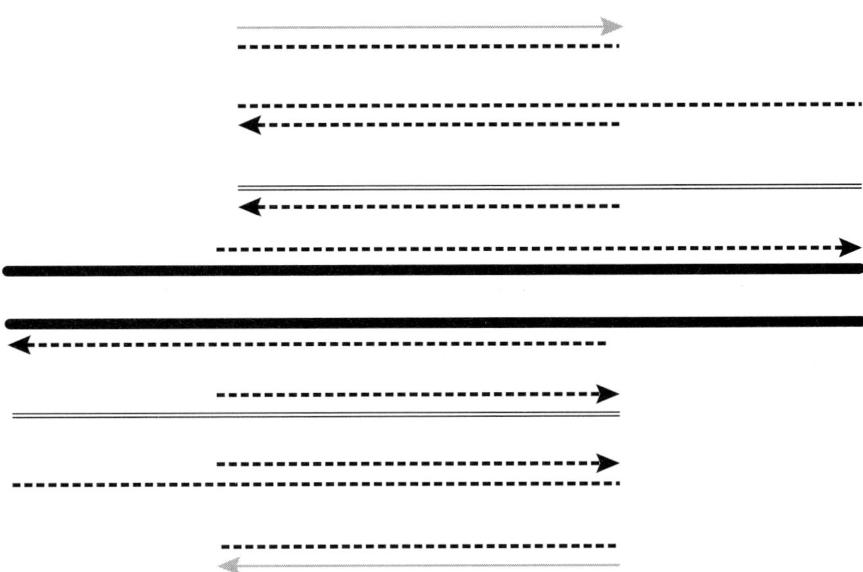

Figure 2.4 Products after cycle 3 of the PCR.

2.4   PCR set-up determinants of signal/noise

2.4.1   *Primer design*

The single most critical determinant of a successful PCR is the oligonucleotide primer pair and their characteristics. No other element of the process, except for contamination, has such a dramatic impact on the outcome of the experiments. For this reason considerable care should be exercised in the selection of the region amplified (the amplicon) and the design of the primer pair.

The first step in developing a primer pair is examining the genetic target to be amplified. In some cases the goal of a PCR experiment is to analyze a specific region of a gene, and this dictates a minimum region to be amplified. In the case of reverse transcription PCR, a portion of an RNA molecule is amplified and any region of that molecule may be a suitable target. In both cases the possible target region should be checked for intra-strand complementarity, i.e. the ability to form stable hairpin structures that would interfere with the progress of the DNA polymerase. The target gene and the amplicon should also be examined for GC content and denaturation temperature (T_m), as a high T_m (>94°C) for either may require a denaturation step more aggressive than the standard 1 minute at 94°C (Drummond *et al.*, 1992; Dutton *et al.*, 1993). Note also that the presence of Mg^{2+} in the PCR increases the T_m of DNA more efficiently than Na^+, so that in calculations of T_m for the target one should consider the Na^+ to be effectively 1 M (Dove and Davidson, 1962). Since *Taq* DNA polymerase is only modestly thermostable above 95°C (Gelfand, 1989), highly thermostable DNA polymerases (e.g. *Pfu*, Vent, Deep Vent, etc.) should be considered along with more aggressive denaturation.

As for the primers themselves, some general rules have become standard:

- primers should be 15–30 bases long
- primers should have G+C content of 40–60%
- each primer should have a T_m in the range of 55 to 80°C, with the two T_ms within 5°C of each other
- the two primers should avoid inter-primer complementarity, particularly at the 3′ ends (this leads to the "primer dimer" effect)
- each primer should avoid self-complementarity and/or internal secondary structure (hairpins)
- long stretches of any one base (runs) should be avoided
- each primer's 3′ end should be highly specific for its target
- the stability of each 3′ terminus should be lower than that of the rest of the primer

A few of these points should be further explained. Intra- or inter-primer complementarity can have several negative consequences. Primer–primer hybridization will compete with primer–target hybridization and reduce the efficiency of the PCR. If these complementary regions are at the 3′ ends of the primers, DNA polymerase will extend the complementary region to create a copy of the primers, called a "primer dimer". The primers will be tied up in such products in just a few cycles. This ability of DNA polymerase to make use of just the 3′ end of the primer also requires that the sequence of each primer's 3′ end be unique to the target. Thus, it is a good practice to check the terminal 7 bases at the 3′ end for other sites of homology on the target; if there are such sites, arranged so that a PCR product could form, choose a different sequence for the primer. Likewise, the 3′ terminus of each primer should be the *least* stable portion of the oligonucleotide: at stringent annealing temperatures the internal portion will form a duplex first, with the 3′ terminus forming a duplex (and hence priming) only on the specific sequence.

Obviously, the process of designing DNA sequences for target and primer sequences with the desired characteristics is ideally suited to computer software. Commercial software packages include Oligo™ (for Windows and Mac, NBI), Primer Premier™ (for Windows and Mac, Clontech), Primer Designer (for Windows, Scientific & Educational Software), and RightPrimer (for Mac, BioDisk). A free software package, Primer 0.5, was developed by Lincoln, Daly, and Lander of the Whitehead Institute for several operating systems (available at http://www-genome.wi.mit.edu/ftp/distribution/software/). Furthermore, Primer 3.0 is a version of this program that can be accessed and executed on the World Wide Web (http://www-genome.wi.mit.edu/cgi-bin/primer/). Not all software packages have equivalent capabilities. In particular, some do not check for uniqueness or stability of the 3′ terminus (the Oligo™ package does do this) and most do not automatically check the target or amplicon for GC content or potential secondary structure. These steps sometimes can be done either manually, through special settings of the software, or through the use of other sequence-analysis software (e.g. GCG). It is highly recommended that some software is used in primer design, as all of these packages do carry out the essential steps.

2.4.2 Cycle number

The number of denaturation–annealing–elongation cycles carried out for any given PCR experiment obviously determines how much of the final, short DNA product is formed. The number of copies of this product is given by the formula $(A^n - A \times n) \times T$, where A is the amplification factor, n is the number of cycles and T is the number of copies of the original template. $A \times n$ gives the number of primary and secondary extension products with indeterminate length ("long" products); in about 8–10 cycles it becomes an inconsequential number and the formula may be simplified to $(A^n) \times T$.

Not surprisingly, the exponential accumulation of product does not continue indefinitely. Rather, after a certain number of cycles the short product accumulates in a more linear fashion and eventually does not accumulate with subsequent cycles (see Figure 2.5). This point is referred to as the "plateau" and is probably caused by a number of factors. Foremost is the eventual overwhelming excess of template over DNA polymerase. Adding more *Taq* polymerase to the reaction can increase the maximum yield up to a point (Katz and Haff, 1989), but other factors apparently still limit the reaction (Sardelli, 1993). From a practical standpoint, when the number of cycles is such that the plateau has been reached, quantitative differences between reaction mixtures cannot be evaluated (see Figure 2.5).

While the amplification factor is theoretically 2, i.e. two short products are produced during each cycle, this number is rarely found to be this high in the laboratory, and can be as low as 1.6. The "efficiency" of the PCR refers to how

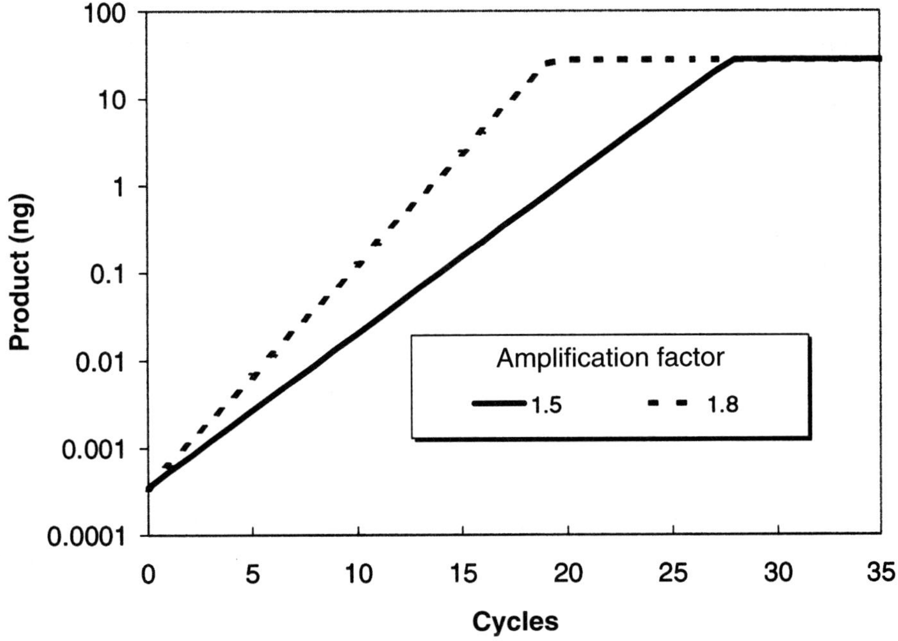

Figure 2.5 Exponential nature of the PCR.

close this number approaches 2, and would thus be 60% if $A = 1.6$. As mentioned above, efficiency (or yield) is a key component in the signal-to-noise ratio, and increasing efficiency is a prime goal of optimization. However, as can be seen in Figure 2.5, two reactions with different efficiencies can still have identical amounts of product if the number of cycles is such that a plateau has been reached. Thus a critical first step *before* optimizing a PCR is determining for a given amount of starting template the range of cycles in which the reaction is still in an exponential phase.

2.4.3   *Thermal profiles and "hot start"*

The PCR process involves a series of incubations at several different temperatures for set periods of time, designed to denature double-stranded DNA, anneal primers to the single strands, and then extend the primers with DNA polymerase (see Figure 2.6). While all of these segments provide opportunities for optimization, with good primer design relatively standard conditions may often be employed with good success. Thus, unless the target or final PCR product is known to have a high T_m (see Section 2.4.1), a denaturation temperature of 94°C is adequate. Likewise, an extension temperature of 72°C is standard for *Taq* DNA polymerase. The times for these incubations are also reasonably standard for any given thermal cycling machine, generally 1 minute for each segment for a machine handling 0.5 ml tubes, but less for those handling smaller tubes.

The desired annealing temperature is obviously the one that gives the greatest amount of desired product with the least amount of extraneous products. Thus one is striving for a temperature that is not so high as to destabilize hybridization to the target, yet high enough to destabilize hybridization of the primers to non-target, but similar, sequences. With good primer design, the annealing temperature can be

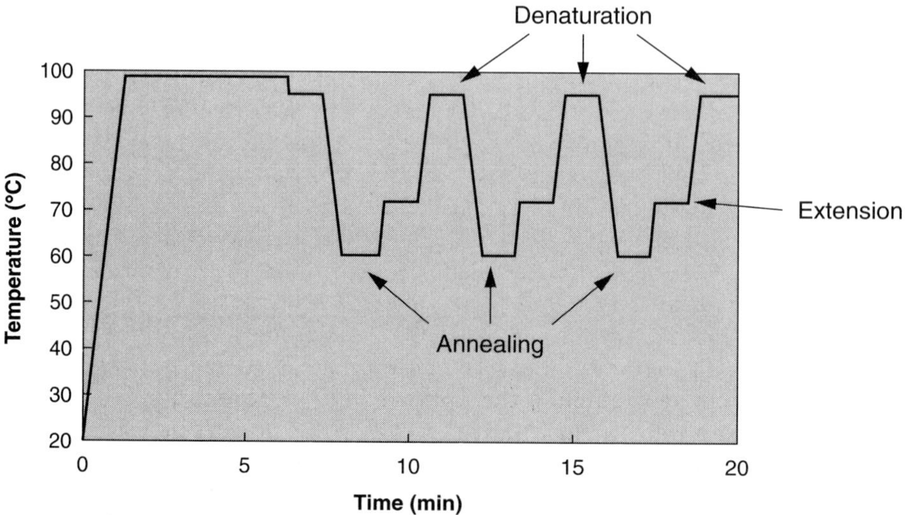

Figure 2.6   PCR thermal profiles.

estimated. According to one empirically derived formula, the "optimal annealing temperature" (T_a^{OPT}) is equal to $0.3 \times T_m(\text{primer}) + 0.7 \times T_m(\text{product}) - 14.9$ (Rychlik *et al.*, 1990) (this is the formula used in the OLIGO program). Alternatively, one can use a formula where T_p is calculated for each primer as $22 + 1.46 \times \{2 \times (\#G \text{ or } C) + (\#A \text{ or } T)\}$ and the lowest value is used for annealing (Wu *et al.*, 1991). The latter formula gives higher annealing temperatures than the former. In any case, if products other than those predicted by the sequence are observed, higher annealing temperatures should be examined.

Ironically, the opportunity for primer annealing to non-target sequences exists in the very first segment of the thermal profile, that of the initial transition from room temperature to the first denaturation cycle. Mis-priming events occurring during these low temperatures can be extended by *Taq* polymerase because this enzyme retains a fair amount of activity even at temperatures as low as 22°C (Gelfand, 1989). Several workers recognized that these mis-priming events unbalanced the PCR right from the start: the myriad of non-target products generated all contain incorporated primer sequences, and hence can be replicated even under stringent conditions later on in the PCR. These products not only show up as extraneous bands upon electrophoresis, but also reduce the yield of the target product through competition for primer and polymerase. The proposed solution was to add a key reagent (e.g. *Taq* polymerase, or deoxynucleotide triphosphates, dNTPs) to the reaction mixture only after a temperature >70°C was reached (D'Aquila *et al.*, 1991; Mullis, 1991), a procedure termed "hot start". This approach was later modified to separate reaction components by a wax barrier that melted only at high temperatures (Chou *et al.*, 1992). Another solution is the inclusion in the PCR mixture of a monoclonal antibody (*TaqStart*™ from CLONTECH) that neutralizes *Taq* polymerase at lower temperatures, but denatures at high temperatures to allow DNA synthesis. "Hot start" improves the performance of most PCR designs, and some form of it should be included in initial optimization steps.

2.4.4 *Components*

Many optimization strategies rightly focus on the components in the PCR and their concentrations, and several commercially available kits simplify this process. In particular, the Mg^{2+} concentration plays a key role in the efficiency and specificity of the PCR. The yield of desired PCR product and the appearance of extraneous products can be vastly different at Mg^{2+} concentrations differing by as little as 1 mM. However, these effects are dependent upon "free" Mg^{2+} ion, not that bound to dNTP; thus any optimization of Mg^{2+} concentration must be done so with an eye to the dNTP concentration. Furthermore, the optimal annealing temperature may be affected by the Mg^{2+} concentration. Thus a change in Mg^{2+} may necessitate a re-examination of the annealing temperature, and a change in dNTP concentration would warrant re-examining the optimal Mg^{2+} concentration.

As mentioned above, the fidelity of the PCR (i.e., the sequence accuracy of the DNA product) should be considered in PCR design, especially if the PCR products are to be cloned. While conditions may be chosen so as to enhance the fidelity of *Taq* DNA polymerase (Eckert and Kunkel, 1990; Ling *et al.*, 1991; Cha and Thilly, 1993), an alternative may be the use of a DNA polymerase with intrinsically greater fidelity, such as Vent™ (Jack *et al.*, 1992), *Pfu*, and others (Cline *et al.*, 1996).

Kary Mullis has proposed that the concentrations of components of the PCR should be described in terms of *molecules per microliter* rather than molarity or µg per µl (Mullis, 1991). A PCR with 1 ng of lambda DNA in 100 µl, contains 1.8×10^7 molecules per microliter, and would be denoted as 1.8 u7 (Table 2.1). This is a paradigm with considerable utility for designing and interpreting the PCR. Indeed, there is little value in comparing PCRs in terms of ng DNA input if the input for one reaction is cellular DNA and the input for another is defined plasmid DNA. Rather, a PCR with DNA input derived from 10^6 diploid cells should be compared with one with DNA input of 2×10^6 copies of a control template.

2.5 Experimental determinants of signal/noise

2.5.1 Contamination

While mis-priming at non-target sites can reduce the signal-to-noise ratio in a PCR, contamination of a PCR with final product or with cloned target can exclude any meaningful analysis of an experimental sample. This is particularly true for experiments looking for mutations in a given target – contamination of the reaction with wild-type product will clearly be disastrous for the project. The magnitude of the problem can be particularly appreciated in light of the paradigm presented in Table 2.1. Consider that if there are ~6×10^5 copies of a unique gene per µg human DNA, and a PCR results in ~2×10^{10} copies of amplified product per µl of PCR, one can calculate that 0.00003 µl of PCR has the same number of input target molecules as 1 µg genomic DNA! In fact, Dr John Sninsky has proposed that "PCR contamination be considered as a form of infection" (Dieffenbach and Dveksler, 1993). Following PCR amplification, samples should be treated as if they were highly radioactive or contain a high titer of a stable pathogenic virus. While other sources of contamination exist (e.g. plasmid or probe preparations containing target sequences), the primary suspects should be the solutions and materials that are handled after PCR amplification.

The simplest most critical step to take to control contamination is to isolate pre-PCR from post-PCR operations. PCR-amplified samples should be processed and

Table 2.1 Kary Mullis' quantitative view of PCR

	Before PCR		After PCR	
	Weight	*Molecules/µl*	*Weight*	*Molecules/µl*
Template (48 500 bp)	1 ng	1.86 u5	1 ng	1.81 u5
Target (500 bp)	10 pg	1.81 u5	1 µg	1.81 u10
Primers (25 mers)	1623 ng	1.20 u12	1574 ng	1.17 u12
dNTPs	39 µg	4.82 u14	37 µg	4.64 u14
Magnesium ion	3.6 µg	9.03 u14	3.6 µg	9.03 u14
Taq DNA polymerase	12.5 ng	8.01 u8	12.5 ng	8.01 u8

Assumption: 100 µl reaction with bacteriophage lambda DNA; 250 000 units/mg *Taq* DNA polymerase; 487 Daltons/average dNTP; 325 Daltons/average dNMP; 10^5-fold amplification.

analyzed in an area separate from that where preparations for PCR are carried out, preferably in different rooms. Alternatively, consider enclosures, or laminar flow hoods, for either one or both of these operations. Enclosures designed for PCR, with ultraviolet sterilization built in, are available (e.g. Oncor, USA Scientific, Inc., Coy Laboratory Products, Inc.) but enclosures designed for radioisotope (e.g. Scotlab) or standard cell culture work may work as well. Separate sets of materials for pre-PCR and post-PCR operations should be maintained, including gloves, tubes, tube openers, pipette tips, ice buckets, pipettors, racks, microcentrifuges, and reagents (e.g. EDTA, buffers, or chloroform for extractions). Separate laboratory coats for the different areas might also be considered, and when equipment from one area is to be used for another it should be thoroughly decontaminated.

As mentioned above, PCR contamination can be compared to radioisotope contamination, and as such can also be controlled through procedures common in radioisotope use. Gloves should be worn and changed frequently. Always use aerosol-barrier pipette tips or positive displacement pipettors to avoid contamination of pipettor barrels with subsequent carry-over. Minimize sample handling and open tubes very carefully (preferably with a tube opener that can be easily decontaminated). Cuvettes that are used in determining the UV absorbance of an amplified sample should be thoroughly decontaminated.

Eliminating PCR contamination from surfaces, equipment, and solutions is more complicated than eliminating biological contamination, since DNA by design is quite stable, and resistant to alcohol, autoclave, or simple filter treatments. An early approach toward PCR surface decontamination was sterilization with ultraviolet light (Cone and Fairfax, 1993). A simple and general approach is to use 10% Chlorox® to treat surfaces and equipment for a short period of time (Prince and Andrus, 1992); this treatment is particularly effective for fragile glassware such as quartz cuvettes. Contamination in solutions can be handled by aliquoting all stock reagents used for setting up PCR mixtures. Thus, if contamination is suspected, the aliquots of stocks currently in use may be discarded and replaced. Alternatively, dUTP may be used in the PCR instead of dTTP; treating pre-PCR mixtures with the enzyme uracil deoxyribose N-glycosylase then degrades any contaminating PCR product (i.e. containing the incorporated deoxyuridine residues) (Hartley and Rashtchian, 1993).

2.5.2 Inhibitors and enhancers

It is well known that many components of unpurified samples can inhibit the PCR, including EDTA, phosphate, and heparin (Beutler *et al.*, 1990). The presence of inhibitors in a sample could be inferred by spiking such a sample with purified target DNA and comparing its amplification to that of the target DNA alone. An obvious solution is further purification of the samples, although additions of spermidine (Wan and Wilkins, 1993) or bovine serum albumin (Kreader, 1996) have been reported to improve amplification of crude samples. Additions of tetramethylammonium chloride (Chevet *et al.*, 1995), spermidine (Wan and Wilkins, 1993), dimethylsulfoxide (Pomp and Medrano, 1991) and formamide (Sarkar *et al.*, 1990; Comey *et al.*, 1991) have been reported to improve the performance of unsatisfactory PCR mixtures, although spermine and spermidine have also been found to inhibit some reactions (Ahokas and Erkkila, 1993). Suffice it to say that such additives may be beneficial, but an ideal situation is a reasonably purified sample with a well-designed PCR.

2.6 Reproducibility and quality controls

Reproducibility in PCR experiments can be improved by precautions common to all molecular biology procedures. Nucleases and proteases from small flakes of dead skin can obviously digest and inactivate enzymes and nucleic acids, so **gloves should be worn at all times** when handling these and other PCR reagents. Likewise, solutions should be prepared with the highest quality reagents and water, and ideally autoclaved before use. Enzymes such as reverse transcriptase should be protected from the freeze–thaw defrost cycles in frost-free freezers. This can be accomplished with storage boxes (from Nalgene or Stratagene) containing a coolant and buffering changes in temperature. Handling small reagent and sample tubes with tube openers can also help to prevent contamination of their contents from dirty gloves or other samples.

Another issue in PCR is that of reproducibly delivering small volumes, sometimes less than 1 μl (requiring a pipettor accurate in this range and appropriate tips). A common pipetting technique is to submerge the pipette tip into the sample, yet in doing this, a finite amount of sample will coat the outside of the tip, and the volume delivered to the reaction will be larger than planned. Instead, the pipette tip should just touch the surface of the sample. Obviously, another source of error in such operations is the pipettor itself, and to ensure reproducibility pipettors should be calibrated on a regular basis, at least every 6 months.

To assess the performance of any given experiment, appropriate controls must be included. A negative control (a PCR mixture with no added DNA template) verifies the absence of contaminating template in the PCR reagents. A procedure control (a mock extraction without sample) verifies the absence of contaminating template in the extraction process. A positive control can consist of a limited number of copies (e.g. 10^5–10^6) of target DNA carried through the entire procedure, but this control is often omitted to avoid contamination.

2.7 PCR from paraffin-embedded tissues

The ability to amplify large quantities of DNA by PCR can be combined with sensitive screening and analytical methods, such as allele-specific hybridization, single-strand conformation polymorphism analysis, and DNA sequencing, to allow investigators to utilize archived tissue samples for genetic analyses. Archives of patient material found in many pathology departments in medical centers can serve as a valuable resource for identification of genetic polymorphisms and the association of genetic differences in DNA structure with specific diseases. In addition, the use of archived patient samples permits one to initiate retrospective studies, whereby samples collected over long periods of time can be utilized for the identification and characterization of genetic polymorphisms. Genetic analyses can also be combined with immunohistochemical determinations to correlate changes in gene sequence with potential effects on the expression of proteins regulated by that gene.

These types of studies allow one to determine the effects of genetic polymorphisms on disease outcome, since the pathology of the lesions has often been determined.

The most common method of tissue preservation involves fixation in a 4–10% formalin-based solution followed by embedding in paraffin. The time of fixation can be quite critical for successful PCR analyses. Extended immersion in formalin can result in DNA damage, probably due to the breakdown of formalin to formic acid, that can markedly impair the integrity of the template (Dubeau *et al.*, 1986). In addition, one should be careful to find out exactly how tissues were preserved. Although not commonly used today, many archived tissues may have been fixed with a modification of the formalin solution that could include additives that either inhibit the amplification reaction or damage DNA. Because of the potential for DNA damage during fixation even under optimal conditions, attempts to amplify >650 base pairs of sequence become problematic (Wright and Manos, 1990). Thus, one limitation of this technique is the need to keep the PCR products relatively short. However, using PCR products in the range of 100–350 base pairs, our laboratory has had excellent results amplifying DNA from embedded samples. In our rodent studies under optimal conditions, where the fixation step was limited to 24 hr and the tissues were analyzed within 6 months, we have been able to amplify 95% of the embedded samples (Wessner *et al.*, 1996; Leone-Kabler *et al.*, 1997; Gressani *et al.*, 1998; Rollins *et al.*, 1998). Utilizing archived tissues from the Department of Pathology at Wake Forest University/Baptist Medical Center, we were able to successfully amplify 80–85% of the embedded samples tested (unpublished observations).

Most of the protocols currently in use to isolate DNA from paraffin-embedded tissues are a modification of the technique originally described by Goelz *et al.* (1985). We have found the modification of this procedure described by Wright and Manos (1990) to be relatively simple and easy to use (Wessner *et al.*, 1996). This procedure involves two rounds of deparaffinization for 20–30 min in xylene, followed by ethanol washes to remove the organic solvent. The samples are then digested with proteinase K (200 µg/ml) in 100–200 µl of buffer for several hours at 55°C or overnight at 37°C, then the proteinase is heat inactivated and the samples are ready for PCR. We have found that 2–5 µl of lysate are usually sufficient to provide ample product following amplification. The use of larger volumes results in less efficient amplification reactions, probably due to the presence of inhibitors (i.e. residual paraffin) in the lysate. These samples can be stored at –20°C until use, but should be used within a few weeks to ensure adequate amplification efficiency. Although we have been able to amplify PCR products from tissue lysates stored this way for almost 2 years (Gressani *et al.*, 1998; Rollins *et al.*, 1998), extended storage under these conditions can result in further DNA damage and the risk of losing some samples. The simplicity of this technique and the minimal amounts of extractions and handling help to keep cross-contamination problems to a minimum. The small volumes required for PCR allow for analysis of multiple gene products and reamplification and verification of any results from one tissue preparation.

In the event that some of the deparaffinized samples prove recalcitrant to attempts to amplify DNA, there are some modifications of the above technique that can improve the ability to recover DNA from the sample. A method referred to as "heat-soaked PCR", originally developed for use in forensic analyses (Ruano *et al.*, 1992), has been modified for use with paraffin-embedded materials (Burns *et al.*, 1997). In this procedure, the deparaffinization step is omitted and embedded sections cut from a microtome are placed in a tube and treated with 1% Triton X-100 at 95°C for 20 min. This material is then used directly in the PCR reaction. Although this method is even faster and simpler than the xylene extraction procedure described above, we have found that a disadvantage of this method is that greater proportions of the starting material are required in the PCR reaction. One modification that we have started to employ in our laboratory is to use the xylene extraction protocol and then heat the samples for 30 min at 95°C if they prove difficult to amplify. If this does not work, Triton X-100 can be added to the lysate and the heating step repeated before deciding that amplification will not be possible from a given sample. We have also noticed that when multiple PCR products are amplified from the same lysate, occasionally specific primer pairs will not work for one gene while other primers for other genes will be effective. Thus, some empirical experimentation may be required to find the primers that work best with your tissue samples. In some instances, a secondary PCR of the first amplification step may be necessary to obtain sufficient yields of product for subsequent analytical steps.

Because of the decreased efficiency of amplification from embedded tissue samples, modification of the standard cycling conditions is often necessary. We have generally found that a doubling of the annealing and extension times will provide a good yield of PCR product, as suggested previously (Wright and Manos, 1990). If your normal cycling conditions use a denaturing time of ≤30 sec, it would also be beneficial to increase the denaturation step to 1 min. In addition, it may also be necessary to increase the number of cycles. If, for example, you would normally use 30 cycles to amplify a particular gene sequence by PCR, increasing the number of cycles to 40 is recommended. However, increasing the cycle number much beyond 45 will probably not increase your yield that much further. Because of the extended cycling times, there is a greater tendency to produce smaller, non-specific PCR products resulting from mispriming events during the amplification reactions. For further analyses, such as restriction fragment length polymorphisms, single-strand conformation polymorphism analysis, or gene sequencing, it may become necessary to gel purify the desired fragment before proceeding to the next step.

One of the problems encountered by amplification of paraffin-embedded tissue is the limited amount of tissue available for analysis due to the small sample size usually obtained from pathology archives. In addition, when looking at cancerous or certain other diseased tissue samples, heterogeneous mixtures of different cell types may be present, making interpretation of the results obtained extremely difficult. While microdissection techniques can aid in removing normal tissue and increasing the specificity of the analysis (Shibata *et al.*, 1992; Emmert-Buck *et al.*, 1996; Going and Lamb, 1996), this usually results in a further reduction in the amount of tissue that is available. A recent variation of the amplification strategy has been developed by Duddy *et al.* (1998) that utilizes whole genome amplification to increase the amount of target DNA available for analysis.

In this procedure, an initial primer extension preamplification (PEP) step (Zhang *et al.*, 1992) is performed to increase the number of genetic loci that can be analyzed by subsequent PCR methods. Paraffin-embedded tissues are deparaffinized and digested with 400 μg/ml of proteinase K in buffer containing 1% Tween 20 (Duddy *et al.*, 1998). The tissue lysates are then purified by adding an equal volume of InstaGene Matrix (Bio-Rad), briefly vortexing the samples and incubating them at 54°C for 30 min, vortexing again for 10–15 sec, heating at 100°C for 8 min, and then centrifuging the samples at ≥12 000 rpm for 2–3 min. As the InstaGene purification step removes inhibitors of *Taq* polymerase, the supernatant can be used directly in the PEP step.

PEP is carried out as described (Zhang *et al.*, 1992; Duddy *et al.*, 1998) using a 50 μM concentration of 15-base random primer sequences (Operon Technologies). This results in the amplification of the entire genome, thus the products from this initial randomly primed amplification reaction can now serve as template in subsequent PCR analyses. Through the use of hemi-nested or nested locus-specific primers, Duddy *et al.* (1998) reported that 10 to greater than 100 genetic loci can be analyzed from a 1 mm^2 piece of 3-μm-thick embedded tissue by succeeding PCR-based analytical techniques such as allele-specific hybridization, single-strand conformation polymorphism analysis, and gene sequencing.

2.8 Allele-specific oligonucleotide (ASO) hybridization

In many instances involving the identification of genetic polymorphisms, restriction fragment length polymorphism (RFLP) analysis can be employed. With this technique, the PCR product is cut with a restriction enzyme that will produce one of two different products, representing cut and uncut DNA, depending on the sequence present. In many cases, however, convenient restriction sites are not present in the particular sequence of interest, necessitating other approaches. One option is to employ primers that introduce restriction sites in the PCR product. This can be a very quick and simple way to identify a particular polymorphic site, but may become more cumbersome if more than one site needs to be screened. For example, there may be several allelic forms of a given gene at a particular site, or one may be interested in screening for potential mutations in a putative disease-related gene. Gene sequencing is often too laborious and time-consuming, and quick and simple screening assays are needed to determine which samples may require further analyses. Two examples of these types of screening techniques, allele-specific oligonucleotide analysis (ASO) and single-strand conformation polymorphism analysis, will be discussed in the next two sections. Although there are a wide range of different molecular assays that can be employed, these two methods are among the most widely used.

ASO analysis is basically a variation of Southern and slot blot techniques (Verlaan-de Vries, 1986; Lyons, 1990). Once the PCR products have been immobilized on the membrane, they can be screened with short oligonucleotide probes, usually around 20 base pairs in length. These can be end-labeled with either radioactive, such as P^{32}, or non-radioactive chemiluminescent nucleotides, such as biotin. When using new primer pairs for the first time, it is advisable to determine the size of the PCR products on an electrophoretic gel, then transfer the samples to a nylon membrane prior to hybridization to be sure the signal is due to the expected sized band. Once you have determined this to be the case, slot blotting, where the

PCR products can be applied directly to the membrane with a filtration apparatus, is quicker and easier, and allows one to screen more samples at one time. The average electrophoretic gel apparatus allows one to load 20 to 40 samples in one gel, whereas slot and dot blot apparatuses contain anywhere from 48 to 96 individual sample wells.

In setting up the ASO experiment, the investigator is usually attempting to screen one or more regions in which the sequence is either polymorphic or contains mutagenic alterations. The PCR primers are designed so that the region of interest will not be at the ends of the PCR product. An example of this technique from the authors' research is shown with the Ki-*ras* gene, which is mutated in codons 12, 13, or 61 in both human and experimental animal tumor models (Reynolds *et al.*, 1992; Wessner *et al.*, 1996; Leone-Kabler *et al.*, 1997). Since codons 12 and 13 of the Ki-*ras* gene are located in exon 1, an approximately 110 base pair PCR product from this exon is amplified that places these codons 36 to 39 base pairs from the 5′ end and over 60 base pairs from the 3′ end. The PCR products are blotted to membranes and then hybridized with labeled oligonucleotide probes. Wash conditions are utilized that allow binding of only perfectly matched probes to the immobilized DNA on the membrane (Miller *et al.*, 1994).

The oligomers are designed so that the potential area of mismatch is located in the center of the probe. In the oligonucleotide shown below, which is complementary to the coding strand of the Ki-*ras* gene, the arrows indicate areas that are potentially mutated in lung tumors:

codon

12   13

5′ – GGA GCT <u>GGT</u> <u>GGC</u> GTA GGC AA – 3′

↑↑ ↑↑

A mismatched base pair in the center of the probe will result in a more unstable conformation, and increase the likelihood that the mismatched probes will melt off the DNA at temperatures that still allow binding of the perfectly matched oligomers (Figure 2.7). Using this method, one can quickly screen a variety of PCR samples and determine the exact sequence of a particular base in each individual PCR product. A representative ASO experiment is shown in Figure 2.8 – DNA from individual mouse lung tumors was amplified and blotted directly to a nylon membrane. The membrane was then sequentially and individually probed with a panel of oligonucleotides that each differed by one base pair in one of the two guanine residues in codon 12 highlighted above. As can be seen from the figure, each of the tumor DNAs contained a wild-type glycine allele. In addition, individual tumors contained different mutated Ki-*ras* alleles, with the majority of the tumors exhibiting G→T transversions in the two guanine residues of codon 12. Thus, one can quickly determine the sequence variation in a large number of samples in a relatively short time. It is recommended that when establishing a new assay, the results of the ASO analysis are confirmed by gene sequencing. However, once the assay conditions have been standardized, it is usually only necessary to occasionally screen a small percentage of your samples for quality control.

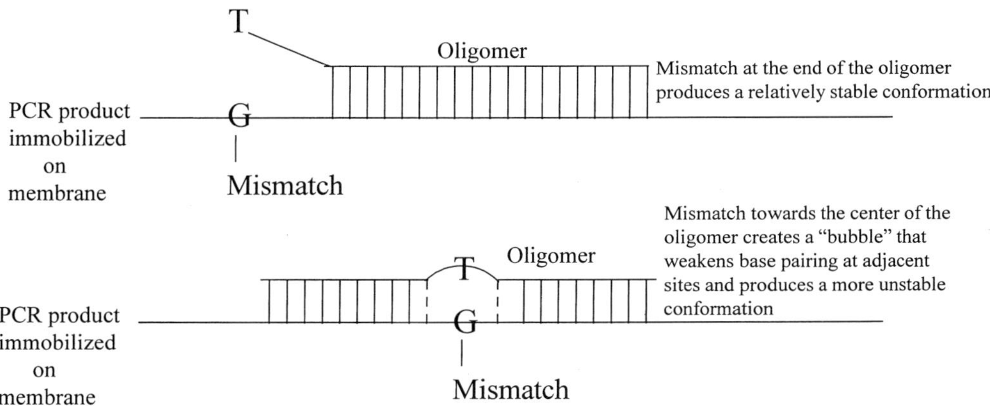

Figure 2.7 Effect of mismatched base pairs on annealing efficiency.

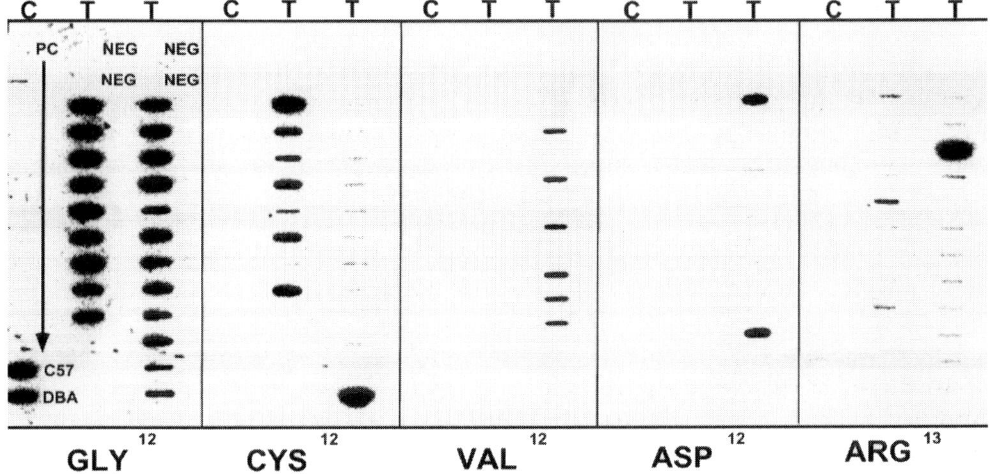

Figure 2.8 Allele-specific oligonucleotide (ASO) analysis of paraffin-embedded tissue for mutations in the Ki-*ras* gene. PCR products were diluted in water and blotted directly onto nylon membranes. The membranes were probed with P[32] end-labeled 20-base-pair oligonucleotides to the 12th codon of the Ki-*ras* gene and visualized on a Molecular Dynamics PhosphorImager 445 SI (Sunnyvale, CA, USA). The first column marked C included all the procedure controls (PC) and normal tissue from C57BL/6 and DBA/2N mice. The next two columns marked T contained the four negative buffer controls for the PCR reaction (NEG) and the tumor DNAs. Both the procedure and negative controls lacked any added template DNA. The figure is reprinted with permission from Leone-Kabler *et al.* (1997).

It should be noted that wash conditions following hybridization will not only vary greatly from one PCR product to the next, but different conditions are often employed by different laboratories to analyze the same PCR products. Again using our past experiences with the Ki-*ras* gene, the blots shown in Figure 2.8 were washed in the final step with $1 \times$ SSC/0.1% SDS at 3°C below the T_m of the oligonucleotide probe (Miller *et al.*, 1994); however, other laboratories have used SSPE (Verlaan-de Vries *et al.*, 1986) or tetramethylammonium chloride-based solutions (Lyons, 1990) to achieve the same results. As the salt concentration decreases and the temperature of the wash increases (increased stringency), the ability to melt off the mismatched probe will increase but the binding of the perfectly matched probe will at the same time be decreased. Thus, a certain amount of empirical testing will need to be done to obtain a reasonable balance of signal to background noise. Finally, the blotting steps should be repeated at least once to be sure that any positive signals were not the result of spurious background binding, particularly when the signal intensity is low.

2.9 Single-strand conformation polymorphism (SSCP) analysis

Although ASO analysis can be a quick and simple way to screen for polymorphisms or mutations in specific gene sequences, a disadvantage and limitation of this method is that it can become very time-consuming if more than one or two bases need to be screened. ASO analysis is not a good choice of techniques when large regions of DNA need to be analyzed for potential sequence alterations or variations. In these instances, a variety of general screening strategies are available that allow one to examine whole PCR products for single base-pair changes relative to a control DNA (i.e. normal) sample. Although several techniques are currently in use for screening small fragments (200 to 400 base pairs) of DNA, the simplest, easiest, and most commonly used is single-strand conformation polymorphism (SSCP) analysis (Orita *et al.*, 1989a,b). Since SSCP allows one to initially screen small fragments of DNA for potential sequence variation without necessarily having to know anything about the sequence, one can rapidly screen a large number of samples for potential sequence alterations. Identification of the few samples with base pair changes thus allows one to concentrate sequencing efforts on the most important specimens.

SSCP analysis basically involves the separation of PCR strands on a non-denaturing polyacrylamide gel, roughly equivalent to a sequencing gel without denaturants (such as urea). As the strands migrate through the gel, they assume their normal conformation and migrate at somewhat different rates, producing two distinct bands. Alterations in base sequence will force the individual strands to assume a somewhat different conformation, and this will result in a diagnostic "band shift", or change in mobility, of one or both fragments. These fragments can then be cut out of the gel and sequenced to identify the changes in base sequence in the DNA. Fragments can be visualized by either end-labeling the primers with P^{32}, adding radioactive dCTP to the PCR mixture, or using sensitive dyes in a radioactive-free "cold SSCP" (Hongyo *et al.*, 1993).

The extent and intensity of the band shift will depend on the nature of the base substitution and percentage of cells that contain an altered DNA sequence in the sample, respectively. Figure 2.9 shows an SSCP gel from a series of mouse liver tumors screened for mutations in the cyclin-dependent kinase inhibitor gene, *Ink4a*.

In this particular case, the tumor DNAs contain single base-pair mutations in this gene not found in normal tissue (Gressani *et al.*, 1998). Tumor 2 contains multiple bands in the tumor that are consistent with the presence of both mutated and normal DNA sequence in the tumor, and the band shifts (see arrows in Figure 2.9) relative to the normal band pattern are clearly visible. In tumor 7, however, all of the tumor DNA contains a mutant sequence. Thus, while only two bands are present in this lane, they are clearly migrating in a very different position relative to DNA isolated from normal mouse tissue (lane 27, marked normal).

It should be pointed out that although SSCP analysis is relatively quick and easy to set up, it does have some drawbacks. First, the false-negative rate with SSCP gels may be as high as 10%, depending on the gene to be analyzed, the primer pairs used, and the quality of the tissue examined (Sheffield *et al.*, 1993). This could result in the investigator missing a significant number of base changes. Second, only small gene segments, ranging from 150 to approximately 300 base pairs in

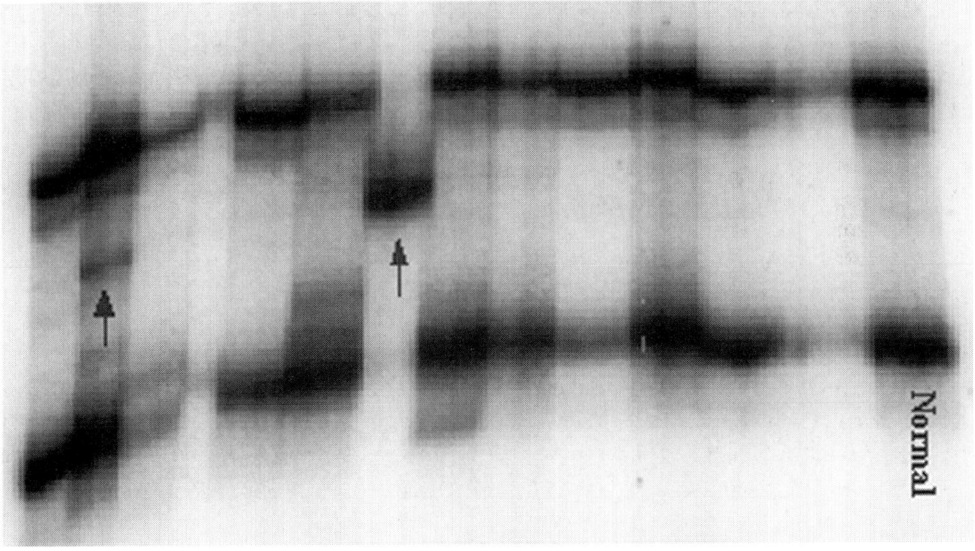

Figure 2.9 SSCP analysis of exon 1 of the *INK4a* gene. Four μl aliquots of labeled PCR products were heat denatured and electrophoresed in a nondenaturing 0.5× MDE gel in the presence of 5% glycerol at 4°C for 9.5 h at 9 W. Lanes 1–13 are tumor samples; lane 27 is normal liver tissue DNA from an untreated mouse. The arrows indicate band shifts. Note that tumor 2 contains multiple bands consistent with the presence of both mutated and normal DNA in the tumor. In tumor 7, however, all of the DNA is apparently mutant sequence, which runs at a different mobility relative to the bands representing wild-type alleles. The figure is reprinted with permission from Gressani *et al.* (1998).

length, can be screened by this technique. Indeed, it has been suggested that 150 base pairs is the optimal length of a PCR fragment for SSCP analyses. For larger genes, this will require multiple sets of primers and PCR products to cover the entire gene. Finally, as discussed by Orita *et al.* (1989b), the sequence context and actual base substitution can markedly influence the sensitivity of the assay, making it difficult to thoroughly standardize the assay for the entire gene sequence. Other techniques, such as denaturing gradient gel electrophoresis (DGGE), conformation sensitive gel electrophoresis (CSGE), and heteroduplex analysis, have the potential to offer increased sensitivity and allow for analysis of larger fragments. In deciding on the appropriate post-PCR analytical techniques for your laboratory, careful attention should be paid to both the advantages and drawbacks of each technique.

2.10   Conclusions

The PCR remains an extremely powerful technique, and a replacement has yet to be found for its ability to amplify genetic targets for subsequent analysis. On the other hand, its apparent simplicity is deceptive, and can lead to an impression that it is not a reliable technique. Like any other biochemical assay, it demands an understanding of its molecular underpinnings and an attention to careful design. Primers must be carefully selected, contamination rigorously controlled, and reactions sometimes optimized to achieve a desirable signal-to-noise ratio. PCR experiments may be automated, but are not inherently high-throughput (particularly if gel analysis of products is required). Post-PCR analyses for gene mutations or polymorphisms likewise are not yet amenable to high-throughput automation. However, the combination of PCR with techniques like ASO and SSCP are among the most powerful tools that can be employed by individual basic and clinical research laboratories to screen for alterations in DNA sequence associated with genetic traits or disease pathogenesis. With the advent of new technologies described in subsequent chapters, PCR will continue to be the basis for a myriad of assays in pharmacogenetics in the future.

Acknowledgments

We would like to thank both past and present members of our laboratories who have contributed significantly to the research efforts of our programs: Manxia Fan, Kiersten Gressani, Sandra Leone-Kabler, Lisa Rollins, and Lisa Wessner.

References

Ahokas, H. and Erkkila, M.J., 1993, Interference of PCR amplification by the polyamines, spermine and spermidine. *PCR Methods Applic.*, 3, 65–68.
Beutler, E., Gelbart, T. and Kuhl, W., 1990, Interference of heparin with the polymerase chain reaction. *BioTechniques*, 9, 166.
Burns, W.C., Liu, Y.S., Dow, C., Thomas, R.J.S. and Phillips, W.A., 1997, Direct PCR from paraffin-embedded tissue. *BioTechniques*, 22, 638–640.
Cha, R.S. and Thilly, W.G., 1993, Specificity, efficiency, and fidelity of PCR. *PCR Methods Applic.*, 3, S18–S29.

Chevet, E., Lemaitre, G. and Katinka, M.D., 1995, Low concentrations of tetramethylammonium chloride increase yield and specificity of PCR. *Nucleic Acids Res.*, **23**, 3343–3344.

Chou, Q., Russell, M., Birch, D.E., Raymond, J. and Bloch, W., 1992, Prevention of pre-PCR mis-priming and primer dimerization improves low-copy-number amplifications. *Nucleic Acids Res.*, **20**, 1717–1723.

Cline, J., Braman, J.C. and Hogrefe, H.H., 1996, PCR fidelity of *pfu* DNA polymerase and other thermostable DNA polymerases. *Nucleic Acids Res.*, **24**, 3546–3551.

Comey, C.T., Jung, J.M. and Budowle, B., 1991, Use of formamide to improve amplification of HLA DQα sequences. *BioTechniques*, **10**, 60–61.

Cone, R.W. and Fairfax, M.R., 1993, Protocol for ultraviolet irradiation of surfaces to reduce PCR contamination. *PCR Methods Applic.*, **3**, S15–S17.

D'Aquila, R.T., Bechtel, L.J., Videler, J.A., Eron, J.J., Gorczyca, P. and Kaplan, J.C., 1991, Maximizing sensitivity and specificity of PCR by pre-amplification heating. *Nucleic Acids Res.*, **19**, 3749.

Dieffenbach, C.W. and Dveksler, G.S., 1993, Setting up a PCR laboratory. *PCR Methods Applic.*, **3**, S2–S7.

Dove, W.F. and Davidson, N., 1962, Cation effects on the denaturation of DNA. *J. Mol. Biol.*, **5**, 467–478.

Drummond, I.A., Madden, S.L., Rohwer-Nutter, P., Bell, G.I., Sukhatme, V.P. and Rauscher, F.J., III, 1992, Repression of the insulin-like growth factor II gene by the Wilms tumor suppressor WT1. *Science*, **257**, 674–678.

Dubeau, L., Chandler, L.A., Gralow, J.R., Nichols, P.W. and Jones, P.A., 1986, Southern blot analysis of DNA extracted from formalin-fixed pathology specimens. *Cancer Res.*, **46**, 2964–2969.

Duddy, S.K., Gorospe, S. and Beavins, M.R., 1998, Genetic analysis of multiple loci in microsamples of fixed paraffin-embedded tissue. *Toxicol. Sci.*, **46**, 317–323.

Dutton, C.M., Paynton, C. and Sommer, S.S., 1993, General method for amplifying regions of very high G+C content. *Nucleic Acids Res.*, **21**, 2953–2954.

Eckert, K.A. and Kunkel, T.A., 1990, High fidelity DNA synthesis by the *Thermus aquaticus* DNA polymerase. *Nucleic Acids Res.*, **18**, 3739–3744.

Emmert-Buck, M.R., Bonner, R.F., Smith, P.D., Chuaqui, R.F., Zhuang, Z., Goldstein, S.R., Weiss, R.A. and Liotta, L.A., 1996, Laser capture microdissection. *Science*, **274**, 998–1001.

Gelfand, D.H., 1989, *Taq* DNA polymerase, in Erlich, H.A. (ed.) *PCR Technology*, pp. 17–22, New York: Stockton Press.

Goelz, S.E., Hamilton, S.R. and Vogelstein, B., 1985, Purification of DNA from formalin fixed and paraffin embedded human tissue. *Biochem. Biophys. Res. Commun.*, **130**, 118–126.

Going, J.J. and Lamb, R.F., 1996, Practical histological microdissection for PCR analysis. *J. Pathol.*, **179**, 121–124.

Gressani, K.M., Rollins, L.A., Leone-Kabler, S., Cline, J.M. and Miller, M.S., 1998, Induction of mutations in Ki-*ras* and *INK4a* in liver tumors of mice exposed *in utero* to 3-methylcholanthrene. *Carcinogenesis*, **19**, 1045–1052.

Hartley, J.L. and Rashtchian, A., 1993, Dealing with contamination: enzymatic control of carryover contamination in PCR. *PCR Methods Applic.*, **3**, S10–S14.

Hongyo, T., Buzard, G.S., Calvert, R.J. and Weghorst, C.M., 1993, 'Cold SSCP': a simple, rapid and non-radioactive method for optimized single-strand conformation polymorphism analyses. *Nucleic Acids Res.*, **21**, 3637–3642.

Jack, W.E., Kucera, R.B. and Kong, H., 1992, Biochemical characterization of Vent(R) and Deep Vent(R) DNA polymerases. *The NEB Transcript*, **4**, 4–5.

Katz, E.D. and Haff, L.A., 1989, Effects of primer concentration and Taq DNA polymerase activity on yield of the PCR process. *Amplifications*, **3**, 8–9.

Kreader, C.A., 1996, Relief of amplification inhibition in PCR with bovine serum albumin or T4 gene 32 protein. *Appl. Environ. Microbiol.*, 62, 1102–1106.

Leone-Kabler, S., Wessner, L.L., McEntee, M.F., D'Agostino, R.B., Jr. and Miller, M.S., 1997, Ki-*ras* mutations are an early event and correlate with tumor stage in transplacentally-induced murine lung tumors. *Carcinogenesis*, 18, 1163–1168.

Ling, L.L., Keohavong, P., Dias, C. and Thilly, W.G., 1991, Optimization of the polymerase chain reaction with regard to fidelity: modified T7, Taq and Vent DNA polymerases. *PCR Methods Applic.*, 1, 63–69.

Lyons, J., 1990, Analysis of *ras* gene point mutations by PCR and oligonucleotide hybridization, in Innis, M.A., Gelfand, D.H., Sninsky, J.J. and White, T.J. (Eds) *PCR Protocols: A Guide to Methods and Applications*, pp. 386–391, San Diego: Academic Press.

Miller, M.S., Baxter, J.L., Moore, J.W., Lewis, J.D. and Schuller, H.M., 1994, Molecular characterization of neuroendocrine lung tumors induced in hamsters by treatment with nitrosamines and hyperoxia. *Intl J. Oncol.*, 4, 5–12.

Mullis, K.B., 1990, The unusual origin of the polymerase chain reaction. *Sci. Am.*, 262, 56–65.

Mullis, K.B., 1991, The polymerase chain reaction in an anemic mode: how to avoid cold oligodeoxyribonuclear fusion. *PCR Methods Applic.*, 1, 1–4.

Orita, M., Iwahana, H., Kanazawa, H., Hayashi, K. and Sekiya, T., 1989a, Detection of polymorphisms of human DNA by gel electrophoresis as single-strand conformation polymorphisms. *Proc. Natl Acad. Sci., USA*, 86, 2766–2770.

Orita, M., Suzuki, Y., Sekiya, T. and Hayashi, K., 1989b, Rapid and sensitive detection of point mutations and DNA polymorphisms using the polymerase chain reaction. *Genomics*, 5, 874–879.

Pomp, D. and Medrano, J.F., 1991, Organic solvents as facilitators of polymerase chain reaction. *BioTechniques*, 10, 58–59.

Prince, A.M. and Andrus, L., 1992, PCR: how to kill unwanted DNA. *BioTechniques*, 12, 358–360.

Reynolds, S.H., Wiest, J.S., Devereux, T.R., Anderson, M.W. and You, M., 1992, Protooncogene activation in spontaneously occurring and chemically induced rodent and human lung tumors, in Klein-Szanto, A.J.P., Anderson, M.W., Barrett, J.C. and Slaga, T.J. (Eds) *Comparative Molecular Carcinogenesis*, pp. 303–320, New York: Wiley-Liss.

Rollins, L.A., Kabler-Leone, S., O'Sullivan, M.G. and Miller, M.S., 1998, Role of tumor suppressor genes in transplacental lung carcinogenesis. *Molec. Carcinogenesis*, 21, 177–184.

Ruano, G., Pagliaro, E.M., Schwartz, T.R., Lamy, K., Messina, D., Gaensslen, R.E. and Lee, H.C., 1992, Heat-soaked PCR: an efficient method for DNA amplification with applications to forensic analysis. *BioTechniques*, 13, 266–274.

Rychlik, W., Spencer, W.J. and Rhoads, R.E., 1990, Optimization of the annealing temperature for DNA amplification *in vitro*. *Nucleic Acids Res.*, 18, 6409–6412.

Saiki, R.K., 1989, The design and optimization of the PCR, in Erlich, H.A. (Ed.) *PCR Technology*, pp. 7–16, NY: Stockton Press.

Sardelli, A.D., 1993, Plateau effect – understanding PCR limitations. *Amplifications*, 9, 1–5.

Sarkar, G., Kapelner, S. and Sommer, S.S., 1990, Formamide can dramatically improve the specificity of PCR. *Nucleic Acids Res.*, 18, 7465.

Sheffield, V.C., Beck, J.S., Kwitek, A.E., Sandstrom, D.W. and Stone, E.M., 1993, The sensitivity of single-strand conformation polymorphism analysis for the detection of single base substitutions. *Genomics*, 16, 325–332.

Shibata, D., Hawes, D., Li, Z.-H., Hernandez, A.M., Spruck, C.H. and Nichols, P.W., 1992, Specific genetic analysis of microscopic tissue after selective ultraviolet radiation fractionation and the polymerase chain reaction. *Am. J. Pathol.*, 141, 539–543.

Verlaan-de Vries, M., Bogaard, M.E., van den Elst, H., van Boom, J.H., van der Eb, A.J. and Bos, J.L., 1986, A dot-blot screening procedure for mutated *ras* oncogenes using synthetic oligonucleotides. *Gene*, **50**, 313–320.

Wan, C.Y. and Wilkins, T.A., 1993, Spermidine facilitates PCR amplification of target DNA. *PCR Methods Applic.*, **3**, 208–210.

Wessner, L.L., Fan M., Schaeffer, D.O., McEntee, M.F. and Miller, M.S., 1996, Mouse lung tumors exhibit specific Ki-*ras* mutations following transplacental exposure to 3-methylcholanthrene. *Carcinogenesis*, **17**, 1519–1526.

Wright, D.K. and Manos, M.M., 1990, Sample preparation from paraffin-embedded tissues, in Innis, M.A., Gelfand, D.H., Sninsky, J.J. and White, T.J. (Ed.) *PCR Protocols: A Guide to Methods and Applications*, pp. 386–391, San Diego: Academic Press.

Wu, D.Y., Ugozzoli, L., Pal, B.J., Qian, J. and Wallace, R.B., 1991, The effect of temperature and oligonucleotide primer length on the specificity and efficiency of amplification by the polymerase chain reaction. *DNA Cell. Biol.*, **10**, 233–238.

Zhang, L., Cui, X., Schmitt, K., Hubert, R., Navidi, W. and Arnheim, N., 1992, Whole genome amplification from a single cell: implications for genetic analysis. *Proc. Natl Acad. Sci., USA*, **89**, 5847–5851.

3

DNA OLIGONUCLEOTIDE MICROARRAYS: APPLICATION TO GENETIC POLYMORPHISM ANALYSIS

Maureen T. Cronin

3.1 Background

The Human Genome Project and related programs, both public and private, have accelerated the pace at which we can accumulate human genetic information and enabled the subsequent increase in our capacity to associate a particular gene sequence with a specific physiological function. The medical genetics and molecular pathology fields arose from access to such information. However, it is clear that single, acutely defined disease phenotypes resulting from a single genetic variant with Mendelian inheritance are the exception rather than the rule among disease of the human population. Similarly, even for diseases associated with a single gene, it is rare within the general, outbred population for a disease phenotype to result from a single mutation. More typically, disease phenotypes present as a continuum between the extremes defined by a healthy, ideal phenotype and a fully involved disease phenotype. The degree of variability observed between the phenotype extremes is recognized to be a consequence of a spectrum of genetic polymorphisms in both the primary candidate gene and all the genes modulating its expression. Together they produce individual disease phenotypes. As our understanding of genetics in both man and animal model systems improves, we have come to accept the idea that many aspects of human disease are genetically determined. In addition, variability across disease phenotypes and variation in individual disease susceptibility are at least partially a consequence of human population genetic variability. As we continue to elucidate specific genotype–phenotype relationships and to integrate increasingly complex genotype–phenotype correlates, our capacity to predict disease susceptibility within human populations should continually improve, and ultimately may be extended to individuals (Lander and Schork, 1994; Thomson, 1995; Dykes, 1996; Lifton, 1996).

Hundreds of disease susceptibility genes have been identified to date. It has become evident that useful genetic profiling depends on having the capability to analyze large numbers of polymorphic sites distributed widely throughout the genome (Schafer and Hawkins, 1998). A number of technical approaches to this

41

type of polymorphic analysis have been developed. These include such techniques as polymerase chain reaction (PCR) with restriction fragment length polymorphism (RFLP) analysis; microsatellite marker PCR with products gel-resolved; allele-specific oligonucleotide (ASO) PCR with the detection specificity built into either a primer or a polymorphism specific hybridizing oligonucleotide; single-strand conformation polymorphism (SSCP) analysis; denaturing gradient gel electrophoresis (DGGE); chemical or enzymatic mismatch cleavage (CCM); and direct Sanger sequencing (Hall *et al.*, 1996; Mifflin, 1996; McKenzie *et al.*, 1998). However, throughput for all of these techniques saturates too quickly to accommodate the high level of parallel analysis required for executing broad genetic profiling efficiently. DNA oligonucleotide microarrays stand out among the numerous technical innovations that have been generated to address this problem. These two-dimensional collections of immobilized DNA sequences offer the option of performing large numbers of genetic analyses in parallel using hybridization-based assays. Their success has made them one of the most popular approaches to performing highly parallel genetic analysis (Marshall and Hodgson, 1998; Ramsay, 1998).

3.2   Essential features of microarray-based assays

3.2.1   *Requirement for surface-bound DNA probes*

DNA microarrays, also commonly called DNA chips, are two-dimensional substrates with collections of DNA probes synthesized *in situ* or immobilized by one of a variety of methods on the substrate surface in defined (addressable) locations. DNA arrays are commonly compared with electronic chip computer microprocessor components, hence the name, DNA chips. This parallel continues in that arrays are composed of individual probe sequences, each confined to a single location called a feature. Each feature has unique X–Y coordinates within the array and, as a result, each feature has an array address that can be associated with its unique nucleotide sequence. Therefore, a target DNA sequence hybridized to any array feature can be decoded since its sequence is the complement of the probe sequence corresponding to that array feature. These aspects of a DNA microarray are illustrated in Figure 3.1.

A wide variety of approaches to DNA array design and fabrication have resulted in a comparably wide variety in array complexity and density. The exact format and probe densities for any array are intimately related to its method of manufacture and its intended analytical purpose. The total number of array features, which are the physical locations on the array where different probe sequences reside, may range from less than 100 to hundreds of thousands per cm^2. Final array dimensions may range from less than 0.25 cm^2 to large membranes with surfaces of 100 cm^2 or more. These characteristics are also intimately dependent on how the array is manufactured and how the array will be used in a hybridization assay (Marshall and Hodgson, 1998).

3.2.2   *Requirement for nucleic acid hybridization target*

The most common method of using microarrays is to label the hybridization target prior to the hybridization reaction, then image the array using a detection

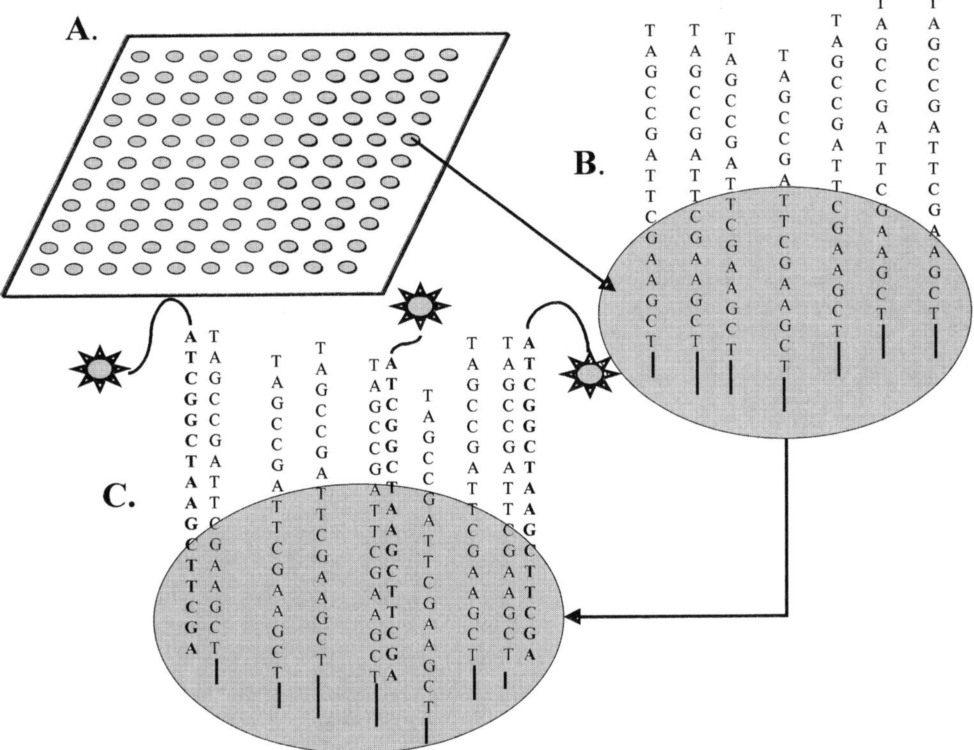

Figure 3.1 Elements of a DNA microarray. **A.** Schematic of a two-dimensional DNA microarray with round features. **B.** Each individual feature is populated with oligonucleotides or cDNA of one sequence. **C.** Hybridization with a labeled target reveals the target's sequence since it is complementary to the oligonucleotides in hybridized features.

instrument compatible with the label that was incorporated into the target. Some current detection formats are isotope based, as were most of the early array hybridization detection strategies, particularly those associated with membranes. Sensitive phosphorimagers with high spatial resolution now substitute for X-ray film to image isotope-based array hybridization results.

DNA microarray detection methods and imaging systems that are currently available do not have sensitivity capable of detecting single copy target hybridization. Therefore, DNA microarray hybridization detection and analysis depend on using one of the alternative amplification strategies available to generate a significant quantity of labeled target. The quantity must be large enough to achieve a hybridization concentration that generates a hybridization signal strong enough to enable imaging hybridized sites on an array. Alternative target amplification methods include the PCR, *in vitro* transcription (IVT), and the oligonucleotide ligation assay (OLA), among others (Nollau and Wagener, 1997).

In general, microarray detection strategies depend on fluorescence from dyes incorporated into the hybridization target either during the target amplification procedure or as a post-amplification processing step (Tamary *et al.*, 1994). Early in array development the fluorescein and rhodamine families of related chemical derivatives that were developed for automated Sanger sequencing were the only fluorescent dyes widely commercially available for fluorescently labeling nucleic acids (Smith *et al.*, 1986). Since the excitation absorbance maxima for these dyes (approximately 490 nm and 525 nm) are spectrally well matched to the principal output lines of argon lasers (488 nm and 514.5 nm), many detection instruments were designed around these dye/laser sets. Evolution of solid-state lasers and a growing variety of alternative nucleic acid compatible dye chemistries have recently given rise to a much more diverse assortment of laser/dye pair options. One example is the cyanine dye CY5 (absorbance maxima 650 nm) paired with a helium–neon laser or a solid-state red diode laser (Zhu *et al.*, 1994).

An alternative to directly incorporating fluorescent dye-conjugated nucleotide substrates into target DNA with enzymes is to substitute biotin- or hapten-conjugated nucleotides. Once target DNA labeled with biotin or a hapten is hybridized to an array, dye-conjugated streptavidin or antibodies can be used to visualize hybridized array features. This strategy offers greater flexibility than using directly dye-labeled nucleotides since nucleotide incorporation can then be managed as a standard, generic protocol, while the dye-staining step may be varied to include a wide variety of dye labels. This labeling method may be extended further to include using streptavidin or antibodies conjugated with an enzyme such as alkaline phosphatase or horseradish peroxidase to target biotin- or hapten-conjugated nucleotides incorporated into the target (Cherry *et al.*, 1994). These labeled sites are then detected with a chromophore or chemiluminescent enzyme substrate and a colorimetric or light-sensitive imaging system (Cano *et al.*, 1992). Using fluorescent labels and sensitive imaging instruments, attomoles of hybridized DNA can be detected routinely and optimization of labeling and detection will permit detection of even smaller quantities of DNA (Mathies and Peck, 1990).

In addition to choosing a labeling–detection strategy for preparing the hybridization target there is a second important consideration to complete the target preparation process. Unless the target is directly generated as a short fragment (approximately 50 bp or less) it is necessary to devise a strategy to fragment the target into that size range. This requirement is due to a number of properties of solution-phase targets and their immobilized complements on microarray surfaces. One factor is that hybridization kinetics are much faster for solution targets that are nearly the same length as the immobilized counterpart probes. In addition, as the length of a target in solution increases, the possible opportunities for the target to form stable sequence-specific, internal structure increases. Any target involved in these structures is essentially unavailable for surface hybridization. Fragmentation minimizes the frequency and stability of secondary target structure and therefore tends to keep a greater percentage of the target available for array hybridization (Chan *et al.*, 1995). A second benefit of target fragmentation is better label-to-fragment distribution. Gross asymmetry in the ratio of the number of labels per target fragment can result in excess label intensity in some regions of an array and loss of signal in others. The ratio is most easily optimized when post-PCR, post-fragmentation end-labeling schemes are employed. The closer to one the label-to-fragment ratio is, the more uniform the hybridization pattern and the more easily the pattern is interpreted.

3.2.3 Requirement for hybridization detection scheme

As pointed out in Section 3.2.2, an essential part of target preparation is to include a target label to make it possible to detect the hybridized target. The label may be isotopic, fluorescent, a colored substrate or chemiluminescent. Regardless of the reporter molecule selected for the detection strategy, a method must be developed to incorporate the label into the hybridization target for it to be detected. This may be done during enzymatic amplification by using amplification primers modified at their 5′ ends with fluorescent dyes, biotin or a hapten recognized by an antibody. Alternatively, deoxynucleotide triphosphates modified with a fluorescent dye or with biotin or a hapten may be included in the amplification reaction mixture for direct incorporation during enzymatic amplification.

Hybridization target labeling can also be carried out as a post-amplification process. This strategy often proves to be the most flexible, economic and efficient, since it is usually problematic to optimize PCR and *in vitro* transcription reactions that include label-modified reagent incorporation. In addition, once these reactions are optimized, they are not easily altered to include new targets or priming strategies as the hybridization assay evolves to incorporate newly emerging genetic information. Post-amplification labeling carried out as a batch process on the entire pool of PCR products intended for a single-array hybridization conserves reagents, improves labeling uniformity and often generates enough labeled material for multiple array hybridization experiments.

There are many alternatives available among detection instruments. These include luminometers to detect chemiluminescence and radiochemiluminescence, spectrophotometric imagers to detect the colored substrates for enzymes such as horseradish peroxidase and alkaline phosphatase, phosphorimagers to detect radioisotope decay and spectrofluorometers to detect fluorescent light emission. Spectrofluorometric imaging can also be modified to be sensitive to fluorescence polarization or variations in fluorescence lifetime. The principal requirement in selecting a detector is that it be well matched to the properties of the reporter used to label the hybridization target. In addition, the impact on hybridization strategy must be considered. Confocal detectors support real-time imaging of hybridization on arrays that are actively interfaced with a hybridization solution while others such as phosphorimagers and fluorimagers may require direct access to a washed and dried hybridized array surface.

3.2.4 Requirement for hybridization analysis algorithms

Automated algorithmic interpretation of the complex hybridization patterns generated by microarray hybridization is a clear necessity to permit full exploitation of the highly parallel analysis that DNA oligonucleotide microarrays provide. However, designing analysis algorithms that provide highly reliable interpretation for array output is not a trivial undertaking. Full normalization of array hybridization signals is generally not completely achievable. This is due to intrinsic variability in enzymatic hybridization target preparation and labeling and to the wide range of intrinsic hybridization optima present on a complex microarray (Chan *et al.*, 1995). As a result, algorithms for interpreting the hybridization data must tolerate a fairly wide range of absolute hybridization signals, and an equally wide range of background signals. Perfect match signal to background mismatch signal ratios will vary across an array even after considerable effort has gone into assay optimization. However, robust algorithms have been derived successfully. In particular, approaches that incorporate internal hybridization controls in the assays and sets of control sequences in the array probes sets have shown some success (Lockhart *et al.*, 1996). In addition, algorithms can be "trained" to recognize a reference sequence hybridization pattern that can be used for global signal normalization across an array. Training requires many separate hybridization experiments using independently prepared and labeled reference samples. Once an algorithm is "trained" it can flag hybridization results that deviate from that "learned" pattern.

Another approach that has proved to be successful is to design the hybridization assay with an internal reference standard that competes with the unknown sample for hybridization sites on the array. If the samples are labeled differentially, for instance with two different fluorescent dyes, differences in sequence between the test sample and the reference sample will appear as areas on the fluorescent image where the spectral ratio of the two dyes deviates from unity (Hacia *et al.*, 1996). These assays are most successful with single-stranded hybridization targets, whereas other approaches permit hybridizations with labeled, double-stranded PCR products to be analyzed.

The high information capacity of microarrays can be exploited to design analyses based on some level of information redundancy (Cronin *et al.*, 1996). Cross-referencing hybridization consistency among multiple related probe sets can improve sequence assignment reliability. However, regardless of the analytical approach chosen, it is essential that the algorithm is tested rigorously on sequenced or otherwise genotyped sample sets different from those used to develop the algorithm. In addition, due to the high potential for intrinsic variability in the target preparation protocols for these assays, it is important to test and retest the algorithm on a variety of sample sets prepared from more than one source.

As arrays and pools of hybridization targets both become more complex, analyzing images of hybridized arrays also becomes increasingly complex. General strategies for both gene expression analysis and genotyping analysis for specific polymorphisms usually involve comparing hybridization results among groups of related oligonucleotide probes. These probes are designed to discriminate differences among related genes or polymorphic variants of individual genes as well as perfectly complementary hybridizations from mismatched hybridizations. Just as with other more traditional types of clinical chemistry measurements, comparative analysis of hybridization intensities among probes matched and mismatched to the intended targets along with non-specific background subtraction depends on having well-designed, well-trained analysis algorithms (Nollau and Wagener, 1997). This type of algorithm development relies on having sets of genotyped or otherwise characterized samples available during hybridization assay development. It is also dependent on having robust, consistent assay conditions to minimize interassay variability. In addition, the assay should be designed with internal controls that the algorithm can use to qualify the results relative to an accepted window of assay performance.

3.3 Two fundamental types of DNA microarrays

3.3.1 *Microarrays constructed using presynthesized DNA*

DNA microarrays can be grouped into two fundamental types based on their method of manufacture. One type of array, commonly used for differential mRNA expression analysis, requires that cDNA probes be synthesized off-line from either cDNA libraries or from isolated mRNA and then purified. Once these libraries of probes have been collected, reproduced on a scale large enough to make many arrays and cataloged, they are "spotted" onto the array surface and immobilized either by passive ionic interactions or with a covalent coupling chemistry (Schena *et al.*, 1995). The choice of substrate material used for the array surface and how it is processed prior to array synthesis direct which immobilization chemistry is best employed. This process also influences the hybridization and detection strategy selections for the final assay format. Substrates most commonly selected as array supports are glass and silica, although plastics, nylon and nitrocellulose membranes have also been used successfully. One significant advantage to making this type of array is that prior knowledge of immobilized probe sequences is not required. Uncharacterized expressed sequence tags (ESTs), sequence tagged sites (STSs) and unsequenced cDNA clone libraries can be used as array probes. Once "hits" arise from differential gene expression studies using these arrays, clones of interest may be isolated and characterized more fully. In addition, the overall probe generation and arraying processes are well suited to interface with high-throughput robotic processes.

Another permutation of this array construction strategy is that developed by Drmanac and colleagues. Their approach requires collecting the target sequences to be analyzed and immobilizing them as an addressable array spotted onto a membrane surface. These sequences are analyzed by hybridizing them sequentially with pools of labeled oligonucleotides whose sequences are known. Each target sequence is reconstructed algorithmically by assembling the overlapping set of oligonucleotide complements to its sequence. This is the set of oligonucleotides that give positive signals during the sequential hybridization process (Drmanac *et al.*, 1990). One limitation of this method is the short target sequence length that can be reconstructed by the algorithms before too many mathematical reconstructions become possible and the target cannot be reconstructed faithfully.

The multiplex allele-specific diagnostic assay (MASDA) is a similar but more directed version of the same approach to array hybridization analysis. MASDA utilizes oligonucleotide hybridization to analyze target DNA sequences. Products of multiplex PCR amplification are immobilized on a solid support and hybridized with a single pool of ASO probes. Sequence-specific band patterns generated by chemical or enzymatic sequencing of the bound ASO probes easily identifies specific mutations present in a given sample (Shuber *et al.*, 1997).

Finally, arrays of presynthesized DNA can be constructed using synthetic oligonucleotides rather than enzymatically generated products (Zhang *et al.*, 1991). This approach also requires that a collection of probes be synthesized, purified and characterized prior to constructing the array. However, the process is more easily managed with oligonucleotides than with cDNA probes because of the easy availability of automated DNA synthesis and parallel automated purification methods. In addition, oligonucleotides can be synthesized with modifications that make chemical coupling to an array surface robust and reliable. An additional advantage is that either the 3' or the 5' end of an oligonucleotide can be used for surface coupling. Strategies that couple the oligonucleotide to the support at the 5' end leave the free 3' end available to participate in DNA polymerase or ligase-directed reactions that support *in situ* dye-terminator sequencing and primer extension reactions (Nikiforov *et al.*, 1994).

3.3.2 *Microarrays constructed using* in situ *DNA synthesis*

The second type of DNA microarray is constructed by synthesizing DNA oligonucleotide probes directly on the substrate surface. Fairly complex arrays can be made by synthesizing and purifying collections of modified oligonucleotides for covalent immobilization on activated membranes or glass surfaces. However, *in situ* oligonucleotide synthesis is a far more efficient method for generating complex arrays, particularly when many different array designs incorporating oligonucleotides of different lengths and composition are required. DNA microarrays for every type of nucleic acid analysis including quantitating specific mRNA expression levels in cells, DNA "resequencing", polymorphism discovery and genotyping can be fabricated using *in situ* DNA oligonucleotide synthesis.

Three different approaches have been developed to synthesize DNA microarrays *in situ*. The first, invented by Southern, adopts the standard phosphoramidite chemistry commonly used on automated DNA synthesizers to synthesize oligonucleotides in a chamber abutting a two-dimensional substrate rather than on controlled pore glass (CPG) beads in a closed column (Southern *et al.*, 1992). The second, invented by Fodor and colleagues employs entirely new nucleotide chemistry. Phosphoramidite monomers prepared with photocleavable protecting groups rather than the common acid-labile protecting groups are used in combination with a photolithographic masking strategy in a light-directed synthesis method (Fodor *et al.*, 1991; Pease *et al.*, 1994). Finally, Brennan and coworkers combine drop-on-demand, standard amidite reagents delivered from piezoelectric nozzles or ink-jet heads with surface tension-defined substrates to make addressable DNA microarrays (Cronin *et al.*, 1998a,b).

3.4 Analyzing genetic polymorphisms using DNA microarrays

3.4.1 *General principles*

Deciphering the human genome is an undertaking far more complex than assembling a primary sequence of its three billion base pairs. Functional genomics seeks to identify, map, sequence and assign function to the estimated 70 000–100 000 genes distributed throughout the genome. This effort is supported by a parallel effort underway to generate a high-resolution physical map of the entire genome. An even more current focus of this effort is to characterize human polymorphic diversity. A full catalog of characterized, human diversity is likely to be the most useful result of the human genome project. Although every human has three billion base pairs of DNA, sequences for any two individuals are estimated to differ from each other, on average, at 1%, or 3 000 000 individual nucleotide positions. Although this is a genome-wide, average value, single-nucleotide polymorphisms (SNPs) are not uniformly distributed throughout the genome (Nickerson *et al.*, 1998). Intergenic region diversity approaches 10% compared with approximately 1% in coding regions. Presumably this reflects differential selective pressure to conserve coding sequences as compared with intergenic sequences.

Nucleotide differences between different human genomes are considered to be polymorphisms rather than mutations when they are maintained in the population at frequencies ≥1%. Primary sequence variability represented by SNPs, nucleotide insertions and nucleotide deletions in human subpopulations is only one important aspect of human variability. Frequencies of individual polymorphisms and groups of polymorphisms that occur together (haplotypes) vary widely among defined segments of the human population. Two copies of a single gene that differ from one another in primary sequence are called alleles of the same gene. These alleles are not uniformly distributed throughout the human population but vary among racial groups and population groups genetically isolated by physical or social mechanisms. Some allele frequency distributions are also associated with a particular environmental selective pressure. An example is the heterozygous form of the sickle-cell trait, which results in a protected phenotype where malaria is endemic (Schafer and Hawkins, 1998).

The ultimate result of genetic variability in primary nucleotide sequence combined with variable allele distribution is tremendous phenotypic variability within the human population. Early studies in medical genetics focused on diversity that could be identified in the context of clearly measurable phenotypes. These were most often monogenetic traits showing Mendelian inheritance such as the sickle-cell trait. Sickle cell is characterized by a single nucleotide change in a single globin gene that results in a clearly defined disease phenotype. This is in sharp contrast to diseases such as the thalassemias where multiple polymorphisms have been defined in multiple globin gene alleles. The resulting diseases, the thalassemias, cover a wide range of phenotypes.

Polygenic diseases are complex traits that result from the combined inheritance of many predisposing genes and, in addition, probably require an environmental exposure to trigger actual manifestation of the disease phenotype. These diseases affect large percentages of the human population, although subpopulations can generally be identified that have unusually high or low incidence of such diseases as diabetes, high blood pressure, asthma, schizophrenia, arteriosclerosis, Alzheimer's and other cardiovascular diseases. Small populations demonstrating high incidence rates for polygenic diseases are being studied to refine phenotypes, collect genetic polymorphism data and study environmental influence on phenotype expression. It is hoped that this integrated, epidemiological approach to studying complex genetic traits will yield a picture of disease predisposition and progression and identify triggering environmental conditions. Ideally this will suggest therapeutic interventions that may prevent or cure such diseases (Dykes, 1996; Schafer and Hawkins, 1998).

A general population-based approach has developed for discovering SNPs associated with a complex trait. It begins with identifying a population with affected individuals displaying the phenotype of interest and unaffected individuals. Genomic samples for both sets of cases are collected from the population and mapped using short tandem repeat or microsatellite DNA markers. Comparing mapping results among affected and unaffected individuals leads to discovery of areas in the genome that are common to the disease phenotype but not common to the unaffected population. These sites are then examined at high resolution for known candidate genes, expressed sequence tagged sites mapped to the area and open reading frames that might define new candidate genes. Finally, candidate genes must be cloned, sequenced and expressed to discover their functions and identify whether any polymorphisms present are likely to be responsible for the observed phenotype variations (Hall *et al.*, 1996).

3.4.2 SNP discovery using microarrays

Today, the most commonly used markers for genomic mapping are variable copy number, short tandem repeat (VNTR) DNA sequences. These are short, repetitive DNA sequences of two (dinucleotide repeats) or more nucleotides that vary with respect to the total copy number of sequence repeats from individual to individual. More than 10 000 VNTR markers, also called microsatellite markers, are known and can easily be used to sort alleles from different individuals. PCR-based assays are used to assess the repeat sites. The products generated using PCR primers designed to prime unique sequences flanking the VNTR sites are resolved on agarose or acrylamide gels.

The frequency and distribution of SNPs makes them prime candidates for high-resolution genomic mapping markers. Consequently, SNP discovery is a high priority for genome mapping as well as for characterizing diversity in medical genetics genotype–phenotype correlation efforts. SNP markers are expected to eventually replace VNTR markers for high-resolution genome mapping. SNPs have the added advantage that they are well suited to the highly parallel hybridization analysis offered by DNA microarrays. VNTR analysis requires gel resolution to interpret PCR products, a method that quickly saturates when there are many sites to analyze in many samples. However, it is anticipated that matrix-assisted laser desorption ionization time-of-flight mass spectrometry (MALDI-TOF MS) can be optimized to provide an effective alternative for analyzing VNTR PCR products (Koster *et al.*, 1996).

Microarray-based SNP discovery is conceptually similar to single-strand conformation polymorphism (SSCP) analysis and other types of heteroduplex detection methods. All of these methods assume that the genomic site under investigation has been characterized well enough to design unique PCR primers that flank the site. In addition, microarray design requires knowing the primary sequence of each site to be analyzed. Alternatively, SSCP heteroduplex analysis depends on having reference samples that form homoduplex and heteroduplex reference standards for the gel analysis. Heteroduplex analysis requires PCR products to be generated from both the unknown samples and reference samples for mixed denaturation, annealing and gel analysis. Samples showing evidence of heteroduplex formation can be further characterized by direct sequencing (Hacia *et al.*, 1996; Winzler *et al.*, 1998).

Microarray-based SNP discovery requires the sequence targeted for analysis to be reconstructed as an array of overlapping oligonucleotide probes. The target sequence is then fragmented, labeled and hybridized to the array. A target perfectly complementary to the oligonucleotide array will result in essentially uniform hybridization across the entire array. A target with SNP sites in its sequence will not be complementary to the subset of oligonucleotide probes overlapping the SNP site. The result will be a loss of hybridization signal at those probe sites in the array. Arrays may be designed to incorporate greater levels of complexity and consequently yield more information from each hybridization analysis. One example is to include alternative probe sets that may be fully complementary to anticipated SNP sequences that may occur in a target. The most complex array possible is a design including hybridization probe sets for all possible nucleotide substitutions in an intended target sequence (Wang *et al.*, 1998).

Generally, once a putative polymorphism has been identified in a target sequence, the site is retested in a collection of genomic DNA samples so that frequencies for each allele can be estimated. It is also generally checked against a kindred pedigree so that the polymorphism inheritance can be ascertained. This can be done using a DNA microarray with probes designed for genotyping that nucleotide position, or it may be confirmed by an alternative method such as Sanger dye terminator sequencing.

3.4.3 SNP genotyping using microarrays

Once the polymorphic sites in a target sequence have been identified and confirmed, hybridization microarrays with probe sets designed specifically to detect the entire set of polymorphisms may be applied to supporting both high-throughput and routine genotyping needs. Although genotyping of this type is more typically done in research settings, examples of clinical applications for polymorphism genotyping are becoming more common. Clinical screens for identifying carriers of mutant forms of such genes as cystic fibrosis (CFTR), the globin genes, and dystrophin are all available. Viral and bacterial gene analysis for drug susceptibility genotype is increasingly becoming incorporated into therapeutic approaches for treating individuals infected with organisms such as HIV and *Mycobacterium tuberculosis*.

DNA microarrays made for detecting characterized mutations and polymorphisms are generally designed very differently from arrays made for discovering genetic variability. Polymorphism discovery demands that arrays be sensitive to a wide range of sequence possibilities with every nucleotide in a target sequence considered to be a candidate site for a polymorphism. In contrast, genotyping arrays contain a smaller set of specific probes since they are designed to test which sequences out of a discrete number of possibilities may be present in a sample. Frequently, additional information, such as the expected frequency for each polymorphism in a defined subpopulation or the population at large and the frequency at which any subset of polymorphisms tends to occur together on the same gene allele, may also be known. All of this information taken together enables array and hybridization assays to be constructed to provide very high-confidence genotyping (Cronin *et al.*, 1996).

3.4.4 Detecting heterozygous polymorphisms using microarray hybridization

DNA microarrays made to provide genetic screening must also support genetic heterozygote discrimination for mutant allele carrier identification. Carrier screening involves identifying individuals with normal phenotypes carrying one mutant allele and one normal allele of a gene. Hybridization to oligonucleotide arrays capable of detecting all possible forms of a gene in parallel is the preferred way to identify heterozygous genotypes. Using this approach, a specific positive hybridization signal will develop for each sequence present in a sample. Sequencing a heterozygous genotype, which is a mixed template, in a standard Sanger reaction may provide only a very subtle indication of when a heterozygous sequence is present. Routine heterozygote discrimination from electropherogram outputs resulting from such mixed template sequencing reactions is often unreliable. Alternatively, hybridization to an array designed to detect both sequence possibilities is unequivocal.

Several probe design approaches have proven to be successful for making microarrays that demonstrate high-quality polymorphism detection in hybridization assays. These may be very simple, such as a single oligonucleotide probe complementary to each of two different polymorphic sequences. Alternatively, they may be complex sets of oligonucleotides designed to detect each known variant gene form on both DNA strands. In any case, it is important to optimize the assay performance for detecting each variant with similar maximum intensity hybridization signals. If this is not done, there is a risk that low-intensity signals generated from weakly hybridized targets which are only partially complementary to the gene alleles present may be misinterpreted as positive heterozygote alleles. This may require some empirical optimization after applying obvious strategies to normalize signals, such as setting an acceptable window of melting temperature (T_{m}) values for each probe in a set and centering the polymorphism detection site in the oligonucleotide probe sequence. Depending on the overall sequence context, a polymorphism may be most efficiently detected when it is positioned somewhat offset from the center of an oligonucleotide probe. Just as it is important to optimize the hybridization signal intensity within a related probe set for all of the polymorphisms to be detected at a particular sequence site, it is also important to optimize the array globally for signal uniformity using a given set of assay conditions. If this is not the case, the goal of high-confidence, highly parallel genotyping may be compromised (Cronin *et al.*, 1996; Hacia *et al.*, 1996).

3.4.5 Detecting polymorphisms in mixed template samples using microarrays

A problem similar to detecting heterozygote cases in genomic DNA is detecting polymorphisms that occur at low frequency within a heterogeneous genetic population. The two principal areas where this problem occurs are in cancer genetics, where genotype characterization of heterogeneous tumor material may be desired, and in infectious disease, where genetically heterogeneous bacterial or viral isolates may need to be scanned for drug susceptibility genotype (Kozal *et al.*, 1996).

Approaches to providing reliable genotyping under these conditions include stratifying the source material either by cloning or microdissection. Alternatively, modifying the hybridization assay to increase its sensitivity to minor species present in a mixed sample may be successful.

3.4.6 Microsatellites, insertions, deletions: limitations to microarray genetic analysis

Although DNA microarray hybridization is very well suited to SNP discovery and detection, it may not be as well suited to detecting other types of genetic polymorphisms. Insertions, deletions, and repetitive DNA sequences, such as microsatellite sequences, provide targets that often result in ambiguous, difficult-to-interpret patterns when hybridized to simple oligonucleotide microarrays. This is generally because stable duplexes with loop-outs in the target or the probe may form for insertions and deletions under standard hybridization conditions. In addition, repetitive target sequences may form stable duplexes with surface probes in many different off-set conformations that make it impossible to discern the number of repeats present in the original sample. The possibility of heterozygosity for the insertion, deletion or repeat makes the interpretation even more complex and the probability of correctly assigning a genotype even more remote.

Nevertheless, high-density, complex arrays of oligonucleotides have been designed that at least partially obviate the difficulties of discovering insertion and deletion polymorphisms. These arrays rely on dense families of closely related, overlapping probe sequences that represent the target sequence with a high degree of redundancy. Target sequences with even single base insertions and deletions are rarely able to hybridize to all of the complementary probes with the same efficiency as the native sequence. Consequently, comparative hybridization between a native target and a test target with an insertion or a deletion will generally identify a subset of probes that show loss of signal with the mismatched target.

Although discovery of insertion and deletion polymorphisms is problematic using hybridization assays, this is not the case for detecting known insertion and deletion polymorphisms. It is generally possible to design probe sets for an oligonucleotide array that are capable of detecting alleles with known insertions and deletions with very sensitive discrimination. ΔF508, a three-nucleotide deletion in exon 10 of the CFTR gene, and the most common CFTR mutation, is an excellent example. Probe-based hybridization assays to detect ΔF508 have been successful in a number of formats, including microarrays (Cronin *et al.*, 1996; Hacia *et al.*, 1996).

3.4.7 High-throughput genotyping applications using microarrays

Once polymorphism discovery has been carried out in any particular target sequence, a high-throughput genotyping array can be designed to carry out routine genotyping for the polymorphisms of particular interest in large sample sets. As described in previous sections, arrays designed for polymorphism discovery are generally more redundant than genotyping arrays. Genotyping requires only the particular subset of probes that enables high-confidence assignment of specific polymorphism sets that define an allele for the gene of interest. These arrays may be combined with robotic sample isolation, high-throughput PCR target preparation and fluorescent labeling to enable high-throughput genotyping capability. In the ideal case, introducing the hybridization target to an array, post-hybridization washing, fluorescent image acquisition and final interpretation of the hybridization pattern would also be automated steps (Hall *et al.*, 1996; Wang *et al.*, 1998).

3.5 Variety of technical approaches to fabricating microarrays

3.5.1 *Repetitive physical masking*

Dr Edwin Southern is well known for developing the Southern blot, a method for carrying out molecular genetic analysis on immobilized, restricted, genomic DNA (Southern, 1975). The precursor of the DNA microarray really is the Southern blot, which brought the realization that complex DNA is more easily characterized when it is reduced to simpler components and immobilized on a surface for hybridization analysis. This method gave way to the dot blot, where DNA was attached to a support for hybridization analysis without prior gel resolution. The "reverse dot blot" represents the first simple DNA microarray. In this method, instead of immobilizing DNA on a support for hybridization analysis, an oligonucleotide probe set is linked to the support and target DNA is hybridized to the probe set to test whether or not complementary sequences are present (Zhang *et al.*, 1991). Other innovations on this basic theme have been developed to make simple test systems to screen for specific polymorphisms such as cystic fibrosis and beta globin mutations or to do human leukocyte antigen (HLA) genotyping (Nikiforov *et al.*, 1994; Wehnert *et al.*, 1994).

In a subsequent basic contribution to molecular biological analytical techniques, Southern has invented *in situ* synthesized DNA microarrays. In a series of publications beginning in 1992 and continuing to the present, he describes a method for synthesizing oligonucleotides by using standard phosphoramidite chemistry directly on a support to be used in hybridization. In his method, he restricts the area of active chemistry onto a specific site on the substrate surface using a Teflon chamber. This he drills with openings to inject and withdraw reagents and flush with argon (Maskos and Southern, 1993a,b). This vessel, which contains the active, anhydrous chemistry of DNA oligonucleotide synthesis, is physically moved along the substrate surface after each cycle of chemical coupling by a predetermined offset.

After a complete synthesis, along the centerline of the cell pathway, the lengths of oligonucleotides synthesized are equal to the diameter of the cell divided by the offset at that position. In addition to the full-length oligonucleotides along the centerline of the array, there are peripheral cells formed by the chamber that contain the full set of "short-mer" probes that are a subset of the full-length sequence. Figure 3.2 illustrates a layout for a microarray made by using repetitive physical masking. Southern has exploited these "side syntheses" to analyze basic characteristics of immobilized probe hybridization (Southern *et al.*, 1992, 1994; Southern, 1996).

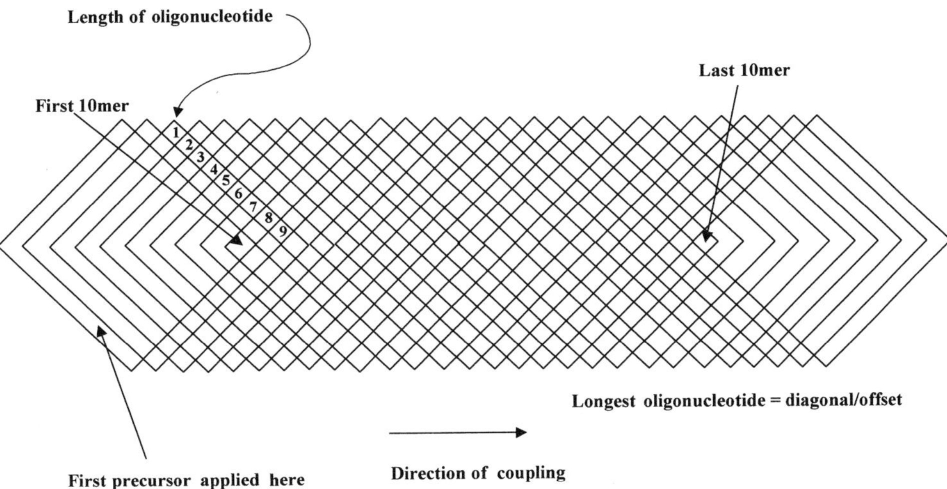

Figure 3.2 Making a scanning array. Arrays of oligonucleotides corresponding to a full set of complements of a known sequence can be made in a single series of base couplings in which each base in the complement is added in turn. Coupling is carried out using a device that applies reagents in a defined area, which is diamond-shaped in the figure. The device is displaced by a fixed movement after each coupling reaction so that consecutive couplings overlap only a portion of previous ones. The amount by which it is displaced at each step determines the length of the oligonucleotide. The diamond shape creates arrays of all oligonucleotides from mononucleotides to the maximum length in a single series of couplings. (Reprinted from Southern, 1996, with permission from Elsevier Science.)

3.5.2 *Microchip arrays assembled in three-dimensional gel elements*

In a different approach to microarray fabrication pioneered by Mirzabekov and coworkers, microarrays are prepared by covalently coupling DNA oligonucleotides within ultrathin polyacrylamide gel elements fixed to a substrate surface (Yershov *et al.*, 1996). The matrix of glass-immobilized gel elements is prepared by coating a glass substrate with a 20-μm-thin 8% polyacrylamide gel and then using a robotic scribe to remove gel strips leaving gel elements as small as 40 μm × 40 μm. The polyacrylamide matrix is activated with a hydrazine-hydrate treatment. Oligonucleotides with a 3′ terminal methyluridine group activated with $NaIO_4$ are coupled to the hydrazine gel matrix using pin-robot transfer. The advantage of this method is a loading capacity more than 100 times greater than that of a typical two-dimensional substrate. The gel elements have a calculated maximal loading capacity of about 50 mM with 70% or more of the immobilized probes being available for hybridization since they are well spaced from each other. Close probe packing on two-dimensional substrates has been identified as a potential limitation to hybridization due to stearic hindrance. This method has been applied to detecting SNPs in human DNA and to *de novo* sequencing by hybridization using complete arrays of octanucleotides and decanucleotides extended by pentanucleotide stacking against the hybridized target DNA (Parinov *et al.*, 1996).

3.5.3 *DNA microarrays synthesized using photolithography*

In 1991 Fodor and coworkers introduced a completely new method for making DNA microarrays by combining photochemistry with a photolithographic masking strategy (Fodor *et al.*, 1991; Pease *et al.*, 1994). This method is based on *in situ* DNA synthesis using novel phosphoramidite nucleotide monomers with photolabile protecting groups rather than the traditional acid-labile protecting groups. Light exposure at the appropriate wavelength removes the protecting group to activate a phosphoramidite, making it competent to participate in a subsequent chemical coupling reaction. By using a photolithographic masking strategy, light is selectively directed to the discrete areas of an array (features) where a particular monomer is intended to extend an oligonucleotide at a particular chemical coupling step. Figure 3.3 illustrates the arrangement of the light source, lithographic mask and synthesis substrate for the photochemical array synthesis process.

The process is very efficient for synthesizing large collections of oligonucleotides *in situ*. The entire array surface is exposed to the coupling chemistry but only the areas that have been selectively exposed to light are competent to react. After each round of four couplings, one each for adenine, cytosine, thymine and guanidine, every oligonucleotide on the surface has been extended by a single nucleotide. The lithographic masking easily permits the standard 1.28 cm × 1.28 cm array substrate

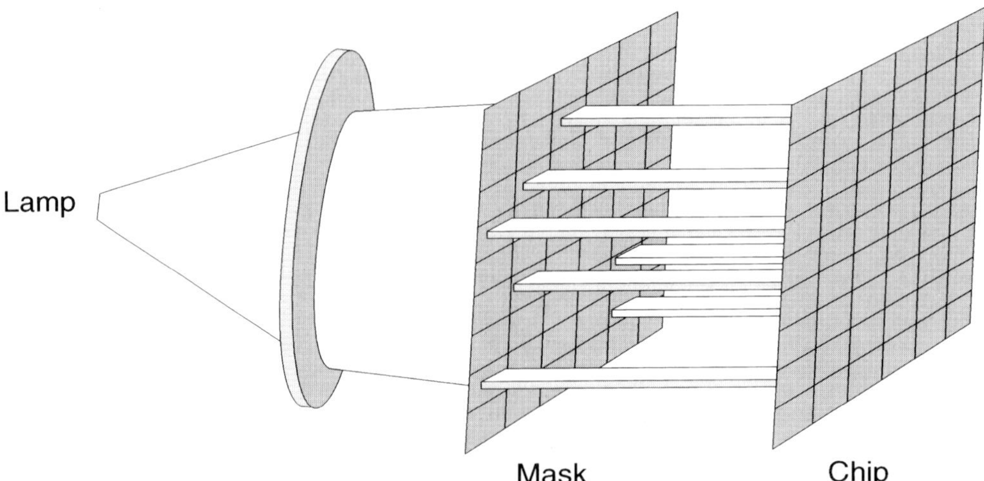

Figure 3.3 Photolithographic microarray synthesis. Light is shown through a series of photolithographic masks to photodeprotect selected sites on the array surface. The light activates exposed sites for chemical coupling by removing a photo-cleavable-protecting group from the growing oligonucleotide polymer. The pattern of photodeprotection, combined with the sequence of monomer additions, determines the sequence within each array feature. (Figure courtesy of Affymetrix, Inc.)

to be divided into 20 μm × 20 μm features, yielding a total oligonucleotide collection of 409 600 sequences. Since each layer of an array of oligonucleotides is synthesized using four coupling steps, four multiplied by the oligonucleotide length used in the array gives the total number of chemical coupling reactions required to synthesize that array's entire collection. An additional facet of economy can be achieved by synthesizing many copies of the same array at once on a single large substrate called a wafer. After the synthesis, the individual arrays are cut or diced out of the wafer for individual use. Figure 3.4 shows the relationship between the packaged microarray, the array surface and a single array feature hybridized with a complementary, labeled target.

Arrays synthesized by this method have been applied to all types of nucleic acid analysis including gene expression monitoring, genotyping, polymorphism discovery, monitoring drug resistance genotypes in HIV, and genome mapping. Figure 3.5 shows images of hybridized photolithographic microarrays for gene expression monitoring and cytochrome P450 genotyping. Since gene expression monitoring by

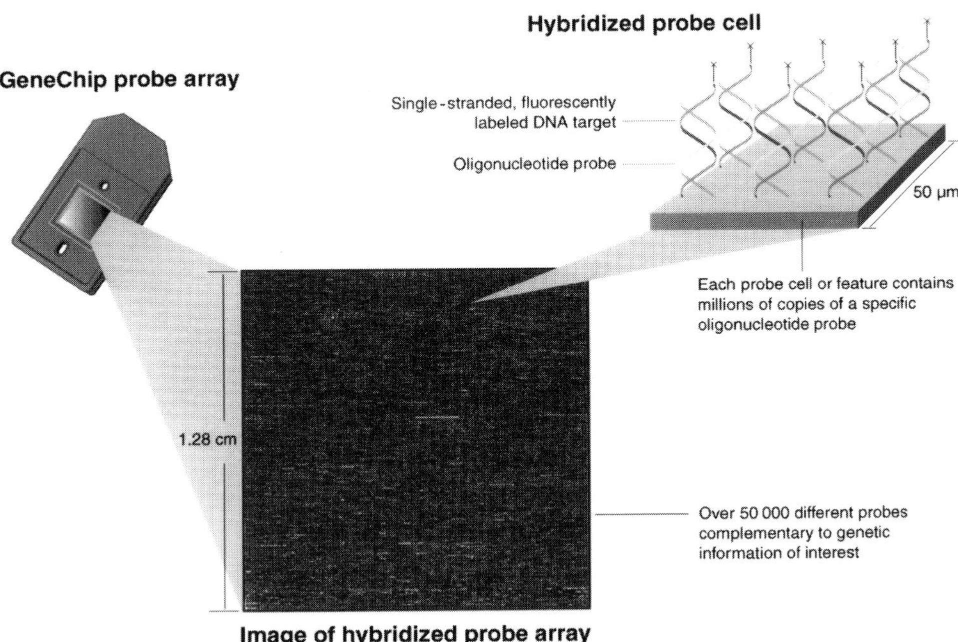

Figure 3.4 The Affymetrix GeneChip® technology. Each Affymetrix GeneChip® probe array is made up of thousands of individual features, each with a different nucleotide sequence. Within each feature are millions of copies of that specific oligonucleotide sequence. The microarrays are synthesized on a glass substrate and packaged into individual plastic flow cells designed to interface with automated fluid-handling stations. Hybridized arrays are scanned with an image acquisition system for assay analysis. (Figure courtesy of Affymetrix, Inc.)

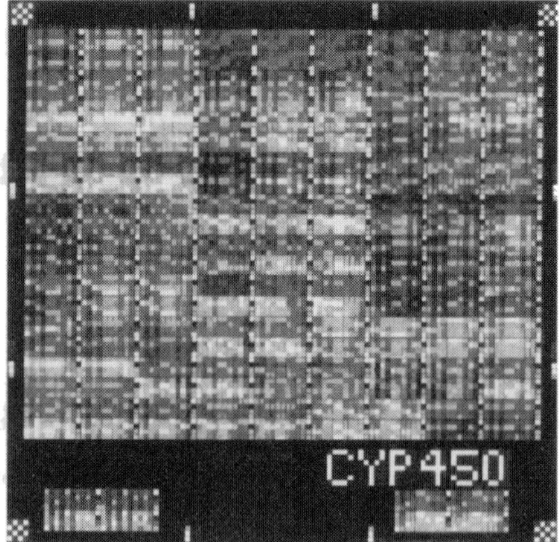

Human 6.8K   CYP450

Figure 3.5 Images of two types of hybridized Affymetrix GeneChip® probe arrays. Affymetrix microarrays have been designed to support both gene expression analysis and genotyping applications. The image on the left shows cRNA hybridization to an array for monitoring 6800 human expressed gene sequences. The image on the right shows the result of a genotyping assay for human cytochromes P450 2D6 and 2C19. (Figure courtesy of Affymetrix, Inc.)

array hybridization was introduced, it has been applied to monitor expression of thousands of human genes as well as the entire collection of yeast genome open reading frame sequences (Lockhart *et al.*, 1996; Wodicka *et al.*, 1997). In yeast, array hybridization has also been applied to genomic mapping (Sapolsky and Lipshutz, 1996). Photolithographic array hybridization has also proven to be useful in a number of human disease analysis applications. Arrays have been designed to genotype mutations that cause cystic fibrosis (Cronin *et al.*, 1996), polymorphic variants of the cytochromes P450 2D6 and 2C19 (Lin *et al.*, 1996), human mitochondrial polymorphisms (Chee *et al.*, 1996), human BRCA1 gene polymorphisms and mutations (Hacia *et al.*, 1996) and drug resistance mutations in HIV (Kozal *et al.*, 1996). Arrays are also being designed with probe sets for genotyping sets of known biallelic markers (Wang *et al.*, 1998). As the set of human SNPs becomes more complete, these arrays are expected to provide genome-wide mapping capability to support disease association case-control studies on a population-wide basis. The hope is that studies of this type will begin to identify candidate genes involved in multigenic disorders such as diabetes, hypertension and schizophrenia.

A more recent version of the process includes modifying the photochemistry to more closely mimic a standard photoresist, photolithographic etching process. Rather than photoprotected phosphoramidites, such a process could be combined with standard, off-the-shelf phosphoramidite chemistry (McGall *et al.*, 1996). This innovation would permit routine synthesis of even smaller array features (<20 μm) and consequently higher density information content in a single microarray. DNA microarrays with oligonucleotide probes at this density have the capacity to interrogate entire genomes.

3.5.4 DNA microarrays synthesized using ink-jet reagent delivery and surface tension-defined substrates

Ink-jet printer heads, piezoelectric nozzles and other similar reagent delivery systems have been conceived to enable reagent delivery for supporting DNA synthesis on a two-dimensional surface. Similarly, robotic systems for delivering presynthesized oligonucleotides or enzymatically generated DNA fragments are becoming increasingly available commercially. All of these delivery systems are plagued by a common problem when they are used to deliver reagents onto a two-dimensional surface. The problem is that the reagents delivered onto a two-dimensional substrate are not constrained in any way, permitting them to spread into irregular, inconsistent, irreproducible shapes on the substrate surface. This is particularly problematic for *in situ* DNA synthesis where the monomer building blocks are delivered sequentially to approximately the same location but each addition may have an area of distribution somewhat different from other prior reagent deliveries after the reagent spreads. The result is an array of ill-defined features that lack homogeneity, reproducibility and a regular, predictable size and shape. This problem defeats effective quantitation of DNA target molecules hybridized to such arrays, which is often the goal of hybridization experiments.

These shortcomings are circumvented by a substrate surface modification recently invented by Thomas Brennan (Cronin *et al.*, 1998a,b). The method involves coating substrate surfaces with a photoresist and then using a generic photomask to define the array features while the resist is exposed to light. Following appropriate photo-exposure, the photoresist is developed and removed, leaving the intended array features protected with photoresist and the remaining surface available for masking chemistry. The features are circular and may be varied in diameter but must be separated by 100 μm or more of intervening substrate to remain compatible with the firing accuracy of the reagent delivery nozzles. The open substrate surface surrounding the protected features is rendered chemically inert and non-wettable by covalently coating it with a perfluoro-octyl silane reagent. The next step in the process is to remove the photoresist protecting the array features to expose the bare substrate surface. These sites are then covalently coupled with a chemically active, wettable linker reagent capable of supporting standard DNA synthesis. Reagents delivered to these array features are now not only constrained by surface tension away from the inert surface into the chemically reactive features but actually self-register and bead up in the wettable sites even when they are delivered less than accurately to the features. Figure 3.6 illustrates the essential elements of this fabrication process and the array characteristics. The surface tension principle exploited for array synthesis applies equally to all reagents delivered to these substrates, whether they are part of *in situ* DNA synthesis or presynthesized DNA reagents intended for covalent immobilization and robotic delivery. This synthesis method provides an elegantly simple but powerful way to make robust, reproducible DNA microarrays.

There is another significant advantage to making arrays by surface tension localization and reagent microdelivery. The lithography used to pattern the substrate surface is a generic process that simply defines the array feature size and distribution. It is completely independent of which sequences will be synthesized or delivered

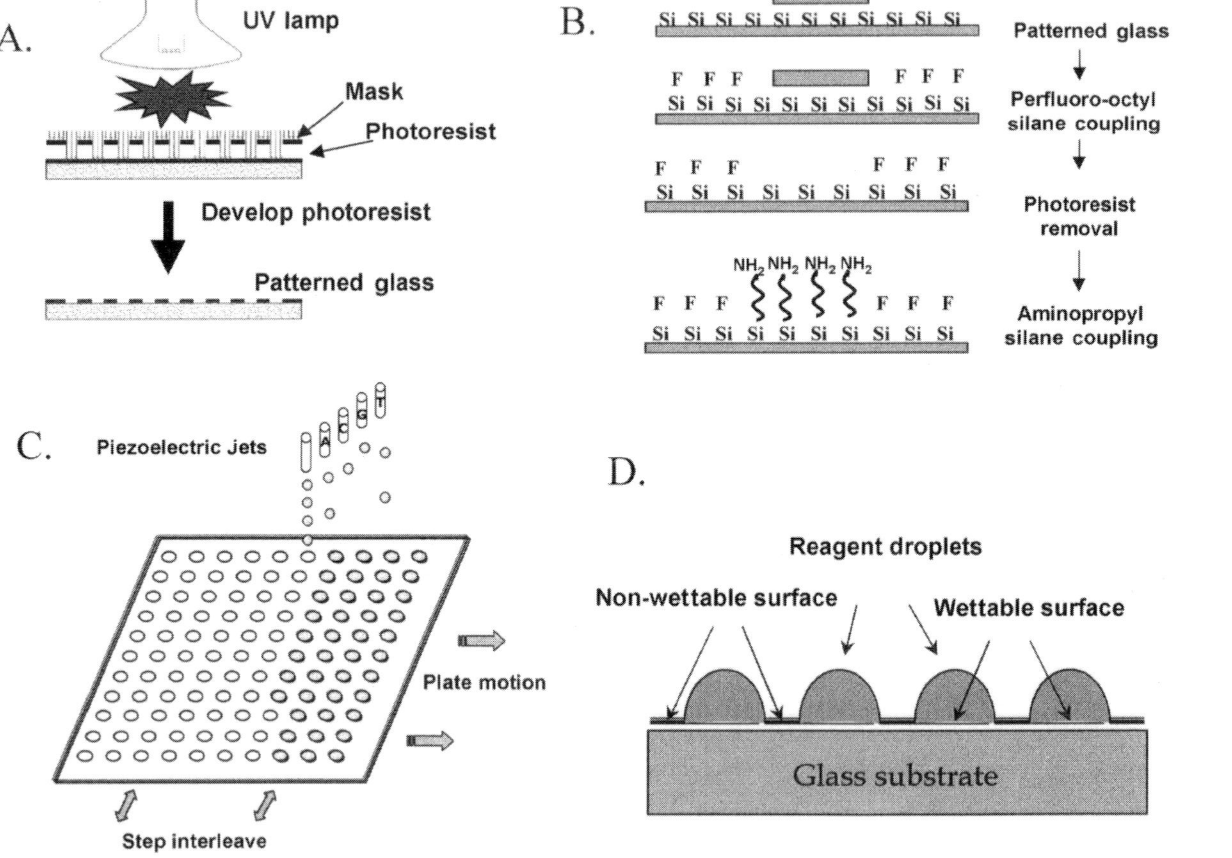

Figure 3.6 Synthesizing arrays on surface tension substrates. **A.** Ultraviolet light is shown through a lithographic patterning mask to expose a pattern onto the substrate surface coated with photoresist. The photoresist is developed leaving the substrate protected with photoresist in a pattern determined by the mask. **B.** The exposed substrate surface is coated with perfluoro-octyl silane to render it chemically inert and non-wettable. The remaining photoresist pattern is removed and the exposed substrate is coupled with a wettable, chemically active linker such as aminopropyl silane. **C.** The patterned substrate is mounted on an X–Y stage that moves under a bank of piezoelectric jets that deliver DNA synthesis reagents for *in situ* oligonucleotide synthesis. **D.** The finished surface tension array maintains its characteristic wettable/non-wettable patterned surface.

at each site. In addition, the chemistry uses standard, not custom, synthesis reagents. The combined result is complete design flexibility with respect to the sequences used in the array, the number and arrangement of array features, and the chemistry used to make them. There is no cost associated with changing the composition of an array once it has been designed. This is an important point since the quality and availability of genetic information is changing so rapidly. Array design is still an art and it is rare that an array design remains optimal in the form it was initially conceived for very long. Experimental results may indicate that probe redesign would improve array performance or new polymorphism or other sequence information may become available that would be useful to incorporate into an existing array design. In addition, DNA synthetic chemistry offers an ever-increasing selection of modified bases and backbones that help to tailor the hybridization performance of complex arrays of oligonucleotides (Nguyen *et al.*, 1995).

3.5.5 *Arrays made by immobilizing DNA targets*

Generally, DNA microarrays are hybridization systems in which the reagent probes are immobilized on the substrate surface, while the DNA to be analyzed is delivered as the labeled hybridization target. In the microarray approach introduced by Drmanac and coworkers, the opposite is the case. Originally invented as a method for *de novo* sequencing by hybridization, this approach is now also being applied to a similar range of human genetic polymorphism analyses as other array systems (Drmanac *et al.*, 1989, 1990, 1998). To execute this approach, the unknown (target) DNA is bound to a substrate as a series of replicate arrays. Each replica array is hybridized with a single labeled oligonucleotide probe added in solution. This is repeated until a universal set of short-length probes has been used to test the unknown samples for complementarity and the final sequence is assembled from the set of short probes that tests positive for each sample (Drmanac *et al.*, 1993).

Limitations to this approach are the need to assemble universal sets of labeled oligonucleotide hybridization probes, the quantity of test sample needed to make the required set of replicate arrays and the time necessary to perform and analyze the required number of probe hybridizations to complete a universal set. This format is more compatible with an analysis service than with product commercialization.

3.5.6 *Arrays of immobilized cDNAs*

Although cDNA "spotted" microarrays are not currently in routine use for discovering or genotyping polymorphisms related to human disease, they play a significant role in microarray optimization and development and are playing a highly visible role in therapeutic discovery. These arrays are intended not for genomic analysis but for monitoring expressed genes. The photolithographic arrays discussed in Section 3.5.3 are used for monitoring expressed genes by designing oligonucleotide probes to known expressed sequences whether they are characterized genes or expressed sequence tagged sites (ESTs). In an alternative approach pioneered by Brown and Shalon, full-length cDNAs made from any RNA source are immobilized on a substrate surface and used to monitor the expression of RNA from other related sources (Schena *et al.*, 1995, 1996). No *a priori* sequence information is required to construct these arrays since they are based on full cDNA libraries from a test source. Novel sequences found to be differentially expressed using these arrays in model systems can be recovered, cloned and sequenced to add to fundamental understanding of regulatory pathways (Geng *et al.*, 1998). An additional advantage is that these arrays can be constructed using commonly available cloning reagents and pipetting devices. Consequently the method is available to any laboratory interested in the technique.

3.6   Second-generation microarrays

Despite the innovation and rapid rate of technical improvement in microarray technology, significant limitations remain not only in array fabrication to improve utility and decrease costs but also in the way array information can be accessed. A major difficulty in using DNA microarray technology for highly parallel DNA analysis is in preparing enough labeled target DNA for each assay. This is particularly problematic in gene expression monitoring where quantitation is important and the relative abundance of the starting material must be preserved during amplification and labeling. In addition, as more and more of the genome is accessed for individual assays, the problem of preparing amplified material from all relevant sites in the genome becomes a significant burden. Multiplex PCR amplification is not reliable as a robust method of quantitatively producing target material from thousands of sites at once. In addition, the cost of reagents for such large-scale amplification is staggering.

An ideal method for doing highly parallel genetic analysis would be one with single copy sensitivity that eliminates the need for amplification and does not require that the target be chemically modified to enable detection. Innovations in several areas hold the promise that a new second-generation of microarray technologies will soon become available to begin addressing some of these issues. Bioelectronic chips developed by Heller and coworkers combine surface-immobilized capture probes on a substrate where each feature is separately electronically addressable. This method vastly accelerates target capture rate by overcoming passive diffusion through generating electric fields at each probe substrate site (Cheng *et al.*, 1997; Sosnowski *et al.*, 1998).

A completely different approach that shows considerable promise and very rapid technical development is DNA microarrays combined with MALDI-TOF MS. This method is attractive because detection is by mass only and does not require separate target labeling. Initial limitations to the system were principally related to mass discrimination of large molecules, but progress has been made in devising strategies to mitigate this problem. MALDI-TOF MS systems are commercially available and development of this method is accelerating (Koster *et al.*, 1996; O'Donnell *et al.*, 1997).

Finally, in a completely different approach, fiber optic bundles are being used as an array support for microbeads coupled with DNA capture sequences. Fluorescently labeled hybridization targets are hybridized with the beads, each of which is associated with a single fiber in the optical bundle that acts as its individual reporter/detector, forming an addressable array of hybridization information (Ferguson *et al.*, 1996; Michael *et al.*, 1998).

References

Cano, R.J., Torres, M.J., Klem, R.E. and Palomares, J.C., 1992, DNA hybridization assay using AttoPhos™, a fluorescent substrate for alkaline phosphatase. *Biotechniques*, **12**, 264–269.

Chan, V., Graves, D.J. and McKenzie, S.E., 1995, The biophysics of DNA hybridization with immobilized oligonucleotide probes. *Biophysical Journal*, **69**, 2243–2255.

Chee, M., Yang, R., Hubbell, E., Berno, A., Huang, X.C., Stern, D., Winkler, J., Lockhart, D.J., Morris, M.S. and Fodor, S.P.A., 1996, Accessing genetic information with high-density DNA arrays. *Science*, **274**, 610–613.

Cheng, J., Sheldon, E.L., Wu, L., Uribe, A., Gerrue, L.O., Carrino, J., Heller, M.J. and O'Connell, J.P., 1997, Preparation and hybridization analysis of DNA/RNA from *E. coli* on microfabricated bioelectronic chips. *Nature Biotechnology*, **16**, 541–546.

Cherry, J.L., Young, H., DiSera, L.J., Ferguson, F.M., Kimball, A.W., Dunn, D.M., Gesteland, R.F. and Weiss, R.B., 1994, Enzyme-linked fluorescent detection for automated multiplex DNA sequencing. *Genomics*, **20**, 68–74.

Cronin, M.T., Fucini, R.V., Kim, S.M., Masino, R.S., Wespi, R.M. and Miyada, C.G., 1996, Cystic fibrosis mutation detection by hybridization to light generated DNA probe arrays. *Human Mutation*, **7**, 244–255.

Cronin, M.T., Butler, J., Giddison, G., Walker, J., Moore, D. and Brennan, T.M., 1998a, Tooling-up *in vitro* diagnostics for pharmacogenetics analysis. Presentation at the 50th Meeting of the American Association for Clinical Chemistry, Chicago, Illinois, August.

Cronin, M.T., Pho, M. and Brennan, T.M., 1998b, Applying an array hybridization strategy to high throughput genotyping for drug metabolizing enzymes. Presentation at the 5th International ISSX Meeting, Cairns, Australia, October.

Drmanac, R., Labat, I., Brukner, I. and Crkvenjakov, R., 1989, Sequencing of megabase plus DNA by hybridization: theory of the method. *Genomics*, **4**, 114–128.

Drmanac, R., Strezoska, Z., Labat, I., Drmanac, S. and Crkvenjakov, R., 1990, Reliable hybridization of oligonucleotides as short as six nucleotides. *DNA and Cell Biology*, **9**, 527–534.

Drmanac, R., Drmanac, S., Strezoska, Z., Paunesku, T., Labat, I., Zaremski, M., Snoddy, J., Funkhouser, W.K., Koop, B. and Hood, L.E., 1993, DNA sequence determination by hybridization: a strategy for efficient large-scale sequencing. *Science*, **260**, 1649–1652.

Drmanac, S., Kita, D., Labat, I., Hauser, B., Schmidt, C., Burczak, J.D. and Drmanac, R., 1998, Accurate sequencing by hybridization for DNA diagnostics and individual genomics. *Nature Biotechnology*, **16**, 54–58.

Dykes, C.W., 1996, Genes, disease and medicine. *British Journal of Clinical Pharmacology*, **42**, 683–695.

Ferguson, J.A., Boles, T.C., Adams, C.P. and Walt, D.R., 1996, A fiber-optic DNA biosensor microarray for the analysis of gene expression. *Nature Biotechnology*, **14**, 1681–1684.

Fodor, S.P.A., Read, L.J., Pirrung, M.C., Stryer, L., Lu, A.M. and Solas, D., 1991, Light directed spatially addressable parallel chemical synthesis. *Science*, **251**, 767–773.

Geng, M., Wallrapp, F., Muller-Pillasch, F., Frohme, M., Hoheisel, J.D. and Gress, T.M., 1998, Isolation of differentially expressed genes by combining representational difference analysis (RDA) and cDNA library arrays. *BioTechniques*, **25**, 434–438.

Hacia, J.H., Brody, L.C., Chee, M.S., Fodor, S.P.A. and Collins, F.S., 1996, Detection of heterozygous mutations in BRCA1 using high-density oligonucleotide arrays and two color fluorescence analysis. *Nature Genetics*, **14**, 441–447.

Hall, J.M., LeDuc, C.A., Watson, A.R. and Roter, A.H., 1996, An approach to high-throughput genotyping. *Genome Research*, **6**, 781–790.

Koster, H., Tang, K., Fu, D., Braun, A., vandenBoom, D., Smith, C.L., Cotter, R.J. and Cantor, C.R., 1996, A strategy for rapid and efficient DNA sequencing by mass spectrometry. *Nature Biotechnology*, **14**, 1123–1128.

Kozal, M.J., Shah, N., Shen, N., Yang, R., Fucini, R., Merigan, T.C., Richman, D.D., Morris, D., Hubbell, E., Chee, M. and Gingeras, T.R., 1996, Extensive polymorphism observed in HIV-1 clade B protease gene using high density oligonucleotide arrays. *Nature Medicine*, **2**, 753–759.

Lander, E.S. and Schork, N.J., 1994, Genetic dissection of complex traits. *Science*, **265**, 2037–2048.

Lifton, R., 1996, Molecular genetics of human blood pressure variation. *Science*, **272**, 676–680.

Lin, C.Y., Hahnenberger, K.M., Cronin, M.T., Lee, D., Sampas, N.M. and Kanemoto, R., 1996, A method for genotyping CYP2D6 and CYP2C19 using GeneChip® probe array hybridization. Presentation at the 7th North American ISSX Meeting, San Diego, CA, October.

Lockhart, D.J., Dong, H., Byrne, M.C., Follettie, M.T., Gallo, M.V., Chee, M.S., Mittmann, M., Wang, C., Kobayashi, M., Horton, H. and Brown, E.L., 1996, Expression monitoring by hybridization to high-density oligonucleotide arrays. *Nature Biotechnology*, **14**, 1675–1680.

Marshall, A. and Hodgson, J., 1998, DNA chips: an array of possibilities. *Nature Biotechnology*, **16**, 27–31.

Maskos, U. and Southern, E.M., 1993a, A novel method for the analysis of multiple sequence variants by hybridization to oligonucleotides. *Nucleic Acids Research*, **21**, 2267–2268.

Maskos, U. and Southern, E.M., 1993b, A novel method for the parallel analysis of multiple mutations in multiple samples. *Nucleic Acids Research*, **21**, 2269–2270.

Mathies, R.A. and Peck, K., 1990, Optimization of high-sensitivity fluorescence detection. *Analytical Chemistry*, **62**, 1786–1791.

McGall, G.H., Labadie, J., Brook, P., Wallraff, G., Nguyen, T. and Hinsberg, W., 1996, Light directed synthesis of high-density oligonucleotide arrays using semiconductor photoresists. *Proceedings of the National Academy of Sciences of the United States of America*, **93**, 13555–13560.

McKenzie, S.E., Mansfield, E., Rappaport, E., Surrey, S. and Fortina, P., 1998, Parallel molecular genetic analysis. *European Journal of Human Genetics*, **6**, 417–429.

Michael, K.L., Taylor, L.C., Schultz, S.L. and Walt, D.R., 1998, Randomly ordered addressable high-density optical sensor arrays. *Analytical Chemistry*, **70**, 1242–1248.

Mifflin, T.E., 1996, Recent developments in molecular diagnostic assays for genetic diseases. *Journal of Clinical Ligand Assay*, **19**, 27–42.

Nguyen, H., Auffray, P., Asseline, U., Dupret, D. and Thuong, N.T., 1995, Modification of DNA duplexes to smooth their thermal stability independently of their base content for DNA sequencing by hybridization. *Nucleic Acids Research*, **25**, 3059–3065.

Nickerson, D.A., Taylor, S.L., Weiss, K.M., Clark, A.G., Hutchinson, R.G., Stengard, J., Salomaa, V., Vartiainen, E., Boerwinkle, E. and Sing, C., 1998, DNA sequence diversity in a 9.7-kb region of the human lipoprotein lipase gene. *Nature Genetics*, **19**, 233–240.

Nikiforov, T.T., Rendle, R.B., Goelet, P., Rogers, Y., Kotewicz, M.L., Anderson, S., Trainor, G.L. and Knapp, M.R., 1994, Genetic bit analysis: a solid phase method for typing single nucleotide polymorphisms, *Nucleic Acids Research*, **22**, 4167–4175.

Nollau, P. and Wagener, C., 1997, Methods for detection of point mutations: performance and quality assessment. *Clinical Chemistry*, **43**, 1114–1128.

O'Donnell, M.J., Tang, K., Koster, H., Smith, C.L. and Cantor, C.R., 1997, High-density, covalent attachment of DNA to silicon wafers for analysis by MALDI-TOF mass spectrometry. *Analytical Chemistry*, **69**, 2438–2443.

Parinov, S., Barsky, V., Yershov, G., Kirillov, E., Timofeev, E., Belgovskiy, A. and Mirzabekov, A., 1996, DNA sequencing by hybridization to microchip octa- and decanucleotides extended by stacked pentanucleotides. *Nucleic Acids Research*, **24**, 2998–3004.

Pease, A.C., Solas, D., Sullivan, E.J., Cronin, M.T., Holmes, C.P. and Fodor, S.P.A., 1994, Light-generated oligonucleotide arrays for rapid DNA sequence analysis. *Proceedings of the National Academy of Sciences of the United States of America*, **91**, 5022–5026.

Ramsay, G., 1998, DNA chips: state of the art. *Nature Biotechnology*, **16**, 40–44.

Sapolsky, R.J. and Lipshutz, R.J., 1996, Mapping genomic library clones using oligonucleotide arrays. *Genomics*, **33**, 445–456.

Schafer, A.J. and Hawkins, J.R., 1998, DNA variation and the future of human genetics. *Nature Biotechnology*, **16**, 33–39.

Schena, M., Shalon, D., Davis, R.W. and Brown, P.O., 1995, Quantitative monitoring of gene expression patterns with a complementary DNA microarray. *Science*, **270**, 467–470.

Schena, M., Shalon, D., Heller, R., Chai, A., Brown, P.O. and Davis, R.W., 1996, Parallel human genome analysis: microarray-based expression monitoring of 1000 genes. *Proceedings of the National Academy of Sciences of the United States of America*, **93**, 10614–10619.

Shuber, A.P., Michalowsky, L.A., Nass, G.S., Skoletsky, J., Hire, L.M., Kotsopoulos, S.K., Phipps, M.F., Barberio, D.M. and Klinger, K.W., 1997, High throughput parallel analysis of hundreds of patient samples for more than 100 mutations in multiple disease genes, *Human Molecular Genetics*, **6**, 337–347.

Smith, L.M., Sanders, J.Z., Kaiser, R.J., Hughes, P., Dodd, C., Connell, C.R., Heiner, C., Kent, S.B.H. and Hood, L.E., 1986, Fluorescence detection in automated DNA sequence analysis. *Nature*, **321**, 674–679.

Sosnowski, R.G., Tu, E., Butler, W.F., O'Connell, J.P. and Heller, M.J., 1998, Rapid determination of single base mismatch mutations in DNA hybrids by direct electric field control. *Proceedings of the National Academy of Sciences of the United States of America*, **94**, 1119–1123.

Southern, E.M., 1975, Detection of specific sequences among DNA fragments separated by gel electrophoresis. *Journal of Molecular Biology*, **98**, 503.

Southern, E.M., 1996, DNA chips: analyzing sequence by hybridization to oligonucleotides on a large scale. *Trends in Genetics*, **12**, 110–115.

Southern, E.M., Maskos, U. and Elder, J.K., 1992, Analyzing and comparing nucleic acid sequences by hybridization to arrays of oligonucleotides: evaluation using experimental models. *Genomics*, **13**, 1008–1017.

Southern, E.M., Case-Green, S.C., Elder, J.K., Johnson, M., Mir, K.U., Wang, L. and Williams, J.C., 1994, Arrays of complementary oligonucleotides for analyzing the hybridization behavior of nucleic acids. *Nucleic Acids Research*, **22**, 1368–1373.

Tamary, H., Surrey, S., Kirschmann, H., Shalmon, L., Zaizov, R., Schwartz, E. and Rappaport, E.F., 1994, Systematic use of automated fluorescence-based sequence analysis of amplified genomic DNA for rapid detection of point mutations. *American Journal of Hematology*, **46**, 127–133.

Thomson, G., 1995, HLA disease associations: models for the study of complex human genetic disorders. *Critical Reviews in Clinical Laboratory Science*, **32**, 183–219.

Wang, D.G., Fan, J., Siao, C., Berno, A., Young, P., Sapolsky, R., Ghandour, G., Perkins, N., Winchester, E., Spencer, J., Kruglyak, L., Stein, L., Hsie, L., Topaloglou, T., Hubbell, E., Robinson, E., Mittmann, M., Morris, M.S., Shen, N., Kilburn, D., Rioux, J., Nusbaum, C., Rozen, S., Hudson, T.J., Lipshutz, R., Chee, M. and Lander, E.S., 1998, Large-scale identification, mapping and genotyping of single-nucleotide polymorphisms in the human genome. *Science*, **280**, 1077–1082.

Wehnert, M.S., Matson, R.S., Rampal, J.B., Coassin, P.J. and Caskey, C.T., 1994, A rapid scanning strip for tri- and dinucleotide short tandem repeats. *Nucleic Acids Research*, **22**, 1701–1704.

Winzler, E.A., Richards, D.R., Conway, A.R., Goldstein, A.L., Kalman, S., McCullough, M.J., McCusker, J.H., Stevens, D.A., Wodicka, L., Lockhart, D.J. and Davis, R.W., 1998, Direct allelic variation scanning of the yeast genome. *Science*, 281, 1194–1197.

Wodicka, L., Dong, H., Mittmann, M., Ho, M. and Lockhart, D.J., 1997, Genome-wide expression monitoring in *Saccharomyces cerevisiae*. *Nature Biotechnology*, 15, 1359–1367.

Yershov, G., Barsky, V., Belgovskiy, A., Kirillov, E., Kreindlin, E., Ivanov, I., Parinov, S., Guschin, D., Drobishev, A., Dubiley, S. and Mirzabekov, A., 1996, DNA analysis and diagnostics on oligonucleotide microchips. *Proceedings of the National Academy of Sciences of the United States of America*, **93**, 4913–4918.

Zhang, Y., Coyne, M.Y., Will, S.G., Levenson, C.H. and Kawasaki, E.S., 1991, Single-base mutational analysis of cancer and genetic diseases using membrane bound modified oligonucleotides. *Nucleic Acids Research*, **19**, 3929–3933.

Zhu, A., Chao, J., Yu, H. and Waggoner, A.S., 1994, Directly labeled DNA probes using fluorescent nucleotides with different length linkers. *Nucleic Acids Research*, **22**, 3418–3422.

4

BIOINFORMATICS – GENETICS SITES ON THE WORLD WIDE WEB

Mark O. Lively

4.1 Introduction

The incredible growth of the biological databases presents each scientist with enormous opportunities for advancement of science as well as a daunting challenge to keep up with the flow of information critical to ongoing research. In the recent past the major information challenge was to follow and read the literature published in scientific journals. This important need has intensified as the rate of growth of the scientific literature continues unabated. However, in addition to the conventional published literature, continued research is producing a virtual torrent of gene sequences, gene maps, chromosome sequences, structures of biological macromolecules and other biological data that enter publicly available databases each day. Much of this information is inter-related and co-dependent. It is not easily conveyed in a useful form by printed media, so scientists must increasingly rely on computers to access and use this information.

Scientists must be able to examine DNA sequence data in the context of the biological and genetic information from which the data derive. It is insufficient just to know the amino acid sequence of a protein when a model of the three-dimensional structure may also be available. The chromosomal localization of the DNA sequence may be known and should be correlated with the nucleotide sequence record. Functional attributes of the DNA sequence or its gene product may be known. Many computational tools are required to retrieve and use the enormous amount of information available. Fortunately, unified mechanisms for retrieving and analyzing the variety of inter-related data are emerging and the tools are as close and convenient as the nearest computer workstation with access to the Internet. This chapter presents an overview of the informatics tools used to access biological databases and highlights selected sites on the World Wide Web (WWW) that bring some of the most important tools right into the laboratory.

67

4.2   The Internet and the World Wide Web

4.2.1   Creation of the Internet

The Internet is not a single network. It is really a large network of networks linking many computer networks all over the world (Baxevanis, 1998a). This network of networks was created initially by the Advanced Research Projects Agency (ARPA) of the US Department of Defense and was called ARPANET. Its purpose in 1969 was to provide a means of communicating defense-related research data between selected laboratories in the US. The network has expanded explosively since its inception and is now available worldwide to any computer with modem access to a telephone network or with a direct link to a computer network connected to the Internet.

A standard instruction set for communication via the Internet was developed by ARPA to define how different computers could exchange data electronically via the network. The protocols are known as the Transmission Control Protocol (TCP) and Internet Protocol (IP) or simply as TCP/IP. In addition to having a common language for information exchange, each computer device connected to the Internet must have a unique address. This objective is met by the IP address. Like a street address for a location in your city, an IP address uniquely defines a single computer device (it can also define printers and other network devices) for the data to be delivered correctly.

IP addresses have a defined format composed of four numbers separated by periods: XX.YY.ZZ.AA. The first two numbers, XX.YY, specify the domain, the third number, ZZ, designates the subnet and the last number, AA, specifies the computer or network device itself. For example, the IP address of one of the main web servers at the National Center for Biotechnology Information (NCBI) at the US National Institutes of Health (NIH) is 130.14.22.107. The domain for the NIH is "130.14" and the subnet "22" specifies the National Library of Medicine (NLM). The number assigned to the computer is "107". Each computer device connected to the Internet must have a unique IP address.

Because many people do not readily remember strings of numbers as easily as names, each IP address can be assigned a fully qualified domain name (FQDN). Computers use the FQDN to find the IP address in a computer that provides a service known as a domain name server (DNS). Each DNS stores records of computer names associated with their corresponding IP addresses. Any network computer that seeks to contact a remote computer on the network by specifying the FQDN must first contact a neighboring DNS to acquire the IP number required to properly route its request to the remote computer. The address of at least one DNS computer must be specified in the network setup of each computer on the Internet. Without access to a DNS to resolve the FQDN into the IP number, the computer will be unable route the message. Fortunately this activity is performed in the background and the average user is usually unaware of the process until their web browser reports an error message such as "unable to locate the server". Failure to locate a server can result because the FQDN is not found in the local DNS, because the FQDN is misspelled, or simply because the requested server is currently not on line on the network.

4.2.2 *The creation of the World Wide Web*

Until creation of the WWW in the early 1990s, data file transfer between computers was difficult for computer users without much experience and understanding of computers. The Internet enables any two computers connected to the network to exchange computer files of any kind. The types of files include text, graphics, sound, movies, executable programs, and many others, even if the two computers use very different types of hardware and operating systems. However, the successful transfer of files between different computer platforms can be difficult if one does not have knowledge of the characteristics and conventions of each computer system. Transferred files can easily be corrupted or cannot be opened if the receiving computer does not have the correct software to read the file. In the early days of the Internet the predominant method of data interchange was by electronic mail (email) but this method requires that a person or program on a remote computer specifically sends a given file to another computer that requests it. Furthermore, different email programs handle attached files differently and can corrupt incoming files so that they cannot be used. Development of the WWW provided a common mechanism for data interchange that allows each network computer to handle file transfer tasks in the background so that users do not need to understand the computing details. This development has revolutionized the way information is shared between computers.

The WWW began in 1990 at the European Nuclear Research Council (CERN) when Tim Berners-Lee wrote the first web browser (www.w3.org/People/Berners-Lee/Overview.html). A web browser is a computer program that runs on a local client computer and takes care of all of the details of file transfer that must occur for one computer to request, receive and display a file from a remote computer. If the file is publicly available then the user is not required to have a previously assigned user identification and password to gain access to the remote computer. A web browser is automatically granted access to public files on the remote computer that it requests and those files are transferred to the client computer. If necessary, secured web sites can require user identification and passwords so that access can be controlled if necessary. The communications connection between the host and client computers is established only for the time needed to complete the file transfer. Older file transfer methods require an interactive login session that remains active until it is terminated by one of the computers.

4.2.3 *Web pages and hypertext markup language*

WWW documents are called web pages and a collection of web pages that are linked together and provided by a given computer is called a web site. Each web page is typically composed of more than one file including text and graphics images but sound, movies and many other types of files can also be included. Significantly, navigation of the web does not require a user to know anything about the location of a remote computer that stores a desired file. Each web page is designated by an address known as a uniform resource locator (URL) and this address is designated in a way that directs the client computer to the specific set of files defined on the remote computer where the web page is located. The file elements of the web page may be stored on that same computer or they may be stored on one or more computers located anywhere on the Internet.

The general format of a URL is protocol://computer.domain. The protocol designates the type of language used to transfer the file from the host web server to the client computer. Most sites today use the hypertext transfer protocol (http) to transfer files but other file transfer methods include file transfer protocol (ftp) and gopher. The computer.domain portion designates the FQDN of the remote computer. For example, the URL for the NCBI web site is designated as www.ncbi.nlm.nih.gov. Web sites given in this chapter will omit the "http://" since most web browsers assume that as the default protocol.

Web pages that use http are written in a format known as hypertext markup language, also known as HTML code. HTML files are simple text files that are interpreted and executed by computer. HTML is independent of the hardware platform and can be interpreted by any web browser written for a specific client computer system. Currently there are two major web browsers in common use, Netscape Communicator provided by Netscape Communications Corporation (www.netscape.com) and Microsoft Internet Explorer (www.microsoft.com). Most WWW sites, especially those of importance to scientists, are compatible with both browsers so the choice of which to use is largely a matter of personal preference. Either browser will provide full access to useful web sites.

In addition to the web browser program, some web sites may also require helper programs, known as plug-ins in the Netscape browser, to be installed on the computer to enhance the presentation of some pages. For example, the molecular modeling page found at the NCBI (www.ncbi.nlm.nih.gov/Structure/) uses a helper program called Cn3D (Hogue, 1997) to display three-dimensional models of biomolecule structures. These helpers are usually obtained free of charge by downloading the program from another web site. Once installed on the client computer, they are associated with the browser so that they are started whenever a file type recognized by the program is selected from a web site. Such programs significantly enhance the capabilities of web browsers.

4.3   Genetic sequence databases

The exponential growth of genetic sequence information and its daily placement on WWW servers requires that the successful biomolecular scientist become a proficient user of computer network resources. Computational tools once accessible only to experts are now readily available for use by every scientist. The days have long passed when it was possible or even desirable to publish long DNA sequences in a printed format. Instead, such sequences are deposited directly into the computer databases where they are stored in standardized formats that make them readily available to the computational tools available for sequence analysis. As a result of an international cooperative effort there is now a single publicly available DNA sequence database composed of data gathered by three separate organizations.

4.3.1   *WWW access to genetic databases and computational resources*

Effective use of the large amount of data present in genetic databases like GenBank is impossible without the use of computers. Fortunately the WWW makes these large databases readily available to anyone with access to the Internet. Computers anywhere on the Internet can search and retrieve data from any database that provides a public web site. Some web sites provide information solely as documents whereas others provide access to powerful computational tools. Interaction with many web sites allows the user to submit a computational job to a much larger computer, even supercomputers at some sites, and easily retrieve the output in an understandable and useful format. The user supplies information via forms posted on a web page and then submits the form for an analysis such as a sequence database search, a protein secondary structure prediction or an alignment of multiple related sequences. Some web sites simply provide lists known as HTML links that direct the viewer to scientific sites on the network. The CMS Molecular Biology Resource at the San Diego Supercomputer Center (www.sdsc.edu/ResTools/) provides an extensive list of current web sites for molecular biology, biotechnology, molecular evolution, biochemistry and biomolecular modeling. Pedro's Research Tools (www.public.iastate.edu/~pedro/research_tools.html) and the US Geological Survey site (info.er.usgs.gov/network/science/biology/index.html) also provide lists of relevant web sites. Most web sites also provide links to related sites that the user may choose to explore. Such sites are good starting points to locate sites with specialized functions or data.

A cautionary note about web site addresses is appropriate at this point. By its very nature the WWW is a rapidly changing and highly fluid information environment. Change is inevitable and a consequence of that change is that web sites often change URLs or disappear altogether as personnel responsible for site maintenance or funding sources supporting the sites change priorities. Consequently, published URLs in printed media such as this book can quickly become outdated as sites expand and modify their documents. For example, while Pedro's is an excellent, useful site, it apparently has not been updated since 1996 so some of its links are "broken". In many cases outdated web addresses will simply refer you to an updated site with a different URL if the information or services are still available. In some cases web sites simply cease to exist. As the era of electronic publication matures consistent mechanisms for preserving older web documents will be developed and it should become easier to locate old links. Such archival procedures are essential for journals published electronically. In the following sections, the URLs cited are served by computers at major institutions with substantial funding and can be expected to be stable for the foreseeable future. If a given long URL reference fails to return the expected document one can usually shorten the URL back to the next "/" symbol to request the next higher directory. Doing so will usually return a page that may have a link to the documents originally sought.

4.3.2 *Nucleotide sequence databases – GenBank*

There are many databases of nucleotide and protein sequences that are accessible via the WWW (Table 4.1). GenBank is the principal genetic sequence database of publicly available DNA sequences. It is maintained by the NCBI. The database is an annotated collection of all publicly available DNA sequences (Benson *et al.*, 1998) and it is freely accessible on the WWW (www.ncbi.nlm.nih.gov/). Currently the database is released six times per year but updates are posted nightly on the web site so the most current information is available daily. GenBank is a partner in the International Nucleotide Sequence Database Collaboration along with the European Molecular Biology Laboratory (EMBL, see Stoesser *et al.*, 1997) and the DNA Data Bank of Japan (DDBJ, see Tateno and Gojobori, 1997). The URLs for the web sites of these organizations are given in Table 4.1. DNA sequence data collected by each organization is exchanged daily so it is not necessary to search each database separately to assure complete coverage of available data. In addition to the public databases, there are privately held gene sequence databases that are commercially available from companies such as Human Genome Sciences, Inc. (www.hgsi.com/) and Incyte Pharmaceuticals, Inc. (www.incyte.com/). There are also specialized industrial databases accessible only within the companies that maintain them.

The NCBI provides an extremely powerful web site (www.ncbi.nlm.nih.gov/) for searching GenBank and related databases. The mission of NCBI is "to develop new information technologies to aid in the understanding of fundamental molecular and genetic processes that control health and disease" (www.ncbi.nlm. nih.gov/Web/NCBI/progsfs.html). The NCBI is a unit of the United States National Library of Medicine (NLM) at the NIH and is charged with providing computer systems for storing and analyzing information about molecular biology, genetics and biochemistry. NCBI also performs basic research in computational biology, often referred to as "bioinformatics", and provides resources that facilitate the use of genetic databases. Finally, it is their responsibility to coordinate efforts to collect, store and manage biotechnology information worldwide.

Table 4.1 Nucleotide and protein sequence database web sites

Web URL	Site information
Nucleotide sequences	
www.ncbi.nlm.nih.gov/	Information technologies to aid the understanding of fundamental molecular and genetic processes that control health and disease Includes GenBank, database services (BLAST and Entrez) and a variety of genetics resources
www.ddbj.nig.ac.jp/	The DNA Data Bank of Japan
www.ebi.ac.uk/ebi_home.html	The European Bioinformatics Institute (EBI), an outstation of the European Molecular Biology Laboratory (EMBL)
Protein sequences	
expasy.hcuge.ch/	Molecular biology WWW server of the Swiss Institute of Bioinformatics for analysis of protein sequences, structures and 2D PAGE
www.mips.biochem.mpg.de/	MIPS, the Munich Information Center for Protein Sequences
pir.georgetown.edu/	The Protein Information Resource – a comprehensive, annotated, and non-redundant set of protein sequence databases in which entries are classified into family groups and alignments

Since its creation in 1982, GenBank has doubled in size approximately every 14 months. At the current rate of growth there will be more than 5 billion bases in the sequence database by the summer of 2000 (Figure 4.1). GenBank release 111.0 (15 April 1999) contains 2 569 578 208 bases from 3 525 418 reported sequences. This database can be compared to the familiar MEDLINE literature database, also maintained at the NLM, that contains approximately 9 million literature citation records dating back to 1966.

GenBank contains DNA sequences obtained from approximately 25 000 different species. The top five most sequenced species currently are: human (1 940 945 sequences); mouse (475 996 sequences); *Caenorhabditis elegans* (77 145 sequences); *Arabidopsis thaliana* (71 938 sequences); and *Drosophila melanogaster* (95 154 sequences). Complete information about the current release of the GenBank database can be retrieved from the NCBI (ftp://ncbi.nlm.nih.gov/genbank/gbrel.txt). The release notes provide useful information about the data in GenBank and details regarding interpretation of the file format. In addition to the DNA sequences, each GenBank record is annotated with literature citations, known functional attributes of the gene sequence, the protein coding sequences and many other facts relevant to the gene sequence.

DNA sequences in GenBank are separated into different divisions (Ouellette and Boguski, 1997). The organismal divisions of GenBank contain annotated sequences grouped by taxonomic classification. For example, there are divisions for sequences from primates, non-primate mammals, other vertebrates, invertebrates, bacteria, and plants. Sequences in the organismal divisions are typically more highly annotated and are generally considered "finished" sequences. The other major divisions of GenBank include the expressed sequence tag (EST), the high-throughput

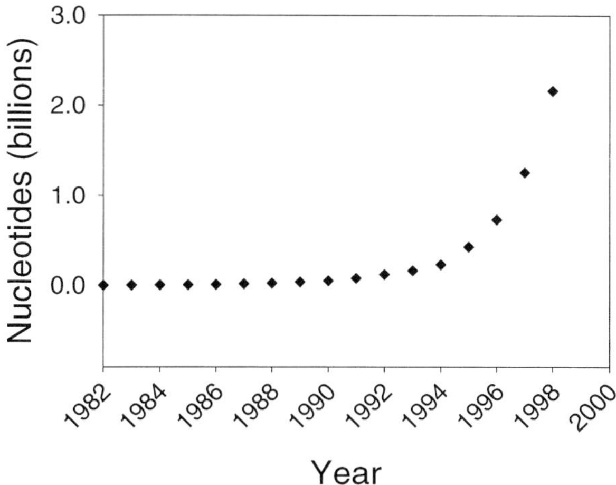

Figure 4.1 Growth of GenBank. The figure shows the increase in numbers of nucleotide bases deposited yearly in GenBank since its creation. The data were obtained from the release notes for GenBank Release 111.0, 15 April 1999 (see ftp://ncbi.nlm.nih.gov/genbank/gbrel.txt for the most current release notes). This release contained 2 569 578 208 bases from 3 525 418 reported sequences.

genome sequence (HTGS), the sequence tagged site (STS) and the genome survey sequence (GSS) divisions. Sequences in the EST, GSS and HTGS divisions serve different purposes and are not considered to be final, finished sequences (www.ncbi.nlm.nih.gov/HTGS/paper.html).

DNA sequences are submitted directly to GenBank by the investigators who produce the data. In the early days of GenBank DNA sequence data were transcribed from published sequences that appeared in printed journals. The NCBI now provides software for direct submission of data to GenBank in the proper format for immediate deposition into the database. Sequin is a standalone program available for all major computer operating systems that allows an investigator to create the submission file on a computer workstation before submitting it by email. Alternatively, Bankit is a WWW interface that allows investigators to directly submit sequence data via the WWW. Direct submission of data significantly speeds up the process of data entry into the database.

Currently the EST division contains the most sequences in GenBank. ESTs are short DNA sequences derived from single sequence determinations of randomly selected clones isolated from selected cDNA libraries (Boguski *et al.*, 1993; Boguski, 1995). GSSs are similar to ESTs except that the sequences originate from genomic DNA. STSs are short sequences having 200 to 500 bases that are unique in the genome. Each STS localizes a specific position on the physical chromosome map (Olson *et al.*, 1989). The EST, GSS, and STS divisions of GenBank are available separately at the NCBI site as searchable databases known as dbEST, dbGSS and dbSTS. In each case the sequences are also present in the main GenBank files. The annotation in the dbGSS (www.ncbi.nlm.nih.gov/dbGSS/index.html) and dbSTS (www.ncbi.nlm.nih.gov/bSTS/index.html) databases is more comprehensive than in GenBank and provides information about contributors, experimental conditions and genetic map locations not found in GenBank. The EST sequences can also be searched separately at the dbEST site (www.ncbi.nlm.nih.gov/dbEST/index.html).

The HTGS division contains "unfinished" genomic sequence data produced by the genome sequencing centers (Ouellette and Boguski, 1997). Upon initial submission to the database the genome sequences are stored in the HTGS division where they are classified as phase I or II, essentially unfinished genomic sequences. Once the DNA sequence fragments have been ordered, oriented and gaps have been closed to yield a contiguous sequence, the finished sequences are reclassified to phase III and are then deposited in the appropriate organismal division of GenBank. It is important to know how the data are divided in GenBank in order to search the proper data set that will achieve a desired objective. Some search programs query the divisions of the database separately so one must be aware of which portions of the database have and have not been searched.

4.3.3 *Protein sequence databases – Swiss-Prot and PIR*

The first sequence databases were collections of amino acid sequences of proteins published in books that appeared in the early 1960s (Dayhoff *et al.*, 1965). The Protein Information Resource (PIR) database (George *et al.*, 1997) traces its roots back to the Dayhoff atlas and today exists as a web site that provides an annotated, non-redundant set of protein sequences that are classified into family groups (pir.georgetown.edu/). The PIR site provides useful information about protein databases and tools for searching protein sequences.

Swiss-Prot is a "curated" protein sequence database (www.expasy.ch/sprot/). ExPASy is the molecular biology WWW server of the Swiss Institute of Bioinformatics. Each entry in the database is provided with a high level of annotation. In addition to the amino acid sequence data, each database entry contains information about the function of the protein, known structural modifications, domain structure, genetic variants, discrepancies in the literature, and other data relevant to the proteins. The TrEMBL database is a supplement to Swiss-Prot that contains the translated sequences from the EMBL nucleotide sequence database. TrEMBL entries are preliminary database records that provide the protein coding information as soon as possible to the database. Once those sequences have been fully annotated they are moved to the Swiss-Prot database. The Swiss-Prot database seeks to minimize redundancy of sequence records and provides a high level of integration with other databases.

The ExPASy web site provides tools to search Swiss-Prot using text or sequence similarity searches. The site also provides other important databases and tools including the PROSITE database of protein families and domains, the SWISS-2DPAGE database of two-dimensional polyacrylamide gel electrophoresis of proteins and databases for protein structure modeling.

4.3.4 Searching GenBank with text – PubMed and Entrez

There are two fundamentally different ways to locate information in sequence databases such as GenBank – either by searching for specific text contained in the annotations or by sequence similarity searching in the sequence data. An in-depth discussion of information retrieval from biological databases has been published by Baxevanis (1998b). The NCBI web site provides powerful tools for searching for text in GenBank, MEDLINE (the bibliographic database provided by the National Library of Medicine) and related databases. PubMed is a database developed jointly by the NCBI and NLM for finding literature citations (www.ncbi.nlm.nih.gov/PubMed/). A major advantage of PubMed is that this database links citations to full-text journals at web sites of participating publishers. If a citation is published in a journal that is available electronically then the PubMed citation will provide a direct hyperlink to the full text version of the journal. PubMed citations also provide links to related articles, to protein or DNA sequences, and to three-dimensional structures associated with the primary citation. PubMed contains all citations in MEDLINE as well as some additional citations not included in that database.

The era of electronic publication is approaching rapidly. As more journals convert to electronic formats the PubMed linkage to on-line journals will further enhance access to the scientific literature. Presently many of the linked journals require a paid subscription for access to the full text of articles. This limits access but protects the important financial interests of the journals. As institutional libraries begin to acquire subscriptions to electronic journals (and perhaps stop subscribing to printed versions) the direct links from PubMed will become available to those investigators with access to the library collections. A plan is being formulated at the NIH to facilitate creation of an electronic publishing site, called "E-biomed". This site would be similar to the existing physics site known as "e-print" (xxx.lanl.gov). The proposed E-biomed would "transmit and maintain, in both permanent on-line and downloaded archives, reports in the many fields that constitute biomedical research, including clinical research, cell and molecular biology, medically-related behavioral research, bioengineering, and other disciplines allied with biology and medicine" (ww.nih.gov/welcome/director.ebiomed/ebiomed.htm).

Another tool provided by NCBI for searching text in biological databases is the Entrez browser (www.ncbi.nlm.nih.gov/Entrez/). Entrez is a powerful integrated information retrieval system that retrieves related molecular biology data and bibliographic citations from five databases. Those databases are the PubMed database, the protein and nucleic acid databases at NCBI, the NCBI three-dimensional structures database and the genomes database. Note that Entrez is not used to search directly for sequence similarities although it does report related DNA and protein sequences. Sequence similarity searches are performed using the BLAST server (see below). The sequence databases searched by Entrez include the GenBank DNA sequence records, protein sequence records from Swiss-Prot, the PIR, the Protein Research Foundation (PRF) in Japan, as well as protein sequences translated from the DNA sequence databases. Entrez also searches genome and chromosome mapping data available at NCBI. This search tool provides a very efficient method for retrieving related information about biological entities.

An Entrez search for "BCR2", the abbreviation for a gene associated with breast carcinomas and other cancers, provides a good example of the extent and types of information that can be obtained using this search engine. A search for this term in early 1999 revealed 301 citations from the Entrez databases. One of the 301 citations near the top of the list was for the mRNA sequence encoded by the human breast cancer 2 (BCR2) gene (Teng *et al.*, 1996). This single citation provided hypertext links to the following sources:

- the GenBank sequence record for this mRNA;
- the sequence of BRCA2 in FASTA format, a simplified sequence format useful for analysis with some programs;
- the ASN.1 report, the fundamental database format for GenBank records;
- a graphical view of the cDNA sequence with annotations;
- a MEDLINE link to the literature source for the sequence which was in turn linked to 356 related MEDLINE citations;
- a protein database entry link with seven links to other protein sources;
- a listing of 11 related nucleotide sequence records.

A single Entrez query for this term provides citations that would take hours or days to gather and correlate if each different database source were searched separately.

4.3.5 Searching GenBank for sequence similarities – BLAST and FASTA

One of the most powerful discovery methods available to biological scientists today is to use alignments of gene sequences to identify distant evolutionary relationships among genes. Pairwise sequence alignment of DNA and protein sequences can reveal important information about evolution of gene structure and function (Schuler, 1998). The most important conclusion one can make from a database search is that the genes encoding two proteins are homologous (Pearson, 1996). If a newly discovered protein sequence is found to be a homologue of another protein whose three-dimensional structure is already known, one can readily use the known structure to build a useful model of the structure of the new protein. Homologous proteins share a common three-dimensional fold and secondary structure.

There are two major programs currently available for performing sequence similarity searches of genetic sequence databases. FASTA was the first widely used sequence similarity search tool (Pearson, 1990). The new versions of this program are available at a number of WWW sites such as at EMBL (www2.ebi.ac.uk/fasta3/). The FASTA search tool is a sensitive method for searching for sequence similarities. The NCBI provides a family of sequence similarity search programs known as BLAST (Basic Local Alignment Search Tool). These programs use pairwise alignment of DNA or amino acid sequences to find related sequences in the databases (Altschul *et al.*, 1990, 1997). BLAST compares single query sequences of DNA or protein against the DNA and protein sequence databases (www.ncbi.nlm.nih.gov/BLAST/). The BLAST algorithm identifies local alignments of sequences to discover statistically significant relationships between sequences that may share only limited regions of similarity.

The BLAST home page (www.ncbi.nlm.nih.gov/BLAST/) provides several useful tools for sequence comparison. There are two implementations of BLAST version 2.0, the basic and advanced. The basic BLAST page requests input of the least amount of information from the user. The program is then executed using default parameters suitable for most queries. The user must first select one of five specific versions of the BLAST program to be run:

- blastp compares an amino acid query sequence against a protein sequence database;
- blastn compares a nucleotide sequence against a nucleotide database;
- blastx compares a nucleotide query sequence translated in all reading frames against a protein sequence database;
- tblastn compares a protein query sequence against a nucleotide sequence database dynamically translated in all reading frames;
- tblastx compares the six-frame translations of a nucleotide query sequence against the six-frame translations of a nucleotide sequence database.

Having selected the proper version of BLAST to run, the user must next select the database to be searched. The default choice is called nr, for non-redundant versions of either the protein or nucleic acid databases. The database selected will depend on the version of BLAST that is chosen. The nr peptide sequence database includes translations of all designated coding sequences from GenBank, sequence entries from the Protein Data Bank (PDB), Swiss-Prot, the PIR and the PRF. The nr nucleic acid database includes all non-redundant sequences from GenBank, EMBL, DDBJ, and PDB. It is critical to note that a search of the nr nucleic acid database does *not* search the EST, STS, GSS or HTGS sequences. These divisions of GenBank must be searched separately. Both versions of the nr databases have a month subsection that contains all new and revised sequences released in the last 30 days. One can use this smaller updated section to perform current awareness searches to identify newly released sequences.

In addition to the non-redundant databases, there are specialized divisions that can be queried. For proteins, those divisions include the last major release of Swiss-Prot, yeast proteins, *E. coli* translations, sequences derived from the PDB structures, the Kabat sequence database of immunological sequences, and translations of selected *alu* sequences. For nucleic acids, the additional choices include dbEST, dbSTS, HTGS, yeast, *E. coli*, PDB, vector sequences, mitochondrial sequences, *alu* sequences, the Eukaryotic Promoter Database and GSS sequences.

Having selected the desired program and database, the user must provide a query sequence that will be searched against the database. The sequence can be directly copied or typed into the sequence input window or an accession number or gene index number referring to a sequence in GenBank can be entered. Once the information has been correctly entered the form is submitted to the BLAST server for analysis. The search results are usually returned within seconds although some searches may take longer depending on the number of other jobs pending and the nature of the search.

The results of the BLAST search are returned to the user as a web page that has three main sections. The top of the page presents a graphical overview of the significant matches aligned with the query sequence. This color-coded view immediately conveys the extent of sequence similarity found. The second section of the report lists the significant alignments ranked by the significance scores. This section provides the one-line descriptions of the database sequences that have significant alignments with the query and each has a hypertext link to the database record. Following the list of matches, the alignment of each region of similarity is shown.

The advanced BLAST page provides users with more control over program variables. In addition to the choices required for the basic version, the user can change the statistical significance threshold for reporting sequences, choose to inactivate the filter function, change the number of descriptions and alignments reported, and modify the format of the sequence alignments reported. The user can select different genetic codes (useful with blastx only) or a different amino acid substitution matrix. For advanced users the page provides a command-line box that permits further modification of program parameters. Explanations of each parameter are provided by hypertext links to assist the user.

In addition to the main BLAST program, there are specialized versions of the program that can be very useful. One can directly compare two sequences to each other using the "BLAST 2 sequences" choice. This alignment tool provides a statistical comparison of the two sequences, a graphical representation of the alignment, and the sequence alignment itself. Position Specific Iterated BLAST (PSI-BLAST) is a program that provides a powerful method to identify patterns of residual similarities among distantly related proteins (Altschul *et al.*, 1997). PSI-BLAST initially searches a single protein sequence against the non-redundant protein sequence database. Sequences with regions with significant local alignments are used to create a multiple sequence alignment. The multiple alignment is then used to create a mathematical profile that describes the aligned proteins. The profile assigns added weight to amino acid positions that are highly conserved among the aligned sequences. The profile is used to search the database again for additional proteins that match the profile but may be too different from the original query sequence to have scored highly in the first search. A new multiple alignment is constructed and a new profile defined. Continued iterations of this process eventually lead to a list of related proteins that fit the profile generated by PSI-BLAST. This process can reveal distant relationships that otherwise would be missed by a single, simple search. The BLAST web page provides a link to the BLAST tutorial that has a number of documents with exercises that demonstrate the BLAST programs in detail.

4.4 Macromolecular structure databases

Three-dimensional structures of proteins and other biological macromolecules are being solved by X-ray crystallography and nuclear magnetic resonance (NMR) methods at an increasingly rapid rate. The Protein Data Bank (PDB) is the repository for publicly available structural data generated worldwide. The PDB was created at the Brookhaven National Laboratories with a mandate to archive and enable access to the extensive data sets that comprise the structural coordinates of 3D structures of macromolecules. The PDB was probably the first true bioinformatics database and it has recently been moved to a new site (see below). Originally, structure data files were distributed by the PDB on magnetic tapes read by large computers. Later, the same data could be obtained on CD-ROM and eventually by either FTP or email from the PDB. Once the data files were obtained, a sophisticated and expensive computer system with tedious software was required to display visual representations of the structures. It was not easy for a non-expert to view molecular models of interest.

Today, repositories of 3D structure data are extremely important scientific resources that are readily available to scientists via the WWW. Analysis of the 3D structures of proteins and other biological molecules complements the genome sequencing projects (Montelione and Anderson, 1999). Knowledge of structure is key to a full understanding of function. Visualization of protein structures is no longer limited to specialized laboratories or to structure experts because the WWW makes available a number of tools to easily examine protein structures (see Table 4.2). It is no longer necessary to have a specialized computer workstation to visualize structure models. Commonly used desktop computer workstations can display and manipulate (magnify, rotate, display in different forms and colors, etc.) models of protein structures using readily available software. More sophisticated modeling software that runs on personal computers can be purchased to enable scientists to display and manipulate structural models.

As the genome sequencing projects continue to produce huge numbers of new protein sequences over the next several years the number of protein structures solved will increase even more rapidly. With as many as 80 000 human gene sequences in addition to thousands of others from different species to be determined in the foreseeable future, the next challenge will be to increase the rate at which structures are solved. With this massive infusion of protein coding sequence data in mind, the crystallography and NMR communities have begun to seriously consider a proposal to fund a Structural Genomics Initiative that would solve structures of

Table 4.2 Structural biology web sites

Web URL	Site information
www.rcsb.org/	International repository for the processing and distribution of 3D structure data of biological macromolecules determined experimentally by X-ray crystallography and NMR
expasy.hcuge.ch/swissmod/ SWISS-MODEL.html	An automated protein modeling server for protein modeling
cmm.info.nih.gov/modeling/	Molecular modeling resources and expertise at the NIH
scop.mrc-lmb.cam.ac.uk/scop/	SCOP database provides detailed and comprehensive description of the structural and evolutionary relationships between proteins of known structure
www.imb-jena.de/IMAGE.html	Jena Image Library of Biological Macromolecules contains information on 3D biopolymer structures
www.ncbi.nlm.nih.gov/Structure/	Database of 3D macromolecular structures with a set of tools for comparing and visualizing structures and sequences

specifically targeted representatives of families of protein folds (Terwilliger *et al.*, 1998). Methods will be developed to increase the rate of throughput of structure determination and even more data will be available for analysis by the scientific community. In the foreseeable future it will be possible to build a useful 3D model of any newly discovered protein based solely on comparison of the deduced amino acid sequence to a known structure of a homologous protein found in the protein data bank.

4.4.1 *Macromolecular structure resources – the PDB*

The PDB is the single international repository for publicly available 3D structure data of macromolecules (www.rcsb.org/pdb/). Unlike the DNA and protein sequence databases where there are many sites that collect and manage databases, there is currently only a single public 3D structure database. Responsibility for the PDB was moved in 1999 from the Brookhaven National Laboratories to the Research Collaboratory for Structural Bioinformatics (RCSB). The RCSB is a non-profit consortium involving three organizations: the Department of Chemistry and the Center for Molecular Biophysics and Biophysical Chemistry at Rutgers, the State University of New Jersey; the San Diego Supercomputer Center (SDSC) at the University of California, San Diego; and the Biotechnology Division of the National Institutes of Standards and Technology. RCSB is supported by funds from the National Science Foundation, the Office of Biology and Environmental Research at the Department of Energy, the National Institute of General Medical Sciences and the National Library of Medicine.

In addition to receiving and managing 3D structure data in the PDB, the RCSB web site also provides a variety of software tools for finding and analyzing 3D structures. If the user already knows the PDB identification code, a four-character alphanumeric name assigned to every database entry, then the structural data can be retrieved directly. Alternatively, one can use keywords or phrases to search the PDB to find structures of interest. For example, a recent keyword search for *oncogene* found 105 structures in the database. The first match on this list was the structure of the H-Ras p21 protein complex with a guanosine 5′-triphosphate analog (Pai *et al.*, 1990). The PDB code for H-ras is 121P.

Each structure matching the search criteria is listed with a summary of details including the PDB code, the classification of the molecule, a brief description of its properties, the date of deposition of the structure, and the experimental method for structure determination. Importantly, each matching structure has a hypertext link to examine the structure further. Choosing the "Explore" button beside an entry leads to the Structure Explorer web page of the PDB site that provides a detailed summary of the selected structure. The primary literature citation is given and it usually has a hypertext link to the NCBI PubMed citation. A summary of the structure data is also provided. Additional hypertext links include:

- view the selected structure
- download or display the primary structure file
- examine information about related structures
- other relevant literature and WWW resources
- details pertaining to the selected structure.

The RCSB site makes it very easy to view a structure model in a variety of formats. Selection of the "View Structure" button leads to a page with several options for viewing representations of the structure on the computer screen. The simplest views of the structure are previously rendered still images that can be directly downloaded for static viewing in a browser window. These images can be saved to the computer workstation and imported by presentation software for inclusion in lectures.

In addition to the option to display previously formatted still images of selected structures, one can choose an interactive 3D display option that allows the user to rotate and modify the structure model in real time. This capability allows the user to select specific regions of a molecule and examine them closely. There are currently four options for interactive viewing of the structures at the RCSB site. Three of the four require the installation of helper software (see above). Helper programs, or plug-ins, can usually be obtained free of charge from supporting web sites. For example, Cosmo Player is a Visual Reality Modeling Language (VRML) viewer that works well with the RCSB "View Structure" page (cosmosoftware.com/). This VRML viewer is similar to the 3D interface used by some computer video games. RASMOL (www.umass.edu/microbio/rasmol/) and the Chime (www.mdli. com/download/) are also programs for viewing structure files found in the PDB. Each program has different features to allow the user to modify the appearance of the molecular model. The VRML viewer has control functions to zoom in or out of a molecule, to rotate it in any direction and to modify its appearance. In addition to those same capabilities, RASMOL allows the user to save image files in a number of useful formats that can be exported to other programs. These web-based tools make it possible for the non-expert to use modeling tools to examine biological molecules with computer resources now common to most laboratories.

4.4.2 Macromolecular structure resources – the Molecular Modeling Database

The NCBI provides another web site to view structure files associated with citations retrieved by Entrez (www.ncbi.nlm.nih.gov/Entrez/). From the Entrez homepage, one can choose "3D structures" to search for known structures using keywords. This search is performed against the Molecular Modeling Database (MMDB) created at the NCBI (Marchler-Bauer *et al.*, 1999). MMDB contains a subset of the structures available from the PDB except that completely theoretical model structures found in PDB are excluded from the MMDB. The goal of the MMDB is to provide structure data in a common format and to facilitate comparative analysis of 3D structures (Hogue and Bryant, 1998). Matches retrieved by the keyword search are listed with summaries of each structure and a number of hypertext links. These links include options to view the structure, MEDLINE entries, structural neighbors (molecules with related structures), and protein sequence database entries. Choosing the "Structure Summary" link leads to a page that gives the details of the structure and provides the opportunity to view the model in an interactive window using a structure viewer called Cn3D (www.ncbi.nlm.nih.gov/Structure/CN3D/cn3d.html) or other viewers.

Unlike the PDB database, the Entrez 3D structure browser provides a new tool to identify structurally similar molecules. Protein "structure neighbors" are identified in Entrez by using the Vector Alignment Search Tool (VAST) algorithm. There are currently more than 15 000 protein domains in the MMDB and the structure of each has been compared to the structure of every other domain. This approach identifies proteins with related folding patterns and thereby can establish relationships that would not be discovered by simple comparison of amino acid sequences. Proteins identified as being structurally similar are listed as protein neighbors in the Entrez search results. Cn3D will display the aligned protein sequence domains to allow direct comparison of structural similarities. Structural comparisons of this type can lead to discovery of protein function in proteins of unknown function.

4.5 Genomic databases

Genome sequencing projects are yielding far more information than the complete DNA sequences of the human chromosomes. In addition to the human genome, more than 18 prokaryotic (archea and bacteria) genomes have already been completed and at least 64 more are currently being sequenced (www.tigr.org/tdb/mdb/mdb.html). A number of eukaryotic genomes other than human are also being sequenced. As they are completed, all of the DNA sequences from these genome projects will be deposited into GenBank. Along with the raw DNA sequence data, these large-scale projects are producing tremendously useful information about chromosome structure and organization. There are genetic and physical maps, nucleotide polymorphisms, genetic markers, disease phenotypes, and gene expression profiles that are correlated with the DNA sequence data. The chromosome maps range from simple circular prokaryotic maps to representations of complex eukaryotic chromosomes. Chromosome maps are essential tools for identification and localization of genes that cause disease. The WWW is ideally suited for making these types of complex data readily available to the scientific community and there are many web sites that provide up-to-date physical and genetic maps (Gelbart, 1998). Interactive web pages currently may be the best way to accurately and effectively describe updated chromosome maps. Web sites like these are essential tools for correlation of the full range of information required to understand the genetic bases of diseases (Rowen *et al.*, 1997; Waterston and Sulston, 1998).

4.5.1 *NCBI Human Genome Web Resource*

The Human Genome Resources page at the NCBI (www.ncbi.nlm.nih.gov/genome/ guide/) is currently the most comprehensive resource for physical mapping data for the human genome. The opening page at this site presents a set of ideograms (cartoons of chromosomes) of the 24 human chromosomes (22 autosomes plus the X and Y sex chromosomes). Clicking on any one of the ideograms produces a page with a detailed radiation hybrid map of that chromosome. Accompanying text lists information about locations of genes on that chromosome as well as information about known disease associations. Microsatellite markers, ESTs, known genes, and currently open reading frames of unknown function are indicated. Each point on the map has a hypertext link that gives additional details relevant to the marker or gene position on the map. Those links lead in turn to citations, DNA sequences, radiation hybrid mapping data, and other important reference information linked to that site. Starting from the highly simplified ideogram of a given chromosome one can quickly locate detailed information about a given region of any human chromosome. This approach appears to be well designed to incorporate new information as it is elaborated.

There is a search box at the top of each chromosome page that queries a new nomenclature cross-referencing tool called LocusLink developed at NCBI. One can search five different databases (LocusLink, MEDLINE, OMIM, GenBank and UniGene) using official gene names, aliases, phenotypes, protein names, sequence accession numbers, database identifiers, UniGene clusters, MIM numbers, EC numbers, or map information. LocusLink is a new tool that will provide a highly organized method to find and retrieve genome information.

The Human Genome Web Resource provides links to related web sites. A link to the Human Genome Sequencing page found on this page returns information and links to the human genome sequencing effort. In addition to this site, each of the genome centers maintains a web site. A link to a list of hypertext links to 23 different genome centers is found on the Human Genome Sequencing front page. From this page one can retrieve data regarding progress toward sequencing of the human genome. If one requires very specific information about a given chromosome it may be helpful to link to the site responsible for that part of the sequencing. For example, one could browse the Chromosome 12 Genome Center (http://paella.med.yale.edu/chr12/Home.html) for information specific to that chromosome. As the genome projects are completed, the information will be incorporated into the NCBI site and will be available in a highly useful, common format available to all investigators via the WWW.

4.5.2 *Entrez Genomes at NCBI*

The Entrez Genomes link (www.ncbi.nlm.nih.gov/Entrez/Genome/org.html) provides direct links to complete genomes of humans as well as other organisms. The Entrez database provides a comprehensive interface to genomic data. The Genomes database integrates genetic and physical maps of chromosomes with the variety of database entries available through Entrez. The Entrez home page has seven different search modes. Selecting the "Genomes" button leads to the Entrez Genomes page that contains a current list of species for which genomic sequences are available. The genomic data includes bacteria, archea and eukaryotes. Selection of one of the hypertext links takes the user to a summary page for the selected organism that includes a chromosome map(s). Clicking on a point on the map will zoom in for a closer view of the region. The maps can be expanded all the way to the point of displaying the DNA sequence with the designated open reading frames (ORFs). Clicking on an ORF will display the DNA sequence and the deduced amino acid sequence. From that page one can search the selected gene against the sequence databases using BLAST. So, starting with a global view of the chromosome structure one can focus the display all the way down to the level of the DNA sequence of the gene.

One can also use the Entrez Genomes page to search for a specific gene using a text search. For example, a search for "p21 Ras" found two entries, the first for human chromosome 5 and the second for the GTPase activating protein ras p21 cited in the UNIGENE database. Selection of the chromosome 5 entry "graphical view" presents seven different chromosome 5 maps produced by different methods (for a more detailed discussion of physical mapping databases, see Stein, 1998). By clicking on the map representations an expanded view of the selected region is returned that shows more detail of gene organization in the region. The citations returned by the search also give MEDLINE links and sequence database links relevant to the query. With this site one can quickly find the chromosomal localizations of known genes as well as related citations.

4.5.3 *Other sources of genomic information*

A number of specialized databases are also found at the NCBI. These include the Online Mendelian Inheritance in Man (OMIM), the Unique Human Gene Sequence Collection (UniGene), Clusters of Orthologous Groups (COGs), the Taxonomy Browser, and the Cancer Genome Anatomy Project (CGAP). Links to these databases (Table 4.3) are found on the NCBI homepage. These sites correlate genetic information with detailed genetic and physical maps of chromosomes. One can locate information about genetic diseases and associated genes. Because there are far too many sites to include in this chapter, only a few of particular interest will be mentioned briefly below.

OMIM is a catalog of human genes and genetic disorders now maintained at the NCBI. This site was authored and edited by Victor A. McKusick and colleagues at Johns Hopkins University and adapted for the WWW by the NCBI (McKusick, 1998). The site contains a searchable database with text and references to identified genetic disorders in man. The entries contain links to citations in the Entrez databases.

Table 4.3 Selected genomic resources at NCBI

Web URL	*Site information*
www.nhgri.nih.gov/Intramural_research/Lab_transfer/Bic/	Breast Cancer Information Core, an on-line breast cancer mutation database
www.ncbi.nlm.nih.gov//ncicgap/	The Cancer Genome Anatomy Project (CGAP) information and technological tools needed to decipher the molecular anatomy of the cancer cell
www.ncbi.nlm.nih.gov/COG/	A system of gene families deduced from complete genomes
www.ncbi.nlm.nih.gov/Taxonomy/	Taxonomy database containing the names of all organisms represented in the genetic databases
www.ncbi.nlm.nih.gov/UniGene/index.html	An experimental system for automatically partitioning GenBank sequences into a non-redundant set of gene-oriented clusters
www.ncbi.nlm.nih.gov/genemap/	A gene map of the human genome
www.ncbi.nlm.nih.gov/Omim/	A catalog of human genes and genetic disorders

The goal of the CGAP is to produce a comprehensive molecular characterization of normal, precancerous and malignant cells. The CGAP web page provides links to five different initiatives: the Human Tumor Gene Index, a Molecular Fingerprinting project, the Cancer Chromosome Aberration Project, the Genetic Anticipation Initiative and the Mouse Tumor Gene Index. The site provides links to a variety of resources relevant to cancer. These include information about where to find cDNA libraries, cDNA clones, cancer genes, chromosome breakpoint maps and cancer gene expression data.

The COGs site documents a system of gene families established by comparison of complete genome sequences (Tatusov *et al.*, 1997). These authors compared proteins encoded by seven complete genomes from five major phylogenetic lineages. They observed patterns of sequence similarities that formed 720 "clusters of orthologous groups" (COGs). The web site allows a user to compare a new sequence against the COGs to determine whether that sequence is a member of a known protein family. This approach provides information about possible functions of unknown gene products.

4.6 Conclusions

Some have compared the potential impact of the WWW on society to the industrial revolution. Clearly this comparison is not an exaggeration when one considers the positive impact that the web has already had on the dissemination of scientific information. Any effort to document the information available on the web in a traditional printed format is likely to be inadequate by the time the ink dries on the printed page. The web is evolving at a remarkable rate and the challenge to each scientist is to identify and use the available resources that will enhance research efforts. The ability to make effective use of scientific resources available through the Internet is now an essential skill for every successful scientist.

References

Altschul, S.F., Gish, W., Miller, W., Myers, E.W. and Lipman, D.J., 1990, Basic local alignment search tool. *J. Molec. Biol.*, **215**, 403–410.

Altschul, S.F., Madden, T.L., Schaffer, A.A., Zhang, J., Zhang, Z., Miller, W. and Lipman, D.J., 1997, Gapped BLAST and PSI-BLAST: a new generation of protein database search programs. *Nucleic Acids Res.*, **25**, 3389–3402.

Baxevanis, A.D., 1998a, The Internet and the biologist, in Baxevanis, A.D. and Ouellette, B.F.F. (Eds) *Bioinformatics: a Practical Guide to the Analysis of Genes and Proteins*, pp. 1–15, New York: John Wiley & Sons.

Baxevanis, A.D., 1998b, Information retrieval from biological databases, in Baxevanis, A.D. and Ouellette, B.F.F. (Eds) *Bioinformatics: a Practical Guide to the Analysis of Genes and Proteins*, pp. 98–120, New York: John Wiley & Sons.

Benson, D.A., Boguski, M.S., Lipman, D.J., Ostell, J. and Ouellette, B.F., 1998, GenBank. *Nucleic Acids Res.*, **26**, 1–7.

Boguski, M.S., 1995, The turning point in genome research. *Trends Biochem. Sci.*, **20**, 295–296.

Boguski, M.S., Lowe, T.M. and Tolstoshev, C.M., 1993, dbEST – database for "expressed sequence tags". *Nat. Genet.*, **4**, 332–333.

Dayhoff, M.O., Eck, R.V., Chang, M.A. and Sochard, M.R., 1965, *Atlas of Protein Sequence and Structure*, Silver Spring, Maryland: National Biomedical Research Foundation.

Gelbart, W.M., 1998, Databases in genomic research. *Science*, **282**, 659–661.

George, D.G., Dodson, R.J., Garavelli, J.S., Haft, D.H., Hunt, L.T., Marzec, C.R., Orcutt, B.C., Sidman, K.E., Srinivasarao, G.Y., Yeh, L.S.L., Arminski, L.M., Ledley, R.S., Tsugita, A. and Barker, W.C., 1997, The Protein Information Resource (PIR) and the PIR-International Protein Sequence Database. *Nucleic Acids Res.*, **25**, 24–28.

Hogue, C.W.V., 1997, Cn3D: a new generation of three-dimensional molecular structure viewer. *Trends Biochem. Sci.*, **22**, 314–316.

Hogue, C.W.V. and Bryant, S.H., 1998, Structure databases, in Baxevanis, A.D. and Ouellette, B.F.F. (Eds) *Bioinformatics: a Practical Guide to the Analysis of Genes and Proteins*, pp. 46–73, New York: John Wiley & Sons.

Marchler-Bauer, A., Address, K.J., Chappey, C., Geer, L., Madej, T., Matsuo, Y., Wang, Y. and Bryant, S.H., 1999, MMDB: Entrez's 3D structure database. *Nucleic Acids Res.*, **27**, 240–243.

McKusick, V.A., 1998, Mendelian inheritance in man, in *Catalogs of Human Genes and Genetic Disorders*, 12th edn, Baltimore: Johns Hopkins University Press.

Montelione, G.T. and Anderson, S., 1999, Structural genomics: keystone for a human proteome project. *Nature Struct. Biol.*, **6**, 11–12.

Olson, M., Hood, L., Cantor, C. and Botstein, D., 1989, A common language for physical mapping of the human genome. *Science*, **245**, 1434–1435.

Ouellette, B.F. and Boguski, M.S., 1997, Database divisions and homology search files: a guide for the perplexed. *Genome Res.*, **7**, 952–955.

Pai, E.F., Krengel, U., Petsko, G.A., Goody, R.S., Kabsch, W. and Wittinghofer, A., 1990, Refined crystal structure of the triphosphate conformation of H-ras p21 at 1.35 Å resolution: implications for the mechanism of GTP hydrolysis. *EMBO J.*, **9**, 2351–2359.

Pearson, W.R., 1990, Rapid and sensitive sequence comparison with FASTP and FASTA. *Methods Enzymol.*, **183**, 63–98.

Pearson, W.R., 1996, Effective protein sequence comparison. *Methods Enzymol.*, **266**, 227–258.

Rowen, L., Mahairas, G. and Hood, L., 1997, Sequencing the human genome. *Science*, **278**, 605–607.

Schuler, G.D., 1998, Sequence alignment and database searching, in Baxevanis, A.D. and Ouellette, B.F.F. (Eds) *Bioinformatics: a Practical Guide to the Analysis of Genes and Proteins*, pp. 145–171, New York: John Wiley & Sons.

Stein, L.D., 1998, Of mice and men: navigating public physical mapping databases, in Baxevanis, A.D. and Ouellette, B.F.F. (Eds) *Bioinformatics: a Practical Guide to the Analysis of Genes and Proteins*, pp. 268–298, New York: John Wiley & Sons.

Stoesser, G., Sterk, P., Tuli, M.A., Stoehr, P.J. and Cameron, G.N., 1997, The EMBL nucleotide sequence database. *Nucleic Acids Res.*, **25**, 7–14.

Tateno, Y. and Gojobori, T., 1997, DNA data bank of Japan in the age of information biology. *Nucleic Acids Res.*, **25**, 14–17.

Tatusov, R.L., Koonin, E.V. and Lipman, D.J., 1997, A genomic perspective on protein families. *Science*, **278**, 631–637.

Teng, D.H., Bogden, R., Mitchell, J., Baumgard, M., Bell, R., Berry, S., Davis, T., Ha, P.C., Kehrer, R., Jammulapati, S., Chen, Q., Offit, K., Skolnick, M.H., Tavtigian, S.V., Jhanwar, S., Swedlund, B., Wong, A.K. and Kamb, A., 1996, Low incidence of BRCA2 mutations in breast carcinoma and other cancers. *Nat. Genet.*, **13**, 241–244.

Terwilliger, T.C., Waldo, G., Peat, T.S., Newman, J.M., Chu, K. and Berendzen, J., 1998, Class-directed structure determination: foundation for a protein structure initiative. *Protein Sci.*, **7**, 1851–1856.

Waterston, R. and Sulston, J.E., 1998, The Human Genome Project: reaching the finish line. *Science*, **282**, 53–54.

5

CANCER SUSCEPTIBILITY GENES

Keith W. Crawford and Peter G. Shields

5.1 Introduction

The emerging discipline of molecular epidemiology seeks to identify biological risk factors that increase an individual's risk of acquiring cancer. More than 70% of all cancers are considered to be due to carcinogen exposures, suggesting that cancer may be highly preventable (Doll and Peto, 1981). Among the most potent of known carcinogens is tobacco smoke, which increases lung cancer risk by more than 14-fold (Doll and Peto, 1978), but the effects of tobacco can vary widely. For example, only about 10% of heavy smokers ever develop lung cancer. It is plausible that these people possess a combination of genetic susceptibility factors that increase their risk following exposure. In an attempt to understand the risks related to specific individuals, a major area of substantial research relates to genetic susceptibilities and biomarker development.

Two fundamental principles that underlie current studies of molecular epidemiology relate to the complexity of the carcinogenesis process and the fact that individuals vary in response to carcinogenic exposures. For the first, cancer is a multistage process where each stage is caused by several genetic mutations. These mutations are the result of several steps, beginning with an exogenous exposure (e.g. chemicals, radiation, or viruses), endogenous exposure (e.g. oxy-radicals), or enzymatic error (e.g. polymerase or recombinase infidelity). But, following these exposures, several processes must occur or go awry. For example, chemical exposures by themselves are not sufficient to cause cancer. Before the mutation occurs, a potential chemical carcinogen has to be absorbed, undergo metabolic activation (which frequently requires several enzymatic steps), escape detoxification, be transported to a target organ, adduct a critical protooncogene or tumor suppressor gene, escape DNA repair, and escape other mechanisms to control the damage, such as cell death. Thus, the multistage process of carcinogenesis is a very complicated process consisting of multiple steps and pathways. This complexity leads to difficulties in the design and interpretation of studies, and the need to formulate *a priori* hypotheses is critical.

The second fundamental principle of molecular epidemiology is based upon data showing interindividual variation (Harris, 1989) in response to carcinogen exposure and carcinogenic processes. In fact, the population is actually quite heterogeneous. Different ages, gender, race, ethnicity, lifestyle (e.g. tobacco use, alcohol, exercise), diet, occupation, and recreational activities all might contribute to differences in responses to environmental agents, although the relative effects of these are currently unknown. Equally important are inherited differences in the host's response to specific exposures, such as in one's metabolic capacity to activate or detoxify carcinogens, and/or repair DNA damage. We are thus further challenged by the paradigm that a combination of exposures and susceptibilities leads to one type of cancer in one person, but another type in a different person, and that different combinations of exposures and susceptibilities can lead to the same type of cancer in different people.

Our approach to studying carcinogenesis in epidemiology is rapidly changing. We are beginning to categorize our genes differently. Specifically, genes have caretaker functions, while others can be thought of as gatekeepers (Kinzler and Vogelstein, 1997). The former category includes genes that maintain basic cellular and genomic integrity, such as DNA repair, carcinogen metabolism, and DNA replication. The latter category includes genes that regulate cell cycle or govern programmed cell death. Dysfunctional caretaker genes increase the probability of mutations in gatekeeper genes, which are necessary to initiate the molecular pathogenesis of cancer. As we classify genes according to this model, rather than whether genes are protooncogenes or tumor suppressor genes, we will also consider gene–gene interactions for cancer risk from a different perspective.

Genetic variation leads to varying degrees of cancer risk. The frequency for "at-risk" genetic variants ranges from rare to common, as do the potential importance and penetrance (i.e. who is affected). Low-penetrant genes typically affect common sporadic cancers, while high-penetrant genes cause the rare family cancer syndromes. The relative roles for inherited susceptibility and carcinogen exposure can be very different for familial and common sporadic cancers (Table 5.1; Caporaso and Goldstein, 1995). Examples of highly penetrant rare mutations are inheritance of *p53* mutations in Li–Fraumeni syndrome families and BRCA1 mutations in breast cancer families. Nonetheless, these rare mutations account for less than 1% of all human cancer, and carcinogen exposure plays a small role in these syndromes (Fearon, 1997). In contrast, polymorphisms for carcinogen-metabolizing genes are examples of low-penetrance traits with relatively small increased risks for the individual, but have important public health consequences because the risk affects many people (a polymorphism is a genetic trait that occurs in at least 1% of a population). Examples include genetic risk factors for either colon or tobacco-related lung cancer. For these commonly occurring sporadic cancers, carcinogen exposures (exogenous and endogenous) play a critical role, and gene–environment interactions are common.

Inherited susceptibilities to cancer are due to variations in the genetic code that alter protein function or localization. A common approach to studying cancer risk is through the determination of genetic polymorphisms. This difference in sequence might be a single nucleic acid base change that changes the protein coding, resulting in a different amino acid sequence and attendant change in function. The possible variations, however, range from single nucleic acid base changes to whole deletions of a gene. Polymorphic variations occur commonly within the genome, although most are silent because they do not affect amino acid changes or they occur in a non-coding region of the gene. The most interesting polymorphisms to study, however, are those that result in changes in protein quantity or function, because those studies can be based on biologically derived hypotheses.

Table 5.1 Genetics and the environment

	Familial cancers (e.g. p53 mutations and Li–Fraumeni Family Cancer Syndrome)	*Sporadic cancers (e.g. tobacco-related lung cancer)*
Frequency in the population	Rare	Common
Role of inherited susceptibilities	Single gene	Multiple genes
Etiology	Primarily genetic	Multifactorial – genes and environment
Familial pattern (e.g. results of family studies)	Gene segregates with disease	Genes aggregate with disease, or difficult to demonstrate heritability
Environmental effect	Minimal	Critical
Genetic risk	Predetermining	Predisposing

Genetic susceptibility can be assessed either phenotypically (measuring the resultant enzymatic function) or genotypically (determining the genetic code). Phenotypic assays may include determining enzymatic activity by administering probe drugs to individuals and measuring urinary metabolites, assessing carcinogen metabolic capacity in cultured lymphocytes, or establishing the ratios of endogenously produced substances, such as estrogen metabolite ratios. Using a genetic-based assay to assess cancer risk is generally preferable because DNA is easier to obtain and the assays are technically simpler. However, phenotypes result from the effects of several genes, and may not be adequately characterized by only one genetic assay. Therefore, there is a role for both genetic- and phenotype-based assays in research studies. Examples of frequently studied genetic polymorphisms in sporadic cancers are the N-acetyltransferase 2 (*NAT2*), glutathione S-transferase M1 (*GSTM1*), and cytochrome P450 1A1 (*CYP1A1*) genes. The "at-risk" genetic variants for these genes in the United States population (Caucasians and African Americans) are approximately 50%, 50%, and <10%, respectively.

5.2 Low-risk penetrance genes: genetic polymorphisms for carcinogen activation

The most intensively studied areas of polymorphism involve genes that govern carcinogen metabolic activation and detoxification. Carcinogens that enter the body from exogenous sources, or are produced endogenously, are recognized typically as foreign agents that need to be excreted. For most carcinogens, this process involves a series of steps that make the carcinogen water-soluble and/or conjugated to a carrier molecule. The first step is that carcinogens are oxidatively metabolized to reactive electrophilic intermediates by cytochrome P450s (CYP450) or other enzymes (phase I reaction), followed by conjugation reactions (phase II reactions) that detoxify the compound. A polymorphism within a phase I enzyme that results in increased production of an electrophilic intermediate or one in a detoxification enzyme that decreases detoxification can result in elevated levels of reactive intermediates that can damage DNA (Figure 5.1).

CYP450 enzymes are heme-containing proteins, which can play either an activating or a detoxifying role, depending on the CYP450 and the substrate. The CYP450 nomenclature (Nelson *et al.*, 1993) is based on the degree of genetic similarity, chromosomal location, and substrate specificity. For human genes, an Arabic numeral and letter follow the "CYP" designation, indicating the family and subfamily of the gene (e.g. CYP1A). A gene family shares greater than 40% homology, has the same number of exons, and similar intron–exon boundaries. Subfamilies are located in the same gene cluster and are non-segregating (e.g. *CYP1A1, CYP1A2*). Mammalian sequences within the same subfamily are always >55% identical. The last Arabic numeral designation is usually based on the order in which the gene sequence of the particular enzyme became known, and on substrate preferences.

5.2.1 Cytochrome P450 1A1

CYP1A1 plays an important role in the activation of potential human carcinogens and also metabolizes estrogens. CYP1A1 substrates include polycyclic aromatic hydrocarbons (PAHs) such as benzo[*a*]pyrene, a carcinogen in tobacco smoke. The

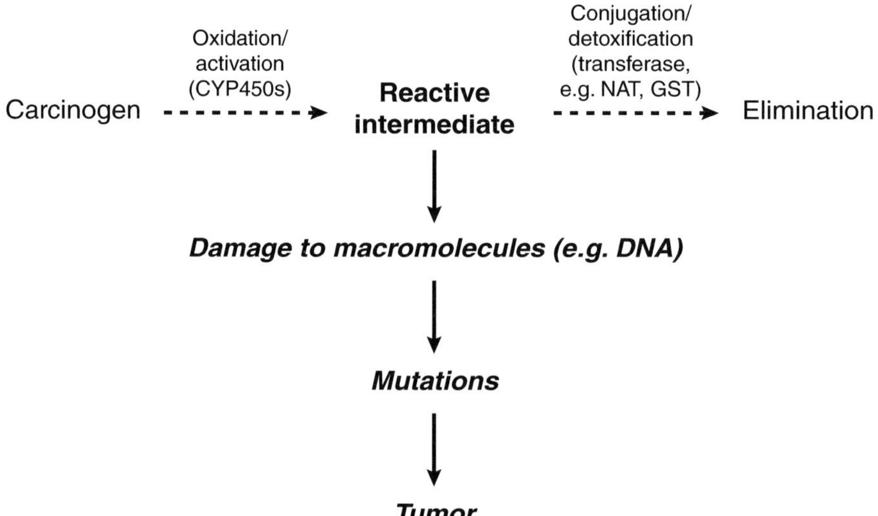

Figure 5.1 This schema illustrates a common pathway where procarcinogens undergo metabolic processing that generally leads to the production of compounds with increased water solubility and enhanced excretion. Enzyme polymorphisms that increase the oxidative pathway (activation of the carcinogen), decrease the conjugation (detoxification), or a combination of both can lead to increased levels of the reactive electrophilic intermediates. These intermediates can damage proteins, lipids, and DNA. The DNA damage is usually repaired, but if the reactive intermediates overwhelm the repair mechanism, a mutation could occur in a gene critical for cell-cycle control, differentiation, or programmed cell death. CYP450, cytochrome P450; NAT, *N*-acetyltransferase; GST, glutathione *S*-transferase.

CYP1A1 gene is primarily expressed in extra-hepatic tissues, such as the lung, implying its importance in activating carcinogens locally within the target organ. The level of CYP1A1 is regulated by the aromatic hydrocarbon receptor (*Ah* receptor), which is stimulated by tobacco smoke, PAHs, organochlorine compounds, steroids, and other compounds. Thus, it is presumed that some carcinogens, through the induction of CYP1A1, can lead to greater activation of those same carcinogens, heightening cancer risk. When an inducer binds to the *Ah* receptor in the cytoplasm, the compound–receptor complex translocates to the nucleus and upregulates the expression of CYP1A1 by binding to the Ah-responsive elements in the 5′ flanking region of the *CYP1A1* gene (Catteau *et al.*, 1995; Fujii-Kuriyama *et al.*, 1995). Genetic variants of this receptor that govern CYP1A1 inducibility have not yet been identified (Kawajiri *et al.*, 1995), although the large degree of interindividual variation for inducibility makes it likely that such genetic variants exist.

CYP1A1 enzyme activity varies greatly among people, more than 100-fold, and this variation is thought to lead to cancer risk. Several *CYP1A1* candidate genetic polymorphisms exist, although none has been conclusively established that can alter the function of the enzyme. One commonly studied genetic polymorphism is found in the 3′ region of the gene, outside the coding area. It occurs at a restriction enzyme sensitive site, so that it is easily identified with a restriction enzyme digestion, specifically *MSP*1. Associations have been shown between *Msp*1 genotype and CYP1A1 activity by some investigators (Landi *et al.*, 1994), but not by others (Cosma *et al.*, 1993). This enzyme polymorphism has been most closely associated with cancer risk, but primarily in Japanese. Kawajiri *et al.* (1990) noted a three-fold higher risk of lung cancer with one variant. Furthermore, it was found that there was an interaction between this polymorphism and smoking for lung cancer risk (Nakachi *et al.*, 1991; Okada *et al.*, 1995), where the combination of the homozygous minor allele and smoking yielded odds ratios similar to having one of the other genotypes and a greater smoking history. The polymorphism increased risk for both squamous cell and adenocarcinomas (Hayashi *et al.*, 1991; Nakachi *et al.*, 1995). One report associates this polymorphism with increased stages of disease (Okada *et al.*, 1995), and recent data suggest that CYP1A1 genetic polymorphisms predict the presence of *p53* mutations (Kawajiri *et al.*, 1996). Also in this study group there was a 4.5-fold increase in risk (95% CI = 1.64, 12.26) for smokers who carried a rare allele for CYP1A1 genetic polymorphisms (Kawajiri *et al.*, 1996). Overall, similar findings have not been found in Western populations (Tefre *et al.*, 1991; Hirvonen *et al.*, 1992; Shields *et al.*, 1993; Alexandrie *et al.*, 1994). One explanation for this discrepancy is that the frequency of the 'at-risk' allele is three times greater in Japanese compared with Westerners, so that this gene might still play a role in Westerners, but our epidemiology studies might be too insensitive. Another possible reason for the differences in findings between different races may be that the *Msp*1 polymorphism is a marker for another genetic variation, which only occurs in Japanese (Wedlund *et al.*, 1994).

A polymorphism exists in exon 7 of *CYP1A1*, which codes for a valine substitution for isoleucine. This site is located in the heme-binding region of the enzyme. Catalytic activity of the variants has been studied, and although the substitution is in a critical area of the gene, there is no difference in catalytic efficiency (Zhang *et al.*, 1996; Persson *et al.*, 1997). This polymorphism has been associated with lung cancer in Japanese (Kawajiri *et al.*, 1993) and in Brazilians (Sugimura *et al.*, 1995), but not in Caucasians (Hirvonen *et al.*, 1992; Alexandrie *et al.*, 1994; Cascorbi *et al.*, 1996). Associations between *CYP1A1* genotype and breast cancer have been reported in lighter smokers (Ambrosone *et al.*, 1995; Ishibe *et al.*, 1997), but these findings are considered preliminary and other studies have been negative (Rebbeck *et al.*, 1996).

The third *CYP1A1* polymorphism exists only in African-Americans (Crofts *et al.*, 1993). It was thought to be associated with adenocarcinoma of the lung (Eliopoulos *et al.*, 1995), but other studies could not confirm this finding (Kelsey *et al.*, 1994; London *et al.*, 1995a; Taioli *et al.*, 1995a). Taoli and Garte (1996) suggest that possible differences in detecting the association of the genotype with histological type may be due to appropriate selection of controls, size of the studies, or effect of adjustment for smoking, but it may also be a spurious finding. This genetic variant might also be a risk factor for breast cancer in African-Americans (Taioli *et al.*, 1995b).

5.2.2 *Glutathione S-transferases*

These enzymes play an important role in detoxification. They conjugate reactive carcinogen and carcinogen metabolites with glutathione, thereby increasing water solubility and enhancing excretion. Five genes for glutathione *S*-transferases have been identified: alpha, kappa, mu, pi, and theta. Of these, genetic polymorphisms in *GSTM1* have been studied extensively, because in approximately one-half of the population the gene is deleted. In these people, there is an inability to detoxify specific carcinogens. These people have decreased DNA damage in lung (Shields *et al.*, 1993; Kato *et al.*, 1995; Rothman *et al.*, 1995) and placenta (Topinka *et al.*, 1997) implicating this gene in teratogenesis or cancer risk. The *GSTM1* null genotype is associated with lung cancer in several studies (Seidegard *et al.*, 1990; Zhong *et al.*, 1991; London *et al.*, 1995b; Nakajima *et al.*, 1995), and a meta-analysis was supportive of increased cancer risk (Rebbeck, 1997). This genotype also predicts the presence of *p53* mutations within those cancers (Ryberg *et al.*, 1994). Separately, the occurrence of the *GSTM1* null genotype is related to bladder cancer risk (Anwar *et al.*, 1996; Brockmoller *et al.*, 1996) and *p53* mutations within those tumors (Brockmöller *et al.*, 1993). GSTM1 also modulates lung GSTM3 levels (Anttila *et al.*, 1995; Nakajima *et al.*, 1995) and CYP1A1 transcription (Vaury *et al.*, 1995), so that the GSTM1-related cancer risk may be due to detoxification and induction of other genes.

There has been increasing attention to polymorphisms of *GSTP1*, where there is a single amino acid (isoleucine/AA or valine/GG) at position 104 in some people. This change alters the conformation of the catalytic site, producing differences in reactivity toward substrates (Hu *et al.*, 1997). This variation has been associated with increased DNA damage in human lung tissues, male lung cancer risk (Ryberg *et al.*, 1997), and oral cavity cancer (Matthias *et al.*, 1998).

GSTT1 is another polymorphism where the entire gene is deleted in some people. This deletion is associated with increased DNA damage in cultured lymphocytes exposed *in vitro* (Norppa *et al.*, 1995) and *in vivo* (Christensen *et al.*, 1998) to 1,3-butadiene metabolites, which are potent lung carcinogens in laboratory animals

and are present in tobacco smoke. Evidence for an association of *GSTT1* null genotype and lung cancer is lacking, but there is some evidence for esophyageal, laryngeal
cancers, and skin cancers (Hung *et al.*, 1997; Lear *et al.*, 1997).

5.2.3 Arylamine transferases

These enzymes are involved in the metabolism of aromatic amines. Aryl aromatic
amines, such as 4-aminobiphenyl, are present in cigarette smoke and the work-
place, and have been implicated in bladder and breast cancers. Heterocyclic aromatic
amines, formed from the overcooking of foods such as meat, chicken, and fish, are
thought to be involved in colon carcinogenesis. Two arylamine transferases have
been implicated in human cancer risk. NAT1 plays a bioactivation role as a phase
I enzyme, while NAT2 can both bioactivate and detoxify, depending on the
substrate.

The NAT2 genetic polymorphism was among the first to be described because
it governs the acetylation of isoniazid for tuberculosis treatment, where slow acetylators have increased chances of hepatic and neurological toxicity. The amount of
NAT2 slow acetylators found among different ethnic or racial groups ranges from
20 to 80%. The slow acetylator phenotype has been linked to occupationally
induced bladder cancer in dye workers exposed to large amounts of *N*-substituted
aryl compounds (Cartwright *et al.*, 1982). This is thought to be related to activation by the liver cytochrome P4501A2 gene, subsequent transport of reactive
metabolites to the bladder via urine, and then decreased detoxification by the variant
NAT2 enzyme. To support this hypothesis in humans, one study (Landi *et al.*,
1996) has shown that rapid oxidizers of CYP1A2 combined with NAT2 slow-
acetylation phenotypes showed the highest level of 4-aminobiphenyl–hemoglobin
adducts. An increased risk for rapid acetylators has been reported in studies of
colon cancer (Lang *et al.*, 1986), where there is a hypothesized relationship to activation of heterocyclic amines by the NAT2 gene in colonic epithelium. Thus, both
CYP1A2 and NAT2 bioactivate these compounds leading to more DNA damage.
For example, persons with higher activity by CYP1A2 and rapid acetylation by
NAT2 had the highest risks for colorectal neoplasia in people who consumed well-
done red meat (Lang *et al.*, 1994). Separately, tobacco smoking was associated with
breast cancer in Caucasian postmenopausal women who were NAT2 slow acetylators, and predictably had a decreased capacity to detoxify aromatic amines
(Ambrosone *et al.*, 1996). Although a prospective study of nurses (Hunter *et al.*,
1997) seemingly provided conflicting data, there were differences in data analysis
and study size. Other studies are nearing completion so that more evidence will
soon be available.

NAT2 and cigarette smoking is a good example of a gene–environment interaction. It was originally shown that the level of 4-aminobiphenyl–hemoglobin adducts,
where 4-aminobiphenyl is a constituent of cigarette smoke, was correlated with
levels of cigarette smoking (Vineis *et al.*, 1990). In fact, it was found that the
adduct levels correlated with the type of cigarette smoked (black versus blond), but
that in all cases, levels were higher in persons who were phenotypically slow acetylators. It was hypothesized that the decreased detoxification by the NAT2 in the
liver led to an increased number of reactive metabolites that were available to

adduct hemoglobin. The data have undergone further analysis (Vineis *et al.*, 1994), where levels of cigarette smoking were considered in study subjects by measuring urinary nicotine and cotinine. The analysis suggested that persons with passive tobacco smoke exposure and who were slow acetylators had a proportional greater risk of forming adducts compared with active tobacco smokers.

The *NAT1* gene polymorphism is located in a polyadenylation site, which was suggested to be associated with higher tissue levels of the enzyme (Badawi *et al.*, 1996), presumably by elevating NAT1 mRNA levels. NAT1 activity in bladder and carcinogen–DNA adduct levels in individuals heterozygous for the **10* allele was twofold higher than what was observed in individuals homozygous for the **4* allele. The highest adduct levels were observed in individuals with NAT1 rapid genotypes (**10* allele) combined with NAT2 slow genotypes. This variant is associated with increased risk for colorectal cancer (Bell *et al.*, 1995).

5.3   High-penetrance genes: familial cancer syndromes

Familial cancer accounts for about 1% of all cancers. These are caused by inheritance of mutated cancer susceptibility genes that are highly penetrant, where the chances of developing cancer in a person with the mutation can be as high as 90%. Although these genes cause only a fraction of total cancers, and so may not have a clear public health importance, they are clearly devastating to those affected cancer families. Research on familial cancers has significantly advanced both basic science and clinical aspects of oncology. In many cases, the same genes are sporadically mutated in common cancers. These genes are typically involved in cell-cycle control, DNA repair, transcription regulation, cell–cell or cell–matrix adhesion, cytoskeletal architecture, and signal transduction.

There are different ways to conceptualize the roles of genes in cancer susceptibility. One way is to consider genes as protooncogenes and tumor suppressor genes. Mutations in protooncogenes are generally inherited in an autosomal dominant fashion, and therefore a mutation in a single allele, whether acquired or inherited, may be adequate to initiate disease. The mutations in oncogenes produce a gain of function, or increase the function of the gene's protein product, which thereby increases the risk of cancer. Mutations in tumor suppressor genes generally act recessively. Mutations in both alleles of a gene need to occur. In the case of a tumor suppressor gene, the mutations result in a loss of function or an abrogation of the activity of the gene's protein product. The second way to conceptualize and categorize the role of genes in cancer was proposed recently (Kinzler and Vogelstein, 1997), whereby we consider them as caretaker and gatekeeper genes. Here, we acknowledge their respective roles as a caretaker in maintaining genomic integrity and as gatekeepers affecting cellular proliferation. Some examples of caretaker genes are those that are involved in DNA repair, while examples of gatekeeper genes are those involved in cell-cycle control and DNA replication. Dysfunctional caretaker genes increase the probability of mutations in gatekeeper genes, which are necessary to initiate the molecular pathogenesis of cancer. It is interesting that the carcinogenic effects of this dysfunction appear to be tissue specific and lead to cancer only in specific organs, even though these genes are expressed in many different organs.

5.3.1 Retinoblastoma

Initial insights into the mechanisms of tumor suppressor genes were gained from early studies of retinoblastoma. Patients without family histories of retinoblastoma usually develop it only in a single eye, whereas in persons with a family history, the cancers occur in both eyes. Also, inherited disease usually occurs much earlier in life than sporadic cancer. Mapping studies for the retinoblastoma susceptibility gene (Rb) revealed that there was a loss of functional gene in both familial and sporadic disease. In most cases, the wild-type allele was lost in the tumor. This observation led Knudson (Cooper, 1995) to propose a "two-hit" hypothesis where a mutation in a tumor suppressor gene either occurs through inheritance or is induced through carcinogen exposure. A second mutation resulting in the loss of the wild-type allele then occurs which obliterates the function of the gene.

5.3.2 The p53 tumor suppressor gene

p53 is a 53-kDa nuclear phosphoprotein tumor suppressor gene that is mutated in more than 50% of sporadic cancer cases. Inherited *p53* germline mutations only account for a few cases of familial cancer, and occur in 60% of Li–Fraumeni syndrome families and about 25% of "Li–Fraumeni-like" syndrome families. The gene is located at chromosome 17p. *p53* functions in cell-cycle control, apoptosis, DNA repair, and transcriptional activation. Li–Fraumeni syndrome is characterized by multiple primary tumors occurring relatively early in life, including early-onset breast cancer, childhood soft-tissue sarcomas, leukemias, adrenocortical carcinomas, and central nervous system tumors (Evans and Lozano, 1997). The classical definition of the tumor requires that a proband be diagnosed with a sarcoma before age 45, a first-degree relative with cancer before age 45, and a first- or second-degree relative with either a sarcoma at any age or any cancer before 45. The distributions of germline mutations in *p53* appear to be similar to the distribution in sporadic cancers (Kleihues *et al.*, 1997). The majority of germline mutations occur within exons 5, 7, and 8 (Varley *et al.*, 1996; Saeki *et al.*, 1997).

5.3.3 Mismatch repair genes hMSH2 and hMLH1

These genes cause hereditary non-polyposis colorectal cancer (HNPCC), also known as the Lynch syndrome, which is characterized by the absence of polyposis, early age of diagnosis, high prevalence of right-sided colon tumors, and susceptibility to other primary malignancies (e.g. uterine, ovarian, stomach, pancreas, and breast). Mutations in hMSH2 and hMLH1 DNA repair genes have been found to account for 65% of the cases of HNPCC (Kinzler and Vogelstein, 1996; Edelmann *et al.*, 1997). Splicing defects producing alternate splicing of exons 9, 10, 15, 16, and 17 in hMLH1 and exon 5 in hMSH2 account for a substantial fraction of the germline mutations in hMSH2 and hMLH1 in HNPCC (Wijnen *et al.*, 1996). Two other mismatch repair genes, hPMS1 and hPMS2 (human post-meiotic segregation 1 and 2) located on chromosomes 2q31–32 and 7p22 may be involved in the remainder of the cases (Nicolaides *et al.*, 1995). Germline mutations in tumor growth factor β (TGF β) II receptor genes also may also play a role in HNPCC (Lu *et al.*, 1998).

There is an interesting gene–dose effect for alterations in MSH2 and MLH1. Replication errors are the hallmark of these tumors, where the effect of defective DNA repair is manifested in the number of repeating nucleotide sequences. Intra-tumor heterogeneity occurs, where there are mild or severe forms, depending on the number of repeats (Habano *et al.*, 1998). The loss of a single MSH2 and MLH1 allele may have only a mild effect on DNA repair capability and result in a mild replication error (RER+) phenotype. However, in tumor cells where both alleles are affected, or one deficient allele occurs with a germline mutation, a severe RER+ phenotype is observed, consistent with the two-hit hypothesis.

Several specific genetic defects have been reported, but alterations in these genes via genetic polymorphisms might also affect sporadic cancer risk. Such candidate polymorphisms include an intronic germline T→C transition in exon 13 of hMSH2, found in 25% of healthy Caucasians and 33% of Japanese. This particular variant may increase the sporadic colorectal cancer risk to 3.2–fold (Goessl *et al.*, 1997).

5.3.4 APC gene

Familial adenomatous polyposis (FAP) syndrome is an autosomal dominantly inherited disease that leads to colon cancer. FAP patients have thousands of adenomatous polyps in the colon by the age of 20, and 90% of these patients develop colon cancer by the fifth decade of life if no treatment is received (Eng and Ponder, 1993). The FAP syndrome is also associated with extra-intestinal manifestations such as congenital hypertrophy of the retinal pigment epithelium (CHRPE) and dental lesions, as well as a high incidence of desmoid tumors in the colon.

Inherited mutations in the adenomatous polyposis coli (APC) gene is considered an early step in FAP, but somatic mutations in sporadic polyps and cancers are also common. APC protein binds to β-catenins, which have been shown to interact with proteins involved in cell–cell adhesion (cadherins). Specific domains in APC are involved in β-catenin binding, and the location of mutations is associated with the phenotypic features of the disease (Fodde and Khan, 1995). Approximately 95% of mutations occur in the 5′ half of the gene, most of which result in protein truncation (Nagase *et al.*, 1992a; van der Luijt *et al.*, 1996). Patients who have frameshift mutations between codons 1250 and 1330 develop more than 5000 polyps, whereas mutations outside this region generally lead to less than 2000 polyps and milder disease (Nagase *et al.*, 1992b; Friedl *et al.*, 1996). Families with mutations located toward the 5′ end of codon 158 are more likely to have heterogeneous features including delayed development of colonic polyposis and colorectal cancer than families with more distal 5′ mutations in the gene (Giardiello *et al.*, 1997). Other mutations such as a frameshift mutation and a termination at codons 1862 and 1987 have been associated with greater variability in the number of colorectal adenomas, and mutations at codon 1309 result in thousands of adenomas at a young age and earlier death (Gayther *et al.*, 1994; Caspari *et al.*, 1995).

5.3.5 *BRCA1 and BRCA2 tumor suppressor genes*

BRCA1 is located on chromosome 17q21. Although its exact function is unknown, it is probably a transcription factor, suggested by the predicted zinc-finger domain. BRCA2 appears to be involved in DNA repair (Patel *et al.*, 1998). The BRCA1 gene contains 22 exons encoding a protein consisting of 1863 amino acids. BRCA2 is located on chromosome 13q12–13. It contains 26 exons and encodes a protein that is 3418 amino acids in length. Germline mutations in BRCA1 account for about 50% of families with a dominant predisposition to breast cancer and up to 90% of families with breast and ovarian cancer (Easton *et al.*, 1995; Szabo and King, 1995). Most germline mutations are frameshift or non-sense mutations that result in a truncated or inactive protein. In these families, women inherit one mutated gene, and the other is lost during life, supporting its role as a tumor suppressor (Shen *et al.*, 1998; Silva *et al.*, 1998). Both a BRCA1 genetic polymorphism (185delAG) and a BRCA2 genetic polymorphism are found in approximately 1% of Ashkenazi Jews, leading to a lifetime breast cancer risk of about 45% (Struewing *et al.*, 1997). In African-American compared with Caucasian women, different BRCA1 mutations are found (Shen *et al.*, 1998).

5.3.6 *RET gene*

The RET gene codes for a tyrosine kinase receptor whose ligand has been identified as glial-cell derived neurotropic factor (Marsh *et al.*, 1997). Germline mutations in *RET* are found in the syndrome of multiple endocrine neoplasia 2 (MEN2) (Borrello *et al.*, 1995; Frank-Raue *et al.*, 1996; Marsh *et al.*, 1996; Uchino *et al.*, 1998). MEN2 is an autosomal dominant inherited cancer syndrome that is subdivided based on which organs are affected. MEN2A is associated with medullary thyroid carcinoma (MTC) of the parafollicular C cells, pheochromocytoma, and hyperparathyroidism. MEN2B is similar to form 2A with the exception that parathyroid disease does not occur. Familial medullary thyroid carcinoma (FMTC) is a form where only MTC is present.

Several types of *RET* mutations have been described, and are associated with different outcomes. Mutations in the extracellular cysteine-rich domain of *RET* are found in the majority of families with MEN2A and FMTC (Mulligan *et al.*, 1995; Blank *et al.*, 1996; Eng *et al.*, 1996; Schuffenecker *et al.*, 1998). People with codon 618 mutations had a lower frequency of pheochromocytoma and parathyroid cancer, and increased life expectancy (Moers *et al.*, 1996). Mutations at codon 768 (exon 13) and 804 (exon 14) in the intracellular domain of the receptor have been associated with FMTC (Eng *et al.*, 1996; Fink *et al.*, 1996; Miyauchi *et al.*, 1997). Mutations in the tyrosine kinase domain lead to oncogenic activation of RET and the development of FMTC (Pasini *et al.*, 1997). A mutation in codon 918 (exon 16) occurs at the substrate recognition site of the receptor (Maeda *et al.*, 1995; Rossel *et al.*, 1995; Blank *et al.*, 1996; Eng *et al.*, 1996;) and has been associated with glaucoma in MEN2B (Mashima *et al.*, 1998).

Mutations in *RET* have also been reported in sporadic cases of MTC (Fink *et al.*, 1996). For example, 29% of tumors contain a somatic mutation at codon 918 that correlates with poor prognosis (Zedenius *et al.*, 1995).

References

Alexandrie, A.-K., Sundberg, M.I., Seidegard, J., Tornling, G. and Rannug, A., 1994, Genetic susceptibility to lung cancer with special emphasis on CYP1A1 and GSTM1: a study on host factors in relation to age at onset, gender and histological cancer types. *Carcinogenesis*, **15**, 1785–1790.

Ambrosone, C.B., Freudenheim, J.L., Graham, S., Marshall, J.R., Vena, J.E., Brasure, J.R., Laughlin, R., Nemoto, T., Michalek, A.M., Harrington, A., Ford, T.D. and Shields, P.G., 1995, Cytochrome P4501A1 and glutathione *S*-transferase (M1) genetic polymorphisms and postmenopausal breast cancer risk. *Cancer Res.*, **55**, 3483–3485.

Ambrosone, C.B., Freudenheim, J.L., Graham, S., Marshall, J.R., Vena, J.E., Brasure, J.R., Michalek, A.M., Laughlin, R., Nemoto, T., Gillenwater, K.A., Harrington, A.M. and Shields, P.G., 1996, Cigarette smoking, *N*-acetyltransferase 2 genetic polymorphisms, and breast cancer risk. *JAMA*, **276**, 1494–1501.

Anttila, S., Luostarinen, L., Hirvonen, A., Elovaara, E., Karjalainen, A., Nurminen, T., Hayes, J.D., Vainio, H. and Ketterer, B., 1995, Pulmonary expression of glutathione *S*-transferase M3 in lung cancer patients: association with GSTM1 polymorphism, smoking, and asbestos exposure. *Cancer Res.*, **55**, 3305–3309.

Anwar, W.A., Abdel-Rahman, S.Z., El-Zein, R.A., Mostafa, H.M. and Au, W.W., 1996, Genetic polymorphism of GSTM1, CYP2E1 and CYP2D6 in Egyptian bladder cancer patients. *Carcinogenesis*, **17**, 1923–1929.

Badawi, A.F., Stern, S.J., Lang, N.P. and Kadlubar, F.F., 1996, Cytochrome P-450 and aceyltransferase expression as biomarkers of carcinogen–DNA adduct levels and human cancer susceptibility, in Walker, C., Groopman, J., Slaga, T.J. and Klein-Szanto, A. (Eds) *Genetics and Cancer Susceptibility: Implications for Risk Assessment*, pp. 109–140, New York: Wiley-Liss.

Bell, D.A., Stephens, E.A., Castranio, T., Umbach, D.M., Watson, M., Deakin, M., Elder, J., Hendrickse, C., Duncan, H. and Strange, R.C., 1995, Polyadenylation polymorphism in the *N*-acetyltransferase 1 gene (NAT1) increases risk of colorectal cancer. *Cancer Res.*, **55**, 3537–3542.

Blank, R.D., Sklar, C.A., Dimich, A.B., LaQuaglia, M.P. and Brennan, M.F., 1996, Clinical presentations and RET protooncogene mutations in seven multiple endocrine neoplasia type 2 kindreds. *Cancer*, **78**, 1996–2003.

Borrello, M.G., Smith, D.P., Pasini, B., Bongarzone, I., Greco, A., Lorenzo, M.J., Arighi, E., Miranda, C., Eng, C. and Alberti, L., 1995, RET activation by germline MEN2A and MEN2B mutations. *Oncogene*, **11**, 2419–2427.

Brockmöller, J., Kerb, R., Drakoulis, N., Nitz, M. and Roots, I., 1993, Genotype and phenotype of glutathione *S*-transferase class mu isoenzymes mu and psi in lung cancer patients and controls. *Cancer Res.*, **53**, 1004–1011.

Brockmöller, J., Cascorbi, I., Kerb, R. and Roots, I., 1996, Combined analysis of inherited polymorphisms in arylamine *N*-acetyltransferase 2, glutathione *S*-transferases M1 and T1, microsomal epoxide hydrolase, and cytochrome P450 enzymes as modulators of bladder cancer risk. *Cancer Res.*, **56**, 3915–3925.

Caporaso, N. and Goldstein, A., 1995, Cancer genes: single and susceptibility. Exposing the difference. *Pharmacogenetics*, **5**, 59–63.

Cartwright, R.A., Glashan, R.W., Rogers, H.J., Ahmad, R.A., Barham-Hall, D., Higgins, E. and Kahn, M.A., 1982, Role of *N*-acetyltransferase phenotypes in bladder carcinogenesis: a pharmacogenetic epidemiological approach to bladder cancer. *Lancet*, **2**, 842–845.

Cascorbi, I., Brockmoller, J. and Roots, I., 1996, A C4887A polymorphism in exon 7 of human CYP1A1: population frequency, mutation linkages, and impact on lung cancer susceptibility. *Cancer Res.*, **56**, 4965–4969.

Caspari, R., Olschwang, S., Friedl, W., Mandl, M., Boisson, C., Boker, T., Augustin, A., Kadmon, M., Moslein, G. and Thomas, G., 1995, Familial adenomatous polyposis: desmoid tumours and lack of ophthalmic lesions (CHRPE) associated with APC mutations beyond codon 1444. *Hum. Mol. Genet.*, **4**, 337–340.

Catteau, A., Douriez, E., Beaune, P., Poisson, N., Bonaiti-Pellie, C. and Laurent, P., 1995, Genetic polymorphism of induction of CYP1A1 (EROD) activity. *Pharmacogenetics*, **5**, 110–119.

Christensen, E.R., Cunningham, J.M., Tester, D.J., Roche, P.C., Burgart, L.J. and Thibodeau, S.N., 1998, Hypermethylation of the hMLH1 promoter in colon cancer with microsatelite instability. *Proc. Am. Assoc. Cancer Res.*, **39**, 460.

Cooper, G.M., 1995, Retinoblastoma and the discovery of tumor suppressor genes, in Jones, J.E. (Ed.) *Oncogenes*, vol. 2, pp. 126–144, Boston: Jones and Bartlett.

Cosma, G., Crofts, F., Currie, D., Wirgin, I., Toniolo, P. and Garte, S.J., 1993, Racial differences in restriction fragment length polymorphisms and messenger RNA inducibility of the human CYP1A1 gene. *Cancer Epidemiol. Biomarkers Prev.*, **2**, 53–57.

Crofts, F., Cosma, G.N., Currie, D., Taioli, E., Toniolo, P. and Garte, S.J., 1993, A novel CYP1A1 gene polymorphism in African-Americans. *Carcinogenesis*, **14**, 1729–1731.

Doll, R. and Peto, R., 1978, Cigarette smoking and bronchial carcinoma: dose and time relationships among regular smokers and lifelong non-smokers. *J. Epidemiol. Community Health*, **32**, 303–313.

Doll, R. and Peto, R., 1981, The causes of cancer: quantitative estimates of avoidable risks of cancer in the United States today. *J. Natl Cancer Inst.*, **66**, 1191–1308.

Easton, D.F., Ford, D. and Bishop, D.T., 1995, Breast and ovarian cancer incidence in BRCA1-mutation carriers. Breast Cancer Linkage Consortium. *Am. J. Hum. Genet.*, **56**, 265–271.

Edelmann, W., Yang, K., Umar, A., Heyer, J., Lau, K., Fan, K., Liedtke, W., Cohen, P.E., Kane, M.F., Lipford, J.R., Yu, N., Crouse, G.F., Pollard, J.W., Kunkel, T., Lipkin, M., Kolodner, R. and Kucherlapati, R., 1997, Mutation in the mismatch repair gene Msh6 causes cancer susceptibility. *Cell*, **91**, 467–477.

Eliopoulos, A.G., Kerr, D.J., Herod, J., Hodgkins, L., Krajewski, S., Reed, J.C. and Young, L.S., 1995, The control of apoptosis and drug resistance in ovarian cancer: influence of *p53* and Bcl-2. *Oncogene*, **11**, 1217–1228.

Eng, C. and Ponder, B.A., 1993, The role of gene mutations in the genesis of familial cancers. *FASEB J.*, **7**, 910–919.

Eng, C., Clayton, D., Schuffenecker, I., Lenoir, G., Cote, G., Gagel, R.F., van Amstel, H.K., Lips, C.J., Nishisho, I., Takai, S.I., Marsh, D.J., Robinson, B.G., Frank-Raue, K., Raue, F., Xue, F., Noll, W.W., Romei, C., Pacini, F., Fink, M., Niederle, B., Zedenius, J., Nordenskjold, M., Komminoth, P., Hendy, G.N. and Mulligan, L.M., 1996, The relationship between specific RET proto-oncogene mutations and disease phenotype in multiple endocrine neoplasia type 2. International RET mutation consortium analysis. *JAMA*, **276**, 1575–1579.

Evans, S.C. and Lozano, G., 1997, The Li–Fraumeni syndrome: an inherited susceptibility to cancer. *Mol. Med. Today*, **3**, 390–395.

Fearon, E.R., 1997, Human cancer syndromes: clues to the origin and nature of cancer. *Science*, **278**, 1043–1050.

Fink, M., Weinhusel, A., Niederle, B. and Haas, O.A., 1996, Distinction between sporadic and hereditary medullary thyroid carcinoma (MTC) by mutation analysis of the RET proto-oncogene. Study Group Multiple Endocrine Neoplasia Austria (SMENA). *Int. J. Cancer*, **69**, 312–316.

Fodde, R. and Khan, P.M., 1995, Genotype–phenotype correlations at the adenomatous polyposis coli (APC) gene. *Crit. Rev. Oncog.*, **6**, 291–303.

Frank-Raue, K., Hoppner, W., Frilling, A., Kotzerke, J., Dralle, H., Haase, R., Mann, K., Seif, F., Kirchner, R., Rendl, J., Deckart, H.F., Ritter, M.M., Hampel, R., Klempa, J., Scholz, G.H. and Raue, F., 1996, Mutations of the ret protooncogene in German multiple endocrine neoplasia families: relation between genotype and phenotype. German Medullary Thyroid Carcinoma Study Group. *J. Clin. Endocrinol. Metab.*, **81**, 1780–1783.

Friedl, W., Meuschel, S., Caspari, R., Lamberti, C., Krieger, S., Sengteller, M. and Propping, P., 1996, Attenuated familial adenomatous polyposis due to a mutation in the 3′ part of the APC gene. A clue for understanding the function of the APC protein. *Hum. Genet.*, **97**, 579–584.

Fujii-Kuriyama, Y., Ema, M., Mimura, J., Matsushita, N. and Sogawa, K., 1995, Polymorphic forms of the *Ah* receptor and induction of the *CYP1A1* gene. *Pharmacogenetics*, **5 Spec No.**, S149–153.

Gayther, S.A., Wells, D., SenGupta, S.B., Chapman, P., Neale, K., Tsioupra, K. and Delhanty, J.D., 1994, Regionally clustered APC mutations are associated with a severe phenotype and occur at a high frequency in new mutation cases of adenomatous polyposis coli. *Hum. Mol. Genet.*, **3**, 53–56.

Giardiello, F.M., Brensinger, J.D., Luce, M.C., Petersen, G.M., Cayouette, M.C., Krush, A.J., Bacon, J.A., Booker, S.V., Bufill, J.A. and Hamilton, S.R., 1997, Phenotypic expression of disease in families that have mutations in the 5′ region of the adenomatous polyposis coli gene. *Ann. Intern. Med.*, **126**, 514–519.

Goessl, C., Plaschke, J., Pistorius, S., Hahn, M., Frank, S., Hampl, M., Gorgens, H., Koch, R., Saeger, H.D. and Schackert, H.K., 1997, An intronic germline transition in the HNPCC gene hMSH2 is associated with sporadic colorectal cancer. *Eur. J. Cancer*, **33**, 1869–1874.

Habano, W., Sugai, T. and Nakamura, S., 1998, Mismatch repair deficiency leads to a unique mode of colorectal tumorigenesis characterized by intratumoral heterogeneity. *Oncogene*, **16**, 1259–1265.

Harris, C.C., 1989, Interindividual variation among humans in carcinogen metabolism, DNA adduct formation and DNA repair. *Carcinogenesis*, **10**, 1563–1566.

Hayashi, S., Watanabe, J., Nakachi, K. and Kawajiri, K., 1991, Genetic linkage of lung cancer-associated *Msp*I polymorphisms with amino acid replacement in the heme binding region of the human cytochrome P450IA1 gene. *J. Biochem. (Tokyo)*, **110**, 407–411.

Hirvonen, A., Husgafvel-Pursiainen, K., Karjalainen, A., Anttila, S. and Vainio, H., 1992, Point-mutational *Msp1* and Ile-Val polymorphisms closely linked in the *CYP1A1* gene: lack of association with susceptibility to lung cancer in a Finnish study population. *CEBP*, **1**, 485–489.

Hu, X., O'Donnell, R., Srivastava, S.K., Xia, H., Zimniak, P., Nanduri, B., Bleicher, R.J., Awasthi, S., Awasthi, Y.C., Ji, X. and Singh, S.V., 1997, Active site architecture of polymorphic forms of human glutathione *S*-transferase P1-1 accounts for their enantio-selectivity and disparate activity in the glutathione conjugation of 7β, 8α-dihydroxy-9α, 10α-epoxy-7,8,9,10–tetrahydrobenzo[*a*]pyrene. *Biochem. Biophys. Res. Commun.*, **235**, 424–428.

Hung, H.C., Chuang, J., Chien, Y.C., Chern, H.D., Chiang, C.P., Kuo, Y.S., Hildesheim, A. and Chen, C.J., 1997, Genetic polymorphisms of CYP2E1, GSTM1, and GSTT1; environmental factors and risk of oral cancer. *Cancer Epidemiol. Biomarkers Prev.*, **6**, 901–905.

Hunter, D.J., Hankinson, S.E., Hough, H., Gertig, D.M., Garcia-Closas, M., Spiegelman, D., Manson, J.E., Colditz, G.A., Willett, W.C., Speizer, F.E. and Kelsey, K. 1997, A prospective study of NAT2 acetylation genotype, cigarette smoking, and risk of breast cancer. *Carcinogenesis*, **18**, 2127–2132.

Ishibe, N., Wiencke, J.K., Zuo, Z.F., McMillan, A., Spitz, M. and Kelsey, K.T., 1997, Susceptibility to lung cancer in light smokers associated with *CYP1A1* polymorphisms in Mexican- and African-Americans. *Cancer Epidemiol. Biomarkers Prev.*, **6**, 1075–1080.

Kato, S., Bowman, E.D., Harrington, A.M., Blomeke, B. and Shields, P.G., 1995, Human lung carcinogen–DNA adduct levels mediated by genetic polymorphisms *in vivo*. *J. Natl Cancer Inst.*, **87**, 902–907.

Kawajiri, K., Nakachi, K., Imai, K., Yoshii, A., Shinoda, N. and Watanabe, J., 1990, Identification of genetically high risk individuals to lung cancer by DNA polymorphisms of the cytochrome P450IA1 gene. *FEBS*, **263**, 131–133.

Kawajiri, K., Nakachi, K., Imai, K., Watanabe, J. and Hayashi, S., 1993, Germ line polymorphisms of *p53* and *CYP1A1* genes involved in human lung cancer. *Carcinogenesis*, **14**, 1085–1089.

Kawajiri, K., Watanabe, J., Eguchi, H., Nakachi, K., Kiyohara, C. and Hayashi, S., 1995, Polymorphisms of human Ah receptor gene are not involved in lung cancer. *Pharmacogenetics*, **5**, 151–158.

Kawajiri, K., Eguchi, H., Nakachi, K., Sekiya, T. and Yamamoto, M., 1996, Association of *CYP1A1* germ line polymorphisms with mutations of the *p53* gene in lung cancer. *Cancer Res.*, **56**, 72–76.

Kelsey, K.T., Wiencke, J.K. and Spitz, M.R., 1994, A race-specific genetic polymorphism in the *CYP1A1* gene is not associated with lung cancer in African Americans. *Carcinogenesis*, **15**, 1121–1124.

Kinzler, K.W. and Vogelstein, B., 1996, Lessons from hereditary colorectal cancer. *Cell*, **87**, 159–170.

Kinzler, K.W. and Vogelstein, B., 1997, Gatekeepers and caretakers. *Nature*, **386**, 761–763.

Kleihues, P., Schauble, B., zur Hausen, A., Esteve, J. and Ohgaki, H., 1997, Tumors associated with *p53* germline mutations: a synopsis of 91 families. *Am. J. Pathol.*, **150**, 1–13.

Landi, M.T., Bertazzi, P.A., Shields, P.G., Clark, G., Lucier, G.W., Garte, S.J., Cosma, G. and Caporaso, N.E., 1994, Association between *CYP1A1* genotype, mRNA expression and enzymatic activity in humans. *Pharmacogenetics*, **4**, 242–246.

Landi, M.T., Zocchetti, C., Bernucci, I., Kadlubar, F.F., Tannenbaum, S., Skipper, P., Bartsch, H., Malaveille, C., Shields, P., Caporaso, N.E. and Vineis, P., 1996, Cytochrome P4501A2: enzyme induction and genetic control in determining 4-aminobiphenyl–hemoglobin adduct levels. *Cancer Epidemiol. Biomarkers Prev.*, **5**, 693–698.

Lang, N.P., Chu, D.Z., Hunter, C.F., Kendall, D.C., Flammang, T.J. and Kadlubar, F.F., 1986, Role of aromatic amine acetyltransferase in human colorectal cancer. *Arch. Surg.*, **121**, 1259–1261.

Lang, N.P., Butler, M.A., Massengill, J., Lawson, M., Stotts, R.C., Hauer-Jensen, M. and Kadlubar, F.F., 1994, Rapid metabolic phenotypes for acetyltransferase and cytochrome P4501A2 and putative exposure to food-borne heterocyclic amines increase the risk for colorectal cancer or polyps. *Cancer Epidemiol. Biomarkers Prev.*, **3**, 675–682.

Lear, J.T., Smith, A.G., Bowers, B., Heagearty, A.H., Jones, P.W., Gilford, J., Alldersea, J., Strange, R.C. and Fryer, A.A., 1997, Truncal tumor site is associated with high risk of multiple basal cell carcinoma and is influenced by glutathione *S*-transferase, GSTT1, and cytochrome P450, *CYP1A1* genotypes, and their interaction. *J. Invest. Dermatol.*, **108**, 519–522.

London, S.J., Daly, A.K., Fairbrother, K.S., Holmes, C., Carpenter, C.L., Navidi, W.C. and Idle, J.R., 1995a, Lung cancer risk in African-Americans in relation to a race-specific *CYP1A1* polymorphism. *Cancer Res.*, **55**, 6035–6037.

London, S.J., Daly, A.K., Cooper, J., Navidi, W.C., Carpenter, C.L. and Idle, J.R., 1995b, Polymorphism of glutathione *S*-transferase M1 and lung cancer risk among African-Americans and Caucasians in Los Angeles County, California. *J. Natl Cancer Inst.*, **87**, 1246–1252.

Lu, S.L., Kawabata, M., Imamura, T., Akiyama, Y., Nomizu, T., Miyazono, K. and Yuasa, Y., 1998, HNPCC associated with germline mutation in the TGF-beta type II receptor gene. *Nat. Genet.*, **19**, 17–18.

Maeda, S., Namba, H., Takamura, N., Tanigawa, K., Takahashi, M., Noguchi, S., Nagataki, S., Kanematsu, T. and Yamashita, S., 1995, A single missense mutation in codon 918 of the RET proto-oncogene in sporadic medullary thyroid carcinomas. *Endocrinol. J.*, **42**, 245–250.

Marsh, D.J., Andrew, S.D., Eng, C., Learoyd, D.L., Capes, A.G., Pojer, R., Richardson, A.L., Houghton, C., Mulligan, L.M., Ponder, B.A. and Robinson, B.G., 1996, Germline and somatic mutations in an oncogene: RET mutations in inherited medullary thyroid carcinoma. *Cancer Res.*, **56**, 1241–1243.

Marsh, D.J., Zheng, Z., Arnold, A., Andrew, S.D., Learoyd, D., Frilling, A., Komminoth, P., Neumann, H.P., Ponder, B.A., Rollins, B.J., Shapiro, G.I., Robinson, B.G., Mulligan, L.M. and Eng, C., 1997, Mutation analysis of glial cell line-derived neurotrophic factor, a ligand for an RET/coreceptor complex, in multiple endocrine neoplasia type 2 and sporadic neuroendocrine tumors. *J. Clin. Endocrinol. Metab.*, **82**, 3025–3028.

Mashima, Y., Konishi, M., Yamada, M., Imamura, Y., Nii, S. and Nakamura, Y., 1998, Multiple endocrine neoplasia 2B with glaucoma associated with codon 918 mutation of the RET proto-oncogene. *Acta Ophthalmol. Scand.*, **76**, 114–116.

Matthias, C., Bockmuhl, U., Jahnke, V., Harries, L.W., Wolf, C.R., Jones, P.W., Alldersea, J., Worrall, S.F., Hand, P., Fryer, A.A. and Strange, R.C., 1998, The glutathione *S*-transferase GSTP1 polymorphism: effects on susceptibility to oral/pharyngeal and laryngeal carcinomas. *Pharmacogenetics*, **8**, 1–6.

Miyauchi, A., Egawa, S., Futami, H., Kuma, K., Obara, T. and Yamaguchi, K., 1997, A novel somatic mutation in the RET proto-oncogene in familial medullary thyroid carcinoma with a germline codon 768 mutation. *Jpn J. Cancer Res.*, **88**, 527–531.

Moers, A.M., Landsvater, R.M., Schaap, C., Jansen-Schillhorn van Veen, J.M., de Valk, I.A., Blijham, G.H., Hoppener, J.W., Vroom, T.M., van Amstel, H.K. and Lips, C.J., 1996, Familial medullary thyroid carcinoma: not a distinct entity? Genotype–phenotype correlation in a large family. *Am. J. Med.*, **101**, 635–641.

Mulligan, L.M., Marsh, D.J., Robinson, B.G., Schuffenecker, I., Zedenius, J., Lips, C.J., Gagel, R.F., Takai, S.I., Noll, W.W. and Fink, M., 1995, Genotype–phenotype correlation in multiple endocrine neoplasia type 2: report of the International RET Mutation Consortium. *J. Intern. Med.*, **238**, 343–346.

Nagase, H., Miyoshi, Y., Horii, A., Aoki, T., Ogawa, M., Utsunomiya, J., Baba, S., Sasazuki, T. and Nakamura, Y., 1992a, Correlation between the location of germ-line mutations in the APC gene and the number of colorectal polyps in familial adenomatous polyposis patients. *Cancer Res.*, **52**, 4055–4057.

Nagase, H., Miyoshi, Y., Horii, A., Aoki, T., Petersen, G.M., Vogelstein, B., Maher, E., Ogawa, M., Maruyama, M. and Utsunomiya, J., 1992b, Screening for germ-line mutations in familial adenomatous polyposis patients: 61 new patients and a summary of 150 unrelated patients. *Hum. Mutat.*, **1**, 467–473.

Nakachi, K., Imai, K., Hayashi, S., Watanabe, J. and Kawajiri, K., 1991, Genetic susceptibility to squamous cell carcinoma of the lung in relation to cigarette smoking dose. *Cancer Res.*, **51**, 5177–5180.

Nakachi, K., Hayashi, S., Kawajiri, K. and Imai, K., 1995, Association of cigarette smoking and *CYP1A1* polymorphisms with adenocarcinoma of the lung by grades of differentiation. *Carcinogenesis*, **16**, 2209–2213.

Nakajima, T., Elovaara, E., Anttila, S., Hirvonen, A., Camus, A.M., Hayes, J.D., Ketterer, B. and Vainio, H., 1995, Expression and polymorphism of glutathione *S*-transferase in human lungs: risk factors in smoking-related lung cancer. *Carcinogenesis*, **16**, 707–711.

Nelson, D.R., Kamataki, T., Waxman, D.J., Guengerich, F.P., Estabrook, R.W., Feyereisen, R., Gonzalez, F.J., Coon, M.J., Gunsalus, I.C. and Gotoh, O., 1993, The P450 super-family: update on new sequences, gene mapping, accession numbers, early trivial names of enzymes, and nomenclature. *DNA Cell Biol.*, **12**, 1–51.

Nicolaides, N.C., Carter, K.C., Shell, B.K., Papadopoulos, N., Vogelstein, B. and Kinzler, K.W., 1995, Genomic organization of the human PMS2 gene family. *Genomics*, **30**, 195–206.

Norppa, H., Hirvonen, A., Jarventaus, H., Uuskula, M., Tasa, G., Ojajarvi, A. and Sorsa, M., 1995, Role of GSTT1 and GSTM1 genotypes in determining individual sensitivity to sister chromatid exchange induction by diepoxybutane in cultured human lymphocytes. *Carcinogenesis*, **16**, 1261–1264.

Okada, T., Kawashima, K., Fukushi, S., Minakuchi, T. and Nishimura, S., 1995, Association between a cytochrome P450 CYPIA1 genotype and incidence of lung cancer. *Pharmacogenetics*, **4**, 333–340.

Pasini, A., Geneste, O., Legrand, P., Schlumberger, M., Rossel, M., Fournier, L., Rudkin, B.B., Schuffenecker, I., Lenoir, G.M. and Billaud, M., 1997, Oncogenic activation of RET by two distinct FMTC mutations affecting the tyrosine kinase domain. *Oncogene*, **15**, 393–402.

Patel, K.J., Yu, V.P.C.C., Lee, H., Corcoran, A., Thistlethwaite, F.C., Evans, M.J., Colledge, W.H., Friedman, L.S., Ponder, B.A. and Venkitaraman, A.R., 1998, Involvement of Brca2 in DNA repair. *Molec. Cell*, **1**, 347–357.

Persson, I., Johansson, I. and Ingelman-Sundberg, M., 1997, *In vitro* kinetics of two human CYP1A1 variant enzymes suggested to be associated with interindividual differences in cancer susceptibility. *Biochem. Biophys. Res. Commun.*, **231**, 227–230.

Rebbeck, T.R., 1997, Molecular epidemiology of the human glutathione S-transferase geno-types GSTM1 and GSTT1 in cancer susceptibility. *Cancer Epidemiol. Biomarkers Prev.*, **6**, 733–743.

Rebbeck, T.R., Godwin, A.K. and Buetow, K.H., 1996, Variability in loss of constitutional heterozygosity across loci and among individuals: association with candidate genes in ductal breast carcinoma. *Molec. Carcinogenesis*, **17**, 117–125.

Rossel, M., Schuffenecker, I., Schlumberger, M., Bonnardel, C., Modigliani, E., Gardet, P., Navarro, J., Luo, Y., Romeo, G. and Lenoir, G., 1995, Detection of a germline mutation at codon 918 of the RET proto-oncogene in French MEN 2B families. *Hum. Genet.*, **95**, 403–406.

Rothman, N., Shields, P.G., Poirier, M.C., Harrington, A.M., Ford, D.P. and Strickland, P.T., 1995, The impact of glutathione S-transferase M1 and cytochrome P450 1A1 geno-types on white-blood-cell polycyclic aromatic hydrocarbon–DNA adduct levels in humans. *Molec. Carcinogenesis*, **14**, 63–68.

Ryberg, D., Kure, E., Lystad, S., Skaug, V., Stangeland, L., Mercy, I., Borresen, A.L. and Haugen, A., 1994, *p53*–mutations in lung tumors. Relationship to putative susceptibility markers for cancer. *Cancer Res.*, **54**, 1551–1555.

Ryberg, D., Skaug, V., Hewer, A., Phillips, D.H., Harries, L.W., Wolf, C.R., Ogreid, D., Ulvik, A., Vu, P. and Haugen, A., 1997, Genotypes of glutathione transferase M1 and P1 and their significance for lung DNA adduct levels and cancer risk. *Carcinogenesis*, **18**, 1285–1289.

Saeki, Y., Tamura, K., Yamamoto, Y., Hatada, T., Furuyama, J. and Utsunomiya, J., 1997, Germline *p53* mutation at codon 133 in a cancer-prone family. *J. Molec. Med.*, **75**, 50–56.

Schuffenecker, I., Virally-Monod, M., Brohet, R., Goldgar, D., Conte-Devolx, B., Leclerc, L., Chabre, O., Boneu, A., Caron, J., Houdent, C., Modigliani, E., Rohmer, V., Schlumberger, M., Eng, C., Guillausseau, P.J. and Lenoir, G.M., 1998, Risk and pene-trance of primary hyperparathyroidism in multiple endocrine neoplasia type 2A families with mutations at codon 634 of the RET proto-oncogene. Groupe D'etude des Tumeurs a Calcitonine. *J. Clin. Endocrinol. Metab.*, **83**, 487–491.

Seidegard, J., Pero, R.W., Markowitz, M.M., Roush, G., Miller, D.G. and Beattie, E.J., 1990, Isoenzyme(s) of glutathione transferase (class Mu) as a marker for the susceptibility to lung cancer: a follow up study. *Carcinogenesis*, **11**, 33–36.

Shen, D., Subbarao, M., Chillar, R. and Vadgama, J.V., 1998, Different patterns of germline BRCA1 gene mutations detected in the young minority breast cancer patients. *Proc. Am. Assoc. Cancer Res.*, **39**, 181.

Shields, P.G., Bowman, E.D., Harrington, A.M., Doan, V.T. and Weston, A. 1993, Polycyclic aromatic hydrocarbon–DNA adducts in human lung and cancer susceptibility genes. *Cancer Res.*, **53**, 3486–3492.

Silva, J.M., Gonzalez, R., Gomendio, B., Munoz, G., Provencio, M., Garcia, J.M., Carretero, L., Espana, P. and Bonilla, F., 1998, Allelic losses in the BRCA1 and BRCA2 regions and high malignancy in breast carcinomas. *Proc. Am. Assoc. Cancer Res.*, **39**, 339.

Struewing, J.P., Hartge, P., Wacholder, S., Baker, S.M., Berlin, M., McAdams, M., Timmerman, M.M., Brody, L.C. and Tucker, M.A., 1997, The risk of cancer associated with specific mutations of BRCA1 and BRCA2 among Ashkenazi Jews. *N. Engl. J. Med.*, **336**, 1401–1408.

Sugimura, H., Hamada, G.S., Suzuki, I., Iwase, T., Kiyokawa, E., Kino, I. and Tsugane, S., 1995, CYP1A1 and CYP2E1 polymorphism and lung cancer, case-control study in Rio de Janeiro, Brazil. *Pharmacogenetics*, **5 Spec No.**, S145–148.

Szabo, C.I. and King, M.C., 1995, Inherited breast and ovarian cancer. *Hum. Molec. Genet.*, **4 Spec No.**, 1811–1817.

Taioli, E., Crofts, F., Trachman, J., Demopoulos, R., Toniolo, P. and Garte, S.J., 1995a, A specific African-American CYP1A1 polymorphism is associated with adenocarcinoma of the lung. *Cancer Res.*, **55**, 472–473.

Taioli, E., Trachman, J., Chen, X., Toniolo, P. and Garte, S.J., 1995b, A CYP1A1 restriction fragment length polymorphism is associated with breast cancer in African-American women. *Cancer Res.*, **55**, 3757–3758.

Taioli, E. and Garte, S.J., 1996, Re: S.J. London *et al.*, Lung cancer risk in African-Americans in relation to a race-specific *CYP1A1* polymorphism. *Cancer Res.*, **55**, 6035–6037, 1995. *Cancer Res.*, **56**, 4275–4277.

Tefre, T., Ryberg, D., Haugen, A., Nebert, D.W., Skaug, V., Brogger, A. and Borresen, A.L., 1991, Human CYP1A1 (cytochrome P1450) gene: lack of association between the *Msp*I restriction fragment length polymorphism and incidence of lung cancer in a Norwegian population. *Pharmacogenetics*, **1**, 20–25.

Topinka, J., Binkova, B., Mrackova, G., Stavkova, Z., Benes, I., Dejmek, J., Lenicek, J. and Sram, R.J., 1997, DNA adducts in human placenta as related to air pollution and to GSTM1 genotype. *Mutat. Res.*, **390**, 59–68.

Uchino, S., Noguchi, S., Adachi, M., Sato, M., Yamashita, H., Watanabe, S., Murakami, T., Toda, M. and Murakami, N., 1998, Novel point mutations and allele loss at the RET locus in sporadic medullary thyroid carcinomas. *Jpn J. Cancer Res.*, **89**, 411–418.

van der Luijt, R.B., Meera Khan, P., Vasen, H.F., Breukel, C., Tops, C.M., Scott, R.J. and Fodde, R., 1996, Germline mutations in the 3′ part of APC exon 15 do not result in truncated proteins and are associated with attenuated adenomatous polyposis coli. *Hum. Genet.*, **98**, 727–734.

Varley, J.M., McGown, G., Thorncroft, M., Cochrane, S., Morrison, P., Woll, P., Kelsey, A.M., Mitchell, E.L., Boyle, J., Birch, J.M. and Evans, D.G., 1996, A previously undescribed mutation within the tetramerisation domain of TP53 in a family with Li–Fraumeni syndrome. *Oncogene*, **12**, 2437–2442.

Vaury, C., Laine, R., Noguiez, P., de Coppet, P., Jaulin, C., Praz, F., Pompon, D. and Amor-Gueret, M., 1995, Human glutathione *S*-transferase M1 null genotype is associated with a high inducibility of cytochrome P450 1A1 gene transcription. *Cancer Res.*, **55**, 5520–5523.

Vineis, P., Caporaso, N., Tannenbaum, S.R., Skipper, P.L., Glogowski, J., Bartsch, H., Coda, M., Talaska, G. and Kadlubar, F., 1990, Acetylation phenotype, carcinogen–hemoglobin adducts, and cigarette smoking. *Cancer Res.*, **50**, 3002–3004.

Vineis, P., Bartsch, H., Caporaso, N., Harrington, A.M., Kadlubar, F.F., Landi, M.T., Malaveille, C., Shields, P.G., Skipper, P., Talaska, G. *et al.*, 1994, Genetically based *N*-acetyltransferase metabolic polymorphism and low-level environmental exposure to carcinogens. *Nature*, **369**, 154–156.

Wedlund, P.J., Kimura, S., Gonzalez, F.J. and Nebert, D.W., 1994, 1462V mutation in the human CYP1A1 gene: lack of correlation with either the *Msp*I 1.9 kb (M2) allele or CYP1A1 inducibility in a three-generation family of east Mediterranean descent. *Pharmacogenetics*, **4**, 21–26.

Wijnen, J., Khan, P.M., Vasen, H., Menko, F., van der Klift, H., van den Broek, M., van Leeuwen-Cornelisse, I., Nagengast, F., Meijers-Heijboer, E.J., Lindhout, D., Griffioen, G., Cats, A., Kleibeuker, J., Varesco, L., Bertario, L., Bisgaard, M.L., Mohr, J., Kolodner, R. and Fodde, R., 1996, Majority of hMLH1 mutations responsible for hereditary nonpolyposis colorectal cancer cluster at the exonic region 15–16. *Am. J. Hum. Genet.*, **58**, 300–307.

Zedenius, J., Larsson, C., Bergholm, U., Bovee, J., Svensson, A., Hallengren, B., Grimelius, L., Backdahl, M., Weber, G. and Wallin, G., 1995, Mutations of codon 918 in the RET proto-oncogene correlate to poor prognosis in sporadic medullary thyroid carcinomas, *J. Clin. Endocrinol. Metab.*, **80**, 3088–3090.

Zhang, Z.Y., Fasco, M.J., Huang, L., Guengerich, F.P. and Kaminsky, L.S., 1996, Characterization of purified human recombinant cytochrome P4501A1-Ile462 and -Val462: assessment of a role for the rare allele in carcinogenesis. *Cancer Res.*, **56**, 3926–3933.

Zhong, S., Howie, A.F., Ketterer, B., Taylor, J., Hayes, J.D., Beckett, G.J., Wathen, C.G., Wolf, C.R. and Spurr, N.K., 1991, Glutathione *S*-transferase mu locus: use of genotyping and phenotyping assays to assess association with lung cancer susceptibility. *Carcinogenesis*, **12**, 1533–1537.

6

STEROID AND NUCLEAR RECEPTOR POLYMORPHISM VARIANTS IN HORMONE RESISTANCE AND HORMONE INDEPENDENCE

Torsten A. Hopp and Suzanne A.W. Fuqua

6.1 Background

Steroid and nuclear receptors (NRs) are an expanding superfamily of ligand-dependent transcription factors important for many aspects of development and maintenance of normal cellular functions (Evans, 1988). This family includes the estrogen receptor (ER), glucocorticoid receptor (GR), androgen receptor (AR), and also receptors for thyroid hormone (TR), vitamin D (VDR), retinoic acid (RAR), and 9-*cis*-retinoic acid (RXR). All the NRs display a modular structure, with five or six distinct domains which show different degrees of evolutionary conservation and are termed domains A through F (Figure 6.1). These domains are involved in transcriptional activation, DNA binding, dimerization, and ligand binding (Kastner *et al.*, 1995; Mangelsdorf *et al.*, 1995). The DNA-binding domain (domain C) is the most highly conserved region within the NR family; characteristic are eight cysteine residues which form two zinc clusters. Zinc ion-coordinated binding is essential for proper folding and DNA binding. The structure of the DNA-binding domains of the GR and ER, respectively, in complex with their response element on the DNA was solved by X-ray crystallography, and almost identical conformations were displayed (Luisi *et al.*, 1991; Schwabe *et al.*, 1993). Two distinct activation functions which flank the DNA-binding domain contribute to transcriptional activity in a cell-type and promoter-specific manner. The N-terminal A/B region exhibits a constitutive transcriptional activation function termed AF-1, while the E domain possesses a multifunctional ligand-binding domain necessary for ligand binding and heterodimerization, and a ligand-dependent transcriptional activation function termed AF-2. Both the AF-1 and AF-2 activation domains appear to act in concert to produce full transcriptional activity on NR-responsive genes, but the contribution of these two activation functions to transcriptional activity of the entire receptor varies distinctly between receptors.

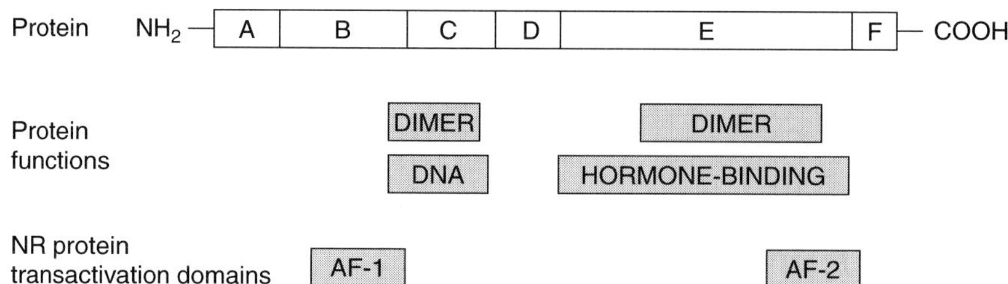

Figure 6.1 Structure of the nuclear receptor (NR) protein. Indicated are functional domains A through E. Also shown are receptor functions such as DNA-binding, hormone-binding, and NR's transcription activation functions (AF-1 and AF-2) in the lower part of the diagram.

The region possessing the activation function AF-2 has been reported to interact directly with the components of the basal transcriptional machinery, but recent studies have identified additional ligand-dependent receptor-interacting proteins, which act as either coactivators or corepressors and play essential roles in mediating NR transcriptional activity (Mangelsdorf *et al.*, 1995; Horwitz *et al.*, 1996; Glass *et al.*, 1997). Resolution of the crystal structure of RXRα (Bourguet *et al.*, 1995), RARγ (Renaud *et al.*, 1995), TRβ (Wagner *et al.*, 1995), ER, and progesterone receptor (PR) (Tanenbaum *et al.*, 1998) revealed that the ligand-binding domain of the NRs share a common structural motif which is composed of 11 or 12 individual α helices. The most carboxyl region, α helix 12, which contains the core of the activation function AF-2 (also termed the AF-2 AD, τC, or τ4 domains, depending on the NR), has been shown to realign over the ligand-binding pocket when associated with agonists. In particular, the hydrophobic residues of helix 12 face toward the ligand-binding pocket, while the charged residues extend into the solvent, possibly mediating protein–protein interactions with coactivators or corepressors (Wurtz *et al.*, 1996; Tanenbaum *et al.*, 1998).

Protein–protein interaction screening has provided several candidate receptor-integrating proteins including CBP/p300 (Chakravarti *et al.*, 1996; Hanstein *et al.*, 1996; Kamei *et al.*, 1996), members of the SRC-1 family of coactivators (SRC-1, TIF-2, and AIB1) (Onate *et al.*, 1995; Voegel *et al.*, 1996; Anzick *et al.*, 1997), as well as nuclear factors RIP-140 and RIP-160 (Cavailles *et al.*, 1994, 1995), ERAP-140 and ERAP-160 (Halachmi *et al.*, 1994), TIF-1 (LeDouarin *et al.*, 1995), TRIP1/SUG1 (Lee *et al.*, 1995; vom Baur *et al.*, 1996), and ARA70 (Yeh and Chang, 1996). However, with the exception of coactivators of the SRC-1 family and CBP/p300, none of these putative coactivators has been shown to unequivocally stimulate AF-2 activity. On the other hand, the silencing activity of unliganded TR, RAR, and VDR has been shown to require corepressor (Baniahmad *et al.*, 1995). Using yeast two-hybrid screening two corepressors have been identified: the nuclear corepressor N-CoR (Horlein *et al.*, 1995; Kurokawa *et al.*, 1995) and the silencing mediator for RAR and TR (called SMRT) (Chen and Evans, 1995; Chen *et al.*, 1996; Sande and Privalsky, 1996). These proteins interact with these receptors in the absence of hormone and inhibit transcription. All of these findings suggest that multiple bridging proteins function to transmit and augment the signal of ligand-induced NR conformational change to the basal transcription machinery.

Thus the mechanism of transcriptional regulation by NRs is a complex network of interactions involving coactivators and corepressors, as well as the basal transcription machinery. Disruption in this intricately regulated circuitry can result in perturbation of NR signaling. Mutations in the NRs or altered receptor expression have been described as the cause for the majority of hormone resistance to date; however, the components of the NR-signaling pathway affected by these mutations are still undefined in many cases. We will focus in this review on genetic mutations and altered expression of the AR, TR, VDR, GR, and ER which are associated with hormone insensitivity. We will also describe possible molecular mechanisms behind hormone resistance and the role of coactivators and/or corepressors in the generation of resistance.

6.2 Androgen receptor

Androgens (testosterone and 5α-dihydrotestosterone) control differentiation of Wolffian ducts and the urogenital sinus during embryogenesis, as well as the masculinization of external genitalia (Wilson *et al.*, 1981; Jost, 1990). Androgens act via the AR which is expressed at relatively high levels in prostate, epididymis, and seminal vesicles, but is low or absent in most other tissues (Ruizeveld de Winter *et al.*, 1991). The AR gene is a phosphoprotein of 910 amino acids with an apparent molecular mass of 110 kDa, has a domain structure typical for steroids and NRs (Bruggenwirth *et al.*, 1998), and resides on the X chromosome at Xq11–q12 (Brown *et al.*, 1989; Kuiper *et al.*, 1989). The amino-terminal domain of the AR also contains a stretch of glutamine residues starting at amino acid 58, and a stretch of glycine residues starting at position 448. The normal function of these amino acid stretches is not yet known. The glutamine stretch is highly polymorphic with a medium length between 11 and 33 amino acids (La Spada *et al.*, 1992; Sleddens *et al.*, 1992), whereas only five different allelic forms of the glycine stretch varying in length from 16 to 24 residues have been found (Sleddens *et al.*, 1993). More than 300 mutations are found in the AR resulting in a broad range of clinical syndromes such as complete and partial androgen insensitivity syndrome (CAIS and PAIS, respectively), spinal and bulbar muscular atrophy (also known as Kennedy disease), and prostate cancer.

Androgen insensitivity syndrome (AIS) is a heterogenous disease with diverse clinical manifestations, ranging from CAIS, with phenotypical females but genotypical males, to mild PAIS with infertility or signs of undervirilization (Quigley *et al.*, 1995; Hiort *et al.*, 1996). More severe forms of PAIS may include hypospadias, cryptorchidism, or ambiguous genitalia. CAIS is also known as testicular feminization (Morris, 1953), whereas various forms of PAIS have been described (infertile male syndrome, Rosewater's syndrome, Reifenstein syndrome, Gilbert-Dreyfus syndrome, and Lubs syndrome) (Reifenstein, 1947; Gilbert-Dreyfus *et al.*, 1957; Lubs *et al.*, 1959; Aiman *et al.*, 1979). All of these aberrations of male differentiation and virilization are associated with AR defects. About two-thirds of all the mutations in the AR have been described in more than 360 patients with AIS, and single base mutations are far more common than all other types of mutations (Tables 6.1 and 6.2)

Table 6.1 AR mutations found in PAIS

Exon	Codon	References
1	Glu 2 Lys	(Choong *et al.*, 1996a)
	Leu 172 STOP	(Holterhus *et al.*, 1997)
2	Gly 568 Trp/Val	(Lobaccaro *et al.*, 1994; Allera *et al.*, 1995)
	Phe 582 Ser/Tyr	(Hiort *et al.*, 1994; Imasaki *et al.*, 1996)
3	Ala 596 Thr	(Gast *et al.*, 1995)
	Ser 597 Gly	(Zoppi *et al.*, 1992)
	Asp 604 Tyr	(Hiort *et al.*, 1994)
	Arg 607 Gln	(Wooster *et al.*, 1992; Hiort *et al.*, 1994; Weidemann *et al.*, 1996)
	Arg 608 Lys	(Saunders *et al.*, 1992; Lobaccaro *et al.*, 1993a,b; Tincello *et al.*, 1997)
	Asn 610 Thr	(Weidemann *et al.*, 1996)
	Arg 615 His/Pro	(Hiort *et al.*, 1996)
	Leu 616 Arg	(De Bellis *et al.*, 1994)
	Arg 617 Pro	(Zoppi *et al.*, 1992)
4	Ala 645 Asp	(Hiort *et al.*, 1994)
	Ile 664 Asn	(Pinsky *et al.*, 1992)
	Pro 671 His	(Hiort *et al.*, 1996)
	Cys 686 Arg	(Hiort *et al.*, 1996)
	Ala 687 Val	(Hiort *et al.*, 1996)
	Asp 695 Asn	(Hiort *et al.*, 1996)
	Ser 703 Gly	(Radmayr *et al.*, 1997)
	Gly 708 Ala	(Hiort *et al.*, 1994)
	Leu 712 Phe	(Hiort *et al.*, 1996)
5	Phe 725 Leu	(Quigley *et al.*, 1995)
	Leu 728 Ser	(McPhaul *et al.*, 1992)
	Gln 733 His	(Hiort *et al.*, 1993)
	Ile 737 Thr	(Quigley *et al.*, 1995)
	Met 742 Val/Ile	(Ris-Stalpers *et al.*, 1994; Bevan *et al.*, 1996)
	Gly 743 Val	(Nakao *et al.*, 1993)
	Leu 744 Phe	(Brinkmann *et al.*, 1995)
	Met 745 Thr	(Ris-Stalpers *et al.*, 1994)
	Val 746 Met	(Hiort *et al.*, 1996)
	Ala 748 Asp	(Marcelli *et al.*, 1994)
	Phe 754 Leu	(Hiort *et al.*, 1996; Weidemann *et al.*, 1996)
	Asn 756 Ser	(Hiort *et al.*, 1996)
	Tyr 763 Cys	(McPhaul *et al.*, 1991)
	Asn 771 His	(Hiort *et al.*, 1994)
	Glu 772 Ala/Glu	(Pinsky *et al.*, 1992; Tincello *et al.*, 1997)
6	Arg 774 His	(Quigley *et al.*, 1995)
	Met 780 Ile	(Pinsky *et al.*, 1992; Brinkmann *et al.*, 1995; Bevan *et al.*, 1996; Rodien *et al.*, 1996)
	Gln 798 Glu	(Quigley *et al.*, 1995; Hiort *et al.*, 1996)
	Cys 806 Tyr	(Brown *et al.*, 1993)
	Ser 814 Asn	(Pinsky *et al.*, 1992)
7	Leu 821 Val	(Pinsky *et al.*, 1992)
	Arg 840 Cys/His	(McPhaul *et al.*, 1992; Hiort *et al.*, 1993; Beitel *et al.*, 1994; Lumbroso *et al.*, 1994; Marcelli *et al.*, 1994; Imai *et al.*, 1995; Bevan *et al.*, 1996; Weidemann *et al.*, 1996)

Table 6.1 continued

Exon	Codon	References
	Ile 841 Ser	(Hiort *et al.*, 1996)
	Ile 842 Thr	(Weidemann *et al.*, 1996)
	Arg 854 Lys	(McPhaul *et al.*, 1992)
	Arg 855 Cys/His	(Batch *et al.*, 1992; Marcelli *et al.*, 1994; Hiort *et al.*, 1996; Weidemann *et al.*, 1996)
	Val 866 Leu/Met	(McPhaul *et al.*, 1992; Saunders *et al.*, 1992; Hiort *et al.*, 1993, 1996; Kazemi-Esfarjani *et al.*, 1993)
8	Ile 869 Met	(Bevan *et al.*, 1996)
	Ala 870 Val	(Hiort *et al.*, 1994)
	Val 903 Met	(McPhaul *et al.*, 1992)
	Gly 909 Leu	(Choong *et al.*, 1996b)

(Sultan *et al.*, 1993; Quigley *et al.*, 1995; Brinkmann *et al.*, 1996; Hiort *et al.*, 1996; Gottlieb *et al.*, 1998). More than 80% of the known mutations are found in the ligand-binding domain, with most of the remaining mutations localized to the DNA-binding domain. Within the ligand-binding domain there are particular sites with unusually high mutations rates including amino acid residues 774, 840, 855, and 866 (Gottlieb *et al.*, 1998). These amino acid residues are probably essential for the formation of the ligand-binding pocket as was predicted by the crystal structure of the domain of the RXR, TR, ER, and PR (Bourguet *et al.*, 1995; Renaud *et al.*, 1995; Wagner *et al.*, 1995; Tanenbaum *et al.*, 1998).

Surprisingly, two AR substitutions located in the DNA-binding domain, Arg607Gln and Arg608Lys, are associated with male breast cancer and PAIS (Poujol *et al.*, 1997). These mutations were observed in two unrelated families: the Arg607Gln substitution in two brothers (Wooster *et al.*, 1992) and the Arg608Lys in 1 out of 13 patients with male breast cancer (Lobaccaro *et al.*, 1993a, b). These mutant ARs exhibit normal androgen-binding affinities but weaker DNA-binding affinity. In cotransfection assays they displayed reduced transactivation efficiencies. Molecular modeling revealed that the Arg residues at codons 607 and 608 were partially surface-exposed and were located in adjacent areas in the AR's DNA-binding domain. The occurrence of two adjacent mutations in patients with PAIS and male breast cancer suggests an association between PAIS and oncogenesis, leading one to suspect a crucial role for the AR DNA-binding domain in the process. However, the same mutations have been found in other patients with PAIS who show no signs of male breast cancer (Saunders *et al.*, 1992; Hiort *et al.*, 1996; Weidemann *et al.*, 1996; Tincello *et al.*, 1997), suggesting that other genes are involved in the development of breast cancer in these patients. Another mutation in the AR DNA-binding domain, an Ala564Asp substitution which is located within the first zinc cluster, results in a transcriptionally inactive receptor as shown by gel retardation and promoter interference assays (Bruggenwirth *et al.*, 1998). Patients carrying this mutation exhibit the CAIS syndrome, demonstrating the importance of DNA binding for AR signaling.

The Arg residue 840 is one of the codons within the AR ligand-binding domain most commonly affected by point mutations. Arg 840 is reported to be the site of

Table 6.2 AR mutations found in CAIS

Exon	Codon	References
1	Gln 60 STOP	(Zoppi *et al.*, 1993)
	Leu 172 STOP	(Hiort *et al.*, 1996)
	Gln 194 Arg	(Komori *et al.*, 1997)
	Gly 371 STOP	(Davies *et al.*, 1995)
	Trp 502 STOP	(Bruggenwirth *et al.*, 1996)
	Tyr 534 STOP	(McPhaul *et al.*, 1991)
2	Cys 559 Tyr	(Zoppi *et al.*, 1992)
	Tyr 571 Cys	(Komori *et al.*, 1998)
	Ala 573 Asp	(Bruggenwirth *et al.*, 1996)
	Cys 576 Arg	(Zoppi *et al.*, 1992)
	Cys 579 Tyr/Phe	(Sultan *et al.*, 1993; Imasaki *et al.*, 1996)
	Val 581 Phe	(Lumbroso *et al.*, 1994)
	Arg 585 Lys	(Sultan *et al.*, 1993)
3	Lys 590 STOP	(Marcelli *et al.*, 1990)
	Cys 601 Phe	(Baldazzi *et al.*, 1994)
	Arg 607 STOP	(Brown *et al.*, 1993)
	Arg 615 His	(Brown *et al.*, 1993; Mowszowicz *et al.*, 1993; Beitel *et al.*, 1994; Ris-Stalpers *et al.*, 1994)
	Leu 616 Pro	(Lobaccarro *et al.*, 1996)
	Arg 617 Pro	(Marcelli *et al.*, 1991; Zoppi *et al.*, 1992)
4	Leu 677 Pro	(Belsham *et al.*, 1995)
	Glu 6812 Lys	(Hiort *et al.*, 1993)
	Gly 688 Glu	(Hiort *et al.*, 1996)
	Asp 695 His/Asn/Val	(Ris-Stalpers *et al.*, 1991; Dork *et al.*, 1998)
	Ser 702 Ala	(Pinsky *et al.*, 1992)
	Asn 705 Ser	(Pinsky *et al.*, 1992; Quigley *et al.*, 1995)
	Leu 707 Arg	(Lumbroso *et al.*, 1996)
	Leu 722 Phe	(Hiort *et al.*, 1996)
5	Asp 732 Asn/Tyr	(Pinsky *et al.*, 1992)
	Trp 741 Arg	(Marcelli *et al.*, 1994)
	Gly 743 Val	(Lobaccaro *et al.*, 1993c)
	Met 749 Val	(Jakubiczka *et al.*, 1997)
	Gly 750 Asp	(Bevan *et al.*, 1997)
	Trp 751 Arg	(Brinkmann *et al.*, 1995)
	Arg 752 STOP/Gln	(Brinkmann *et al.*, 1995; Komori *et al.*, 1998)
	Phe 754 Val	(Pinsky *et al.*, 1992)
	Leu 762 Phe	(Lobaccaro *et al.*, 1993c; Hiort *et al.*, 1996)
	Tyr 763 Cys	(Bevan *et al.*, 1997)
	Phe 764 Leu	(Quigley *et al.*, 1995)
	Ala 765 Thr/Val	(Pinsky *et al.*, 1992; Marcelli *et al.*, 1994; Ris-Stalpers *et al.*, 1994)
	Pro 766 Ser	(Marcelli *et al.*, 1994)
	Glu 772 STOP	(Imasaki *et al.*, 1995)
6	Arg 774 His	(Brown *et al.*, 1990; Marcelli *et al.*, 1991; Batch *et al.*, 1992; Prior *et al.*, 1992; Hiort *et al.*, 1996; Komori *et al.*, 1998)
	Arg 779 Trp	(Hiort *et al.*, 1994; Sinnecker *et al.*, 1997)
	Met 780 Ile	(Rodien *et al.*, 1996; Jakubiczka *et al.*, 1997)
	Arg 786 STOP	(Pinsky *et al.*, 1992)

Table 6.2 continued

Exon	Codon	References
	Met 787 Val	(Nakao *et al.*, 1993)
	Phe 794 Ser	(Hiort *et al.*, 1996; Jakubiczka *et al.*, 1997)
	Trp 796 STOP	(Marcelli *et al.*, 1990)
	Met 807 Val/Arg	(Adeyemo *et al.*, 1993)
7	Gly 820 Ala	(Kasumi *et al.*, 1993)
	Arg 831 STOP/ Gln/Leu	(Brown *et al.*, 1990; De Bellis *et al.*, 1992; McPhaul *et al.*, 1992; Pinsky *et al.*, 1992)
	Tyr 834 Cys	(Wilson, 1992)
	Ile 842 Thr	(Hiort *et al.*, 1993)
	Asn 848 Lys	(Brinkmann *et al.*, 1995)
	Ser 853 STOP	(Wilson, 1992; Jakubiczka *et al.*, 1997)
	Arg 855 Cys/His	(De Bellis *et al.*, 1992; McPhaul *et al.*, 1992; Sultan *et al.*, 1993; Brinkmann *et al.*, 1995; Hiort *et al.*, 1996; Malmgren *et al.*, 1996; Weidemann *et al.*, 1996; Komori *et al.*, 1997)
	Leu 863 Arg	(Brown *et al.*, 1993)
	Asp 864 Asn/Gly	(De Bellis *et al.*, 1992; Bevan *et al.*, 1996)
	Val 866 Met/Glu	(McPhaul *et al.*, 1992; Kazemi-Esfarjani *et al.*, 1993; Weidemann *et al.*, 1996)
8	Ala 870 Val	(Hiort *et al.*, 1994)
	Leu 881 Val	(Davies *et al.*, 1995)
	Lys 883 STOP	(Trifiro *et al.*, 1991)
	Val 889 Met	(Pinsky *et al.*, 1992)
	Ile 898 Thr	(Hiort *et al.*, 1996)
	Pro 904 Ser/His	(McPhaul *et al.*, 1992; Pinsky *et al.*, 1992)
	Leu 907 Phe	(Bevan *et al.*, 1997)
	Phe 916 Leu	(Radmayr *et al.*, 1997)

substitution by either cysteine or histidine (3 and 12 patients, respectively) in multiple patients (Gottlieb *et al.*, 1998). These substitutions are associated with a variety of PAIS phenotypes. In one recently published case, two siblings with the same mutation in codon 840 are described with PAIS ranging from cliteromegaly and labial fusion in one sibling, who was raised as a girl, to micropenis and peno-scrotal hypospadias in the other, who was raised as a boy (Evans *et al.*, 1997). This highlights the complexity of the genotype–phenotype relationship of AIS. The molecular basis of this phenotypic variation has been investigated only in a few families and is still not well understood (Batch *et al.*, 1992).

In an attempt to correlate the degree of androgen resistance with AR binding and thermolability, and/or the nature of the mutation, the biochemical parameters of the AR of 20 patients with various degrees of defects in virilization were analyzed (Weidemann *et al.*, 1996). This study showed that even if there was abnormal binding of androgen and a specific mutation, there was no clear relationship with other parameters of the receptor, or the degree of virilization. However, this study did not analyze the ability of the ligand-binding domain of the defective ARs to interact with any of the recently identified receptor-interacting coactivator and core-pressor proteins. In contrast, numerous studies have described ER mutations within the ligand-binding domain which affect its interaction with coactivator (Weis *et al.*, 1996; Lazennec *et al.*, 1997). It could be hypothesized that AR mutations within the ligand-binding domain will similarly affect its interaction with coactivators or corepressors resulting in decreased transcription of AR-regulated genes. Individual differences in expression of these cofactors might help to explain how the same mutation results in CAIS in one patient and PAIS in another.

X-linked spinal and bulbar muscular atrophy (SMBA), also called Kennedy disease, is a hereditary depletion of primary motor neurons in the anterior horn of the spinal cord, a loss of sensory neurons in the dorsal root ganglia, and the degeneration of brainstem motor nuclei (Kennedy *et al.*, 1968; Nagashima *et al.*, 1988; Sobue *et al.*, 1989). It is characterized by adult onset, slowly progressive bulbospinal atrophy and weakness, hand tremors, painful muscle cramps, androgen insensitivity, and mildly increased serum creatine levels (Kennedy *et al.*, 1968; Harding *et al.*, 1982; Arbizu *et al.*, 1983; Wilde *et al.*, 1987; Sobue *et al.*, 1989). This neurodegenerative disease is caused by an expansion of the glutamine stretch within the first exon of the AR (La Spada *et al.*, 1992). Remarkably, polyglutamine expansion in other proteins such as Huntingtin, atrophin-1, and ataxin-3 results in disorders with numerous similarities to SMBA, although the causative proteins do not share structural homology outside of the glutamine tract (Ross *et al.*, 1997; Warrick *et al.*, 1998). The length of the glutamine stretch is roughly doubled in SMBA, but it is unstable, and can shift in length as it is passed from one generation to the next (La Spada *et al.*, 1991, 1992). The age of onset of SMBA is inversely correlated with the number of glutamine repeats, and the severity of the disease has also been correlated with the length of the glutamine stretch (La Spada *et al.*, 1992). Since the loss of AR function results in CAIS or PAIS, it has been hypothesized that the glutamine repeats cause a toxic gain of AR function which then leads to the degeneration of motor neurons. A recent study has also shown that cellular toxicity associated with the AR is triggered through a mechanism that involves amino-terminal cleavage of the AR by the protease caspase-3 (Wellington *et al.*, 1998). Caspases are a large family of cysteine proteases that become activated as cells initiate programmed cell death (Ellis and Horvitz, 1986). However, it is unclear why and how the caspase pathway is activated and how a truncated AR mechanistically affects motor neurons.

6.3 Thyroid hormone receptor

Thyroid hormones (thyroxine and triiodothyronine) are important regulators of growth, metabolic rate, and myocardial contractility. They are also critical for development of the central nervous system. Synthesis of thyroid hormones is controlled by pituitary thyroid-stimulating hormone (TSH), which in turn is negatively regulated by thyroid hormones. The effects of thyroid hormones are mediated by the TR which activates the transcription of target genes either as a heterodimer with the retinoid X receptor or as homodimer. There are two TR genes: TRα located on chromosome 17 and TRβ on chromosome 3 (Weinberger *et al.*, 1986; Nakai *et al.*, 1990). Both genes undergo alternative splicing resulting in distinct proteins with differing properties and tissue distributions (Sakurai *et al.*, 1989). The TRα gene produces TRα1 and c-erb-A α2, and the TRβ generates two isoforms, TRβ1 and TRβ2. All except c-erb-A α2 bind hormone, and c-erb-A α2 is thought to inhibit the function of both TRα1 and TRβ1 in a dominant-negative fashion (Koenig *et al.*, 1989; Lazar *et al.*, 1989; Nakai *et al.*, 1990). TRα1 is predominantly expressed in brain, heart and skeletal muscle, while TRβ1 is found in liver and kidney, and TRβ2 is most abundant in pituitary and hypothalamus (Lazar, 1993).

Mutations in the TRβ gene are tightly linked to familial cases of resistance to thyroid hormone (RTH), which is characterized by elevated serum-free thyroid hormone levels, failure to suppress TSH secretion, and variable refractoriness to hormone action in peripheral tissues (Usala *et al.*, 1988; Refetoff *et al.*, 1993; Refetoff, 1994). Other clinical features of RTH are goiter, growth retardation, low body mass index, attention-deficit hyperactivity disorder, heart abnormalities, hearing loss, and reduced skeletal bone mineral density (McDermott and Ridgway, 1993; Chatterjee, 1997). Two major forms of RTH are recognized: asymptomatic individuals with generalized resistance (GRTH) and patients with thyrotoxic features due to a selective pituitary resistance to thyroid hormone (PRTH) (Chatterjee, 1997). GRTH is characterized by partial resistance of both the pituitary gland and peripheral tissues to the effects of thyroid hormone. Individuals with GRTH show relatively high levels of thyroid hormones and maintain normal or increased TSH concentrations (McDermott and Ridgway, 1993). On the contrary, PRTH is characterized by thyroid hormone resistance in the pituitary gland while the peripheral tissues retain apparent normal responsiveness. In individuals with PRTH relatively high levels of thyroid hormone are needed to inhibit TSH secretion. These high thyroid hormone levels result in thyrotoxicosis in peripheral tissues. However, GRTH and PRTH can coexist in single families with the same mutation (Chatterjee, 1997), highlighting the variability of clinical features in RTH and supporting previous observations which indicate that generalized resistance and pituitary resistance are differing manifestations of a single genetic disorder.

RTH is inherited in an autosomal dominant fashion, except for one family with complete deletion of both TRβ alleles which results in recessive inheritance of RTH (Refetoff *et al.*, 1967). The majority of RTH-affected individuals express both the mutant and the WT TRβ allele, consistent with the dominant mode of inheritance of this disorder. More than 60 different mutations have been identified in more than 100 families, with most of the known RTH mutations mapping to three distinct areas within the hormone-binding domain (codons 234 to 282, 310 to 353, and 429 to 461) (Refetoff *et al.*, 1993; Beck-Peccoz *et al.*, 1994; Kopp *et al.*, 1996; Chatterjee, 1997). These sites within TRβ convey multiple functions including hormone binding, receptor dimerization and interactions with the transcriptional machinery, as well as coactivator and/or corepressor interactions. Consistent with this location, the ability of mutated receptors to bind hormone is moderately or markedly reduced, whereas their ability to bind to DNA and form heterodimers with RXR is generally not affected (Collingwood *et al.*, 1994). Apparently as a consequence of reduced hormone binding, many TRβ mutants can repress gene transcription of WT TR, but are defective in gene activation alone in response to hormone (Yoh *et al.*, 1997).

General transcriptional repression requires the association of the receptor with a family of corepressor proteins including the nuclear receptor corepressor N-CoR, silencing mediator for retinoid and thyroid hormone receptor (SMRT), retinoid X receptor-interacting protein-13 (RIP-13), and thyroid hormone receptor-associated cofactor (TRAC) (Baniahmad *et al.*, 1992, 1994; Chen and Evans, 1995; Horlein *et al.*, 1995; Kurokawa *et al.*, 1995; Chen *et al.*, 1996; Sande and Privalsky, 1996). Upon binding of thyroid hormone to the TRβ, corepressors dissociate from the receptor and a new set of transcriptional regulators, the coactivators, are then recruited to TRβ leading to the formation of an active transcriptional complex. A recent study analyzing the interaction of corepressors with TRβ mutants has also revealed that these receptors exhibit an impaired ability to dissociate from corepressors in the presence of thyroid hormone (Yoh *et al.*, 1997). This altered corepressor interaction might thus play a critical role in the dominant-negative effect of TRβ mutants and contribute to the variable phenotype seen in RTH. On the other hand, the interaction of coactivators such as receptor-interacting protein-140 (RIP-140) and steroid receptor coactivator-1 (SRC-1) with the TRβ mutant L454V is markedly reduced (Collingwood *et al.*, 1994), suggesting that an alteration in TRβ coactivator-interaction could also contribute to the RTH phenotype. If corepressors and coactivators are expressed in specific tissues, this could help to explain the varying organ-specific effects in RTH. Furthermore, it can be hypothesized that genetic polymorphisms within corepressors and coactivator families might also contribute to the variable resistance of the RTH syndrome in different individuals carrying the same TRβ mutation.

6.4 Vitamin D receptor

1,25-Dihydroxy vitamin D3 [1,25(OH)2D3], an active form of vitamin D, plays a critical role in calcium homeostasis and bone remodeling, but the molecular basis for the action of 1,25(OH)2D3 in bone formation and its role during development are uncertain (Walters, 1992; Studzinski *et al.*, 1993; Bouillon *et al.*, 1995). The actions of 1,25(OH)2D3 are mediated by the VDR, which is highly expressed in bone, kidney, and intestine, but is also present in many other tissues, including activated immune cells such as T cells (Dame *et al.*, 1985; Alroy *et al.*, 1995). Upon addition of 1,25(OH)2D3, the VDR is rapidly phosphorylated in the ligand-binding domain (Brown and DeLuca, 1990). However, the exact functional consequences of this phosphorylation have not yet been determined. Additionally, ligand binding promotes the formation of VDR–RXR heterodimers resulting in the activation of transcription from vitamin D target genes such as osteocalcin and calcium-binding protein (MacDonald *et al.*, 1993). Unlike the TR, the VDR does not bind to N-CoR, which may in part explain why the VDR does not seem to exhibit dominant-negative effects as seen with TR.

Mutations in the VDR gene, which resides on chromosome 12q13–14 (Labuda *et al.*, 1992), have been shown to cause a rare autosomal recessive disorder called hereditary vitamin D-resistant rickets (HVDRR), which is also known as vitamin D-dependent rickets type II. The major clinical features are early-onset rickets, hypocalcemia, elevated serum 1,25(OH)2D3 levels, secondary hyperparathyroidism, and often, alopecia (Rut *et al.*, 1994). The symptoms of HVDRR, with the exception of alopecia, mimic the classic vitamin D-deficiency rickets. Recently, the generation of VDR knockout mice has revealed apparently normal heterozygotes, but severely affected homozygotes in the post-weaning stage (Yoshizawa *et al.*, 1997).

The VDR null mice show no defects in development and growth before weaning, irrespective of the complete absence of VDR gene expression. After weaning, however, these null mice exhibit symptoms typical for HVDRR, thus establishing a critical role for VDR in growth and bone formation in the post-weaning stage.

In cases of HVDRR, genetic defects such as missense mutations, premature termination, splice site mutations that cause exon skipping, or partial gene deletions are all found in the VDR gene (Hughes *et al.*, 1988; Ritchie *et al.*, 1989; Malloy *et al.*, 1990, 1994, 1997; Sone *et al.*, 1990; Saijo *et al.*, 1991; Kristjansson *et al.*, 1993; Wiese *et al.*, 1993; Yagi *et al.*, 1993). Thus far, 10 unique mutations have been identified within the DNA-binding domain, which do not affect the ligand binding properties of the receptor. Most of these mutations are located within the conserved zinc-finger region that is involved in DNA base recognition and interaction with the DNA phosphate backbone (Rastinejad *et al.*, 1995). On the other hand, seven mutations have been located in the ligand-binding domain subsequently affecting both hormone binding and/or heterodimerization with RXR (Haussler *et al.*, 1997). It is believed that these mutated VDRs cause HVDRR due to a decreased affinity for 1,25(OH)2D3, but can be effectively treated with extremely high doses of hormone.

6.5 Glucocorticoid resistance

Glucocorticoids, in particular cortisol, not only regulate glucose and fat metabolism as well as modulate the cardiovascular system and immune functions, but also have additional effects on growth and development, calcium metabolism, connective tissue, bone regulation, and thyroid and gonadal function. An elaborate feedback systems exists that regulates glucocorticoid synthesis. Glucocorticoids are secreted in response to adrenocorticotrophic hormone (ACTH); in turn, glucocorticoids exert negative feedback effects on the pituitary secretion of ACTH. They also inhibit hypothalamic secretion of corticotropin-releasing hormone and vasopressin, as well as negatively regulate the activity of the hypothalamic–pituitary–adrenal (HPA) axis, such as the hippocampus and amygdala. The biological effects of glucocorticoids are mediated by the glucocorticoid receptor (GR), which is expressed as two isoforms: a ligand-binding GRα, and a β isoform that does not bind hormone and is thought to act as a ligand-independent negative regulator of glucocorticoid action (Hollenberg *et al.*, 1985; Bamberger *et al.*, 1995). GRα is expressed in all nucleated normal cells, consistent with the widespread effects of glucocorticoids on metabolism, differentiation, and development. Ligand-bound GRα activates transcription of glucocorticoid responsive genes by a mechanism which represents the classical model of steroid hormone receptor action. In addition, hormone-occupied GRα can inhibit activation of target genes involved in anti-inflammatory and immunosuppressive responses through direct protein–protein interactions with the transcription factors AP-1 and NF-κB (for a review see McEwan *et al.*, 1997). However, this interaction is not dependent on GRα binding to its response element. GR's activity is additionally regulated through heterodimerization with GRβ, hyperphosphorylation, and modulation of the surrounding chromatin to allow the assembly of coactivators and/or corepressors and the basal transcription machinery on the DNA. The complexity of glucocorticoid signal transduction can give rise to a variety of aberrations on several levels, all affecting glucocorticoid sensitivity via different pathways.

Two forms of glucocorticoid resistance can be distinguished. First, a generalized resistance exists in which all tissues are partly resistant (called generalized inherited glucocorticoid resistance or GIGR), and second, an acquired tissue-specific resistance exists that may play a role in the pathogenesis of depression, autoimmune disorders, AIDS, leukemia, and Nelson's syndrome (Werner and Bronnegard, 1996). GIGR is inevitably partial, as complete receptor inactivation results in early fetal death (Cole *et al.*, 1995). Fortunately, GIGR is a rare disease, which has only been described in a few families and is characterized by elevated levels of plasma cortisol and ACTH, but lacking glucocorticoid-induced pathology (Cushing's syndrome) (Lamberts *et al.*, 1996). There is, however, considerable variability in the clinical presentations of GIGR ranging from asymptomatic to isolated chronic fatigue, to hypertension with or without hypokalemic alkalosis or hyperandrogenism, or both. In these families point mutations in the GR at codons 559, 641, and 729 have been identified as the cause of GIGR (Hurley *et al.*, 1991; Malchoff *et al.*, 1993; Karl *et al.*, 1996a). In addition, a 4-base deletion at the 3'-boundary of exon 6, which removes a RNA splicing site, results in a functional knock-out of the allele leading to decreased GR expression (Karl *et al.*, 1993). The long-term effects of GIGR are unknown, but there is some evidence that it may result in the development of hyperplasia of the pituitary corticotroph cells, and eventually the development of pituitary corticotroph tumors (Karl *et al.*, 1996b; Lamberts *et al.*, 1996). Moreover, GR gene polymorphisms are frequently found in healthy subjects and have been thought to contribute to glucocorticoid resistance. A recent study, however, suggests that there is no clear association between specific GR polymorphisms and glucocorticoid resistance, although some individuals carrying an Asp363Ser polymorphism did exhibit a higher sensitivity to exogenously administered glucocorticoids, with respect to both cortisol suppression and insulin response (Koper *et al.*, 1997; Huizenga *et al.*, 1998).

Quantitative and qualitative alteration in GR expression might also cause primary and secondary resistance to glucocorticoids which is found in many leukemia patients (Moalli and Rosen, 1994). Glucocorticoids are routinely used in the treatment of several leukemic and lymphoma disorders, and are thought to induce apoptosis in these hematological malignancies (Thompson, 1994) The underlying molecular mechanism by which glucocorticoids induce apoptosis is poorly understood. Various studies using the human leukemic CCRF-CEM cell line have shown that subthreshold expression of functional GR, as well as various point mutations (substitutions, deletion of single basepairs, and non-sense mutations) can all result in glucocorticoid resistance (Geley *et al.*, 1996; Hala *et al.*, 1996). Whether these mechanisms also occur *in vivo* in the development of glucocorticoid resistance, and if so at what frequency, is presently unknown.

6.6 Estrogen receptor

Estrogens are essential for bone maturation, epiphyseal fusion, pubertal growth spurts, and achievement of normal bone mineral mass in both females and males. They also influence normal female secondary sexual maturation, insulin sensitivity, and lipid homeostasis. However, estrogens do not appear to be essential for fetal survival, placental growth, or female sexual differentiation. These hormones act via the ER which, upon binding of estrogen, homodimerizes and activates target gene expression directly by binding to specific estrogen response elements (Kumar and Chambon, 1988), or indirectly through its association with other DNA-bound proteins such as AP-1, NF-κB, C/EBPβ, and GATA-1 (Blobel *et al.*, 1995; Stein and Yang, 1995; Webb *et al.*, 1995; Elgort *et al.*, 1996). There are two ER genes: ERα resides on chromosome 6q25.1 (Menasce *et al.*, 1993) and the recently identified ERβ gene resides on chromosome 14q22–24 (Enmark *et al.*, 1997). Both genes undergo alternative splicing, resulting in ER mRNA variants with single or multiple exons being skipped (Hopp and Fuqua, 1998; Lu *et al.*, 1998). Furthermore, the formation of ERα and ERβ heterodimers in cells expressing both receptor subtypes introduces another level of complexity in estrogen signaling pathways (Pettersson *et al.*, 1997; Ogawa *et al.*, 1998). ERα mRNA variants are frequently found in both normal and neoplastic estrogen-responsive tissues, although their physiological or pathological functions remain largely unclear (Murphy *et al.*, 1997a). The ERβ gene is strongly expressed in testis and in ovary, but its expression has also been found in lung, spleen, pituitary gland, leukocytes, bone marrow, colon, and uterine tissues (Enmark *et al.*, 1997).

In contrast to the abundance of ER mRNA variants, naturally occurring mutations of the ERα gene appear to be relatively rare, and only one of these mutations with a cytosine to thymidine transition at codon 157 results in estrogen insensitivity in a homozygous carrier (Smith *et al.*, 1994). This patient, a male, was normally masculinized, showed elevated levels of estradiol, estrone, follicle-stimulating hormone, and luteinizing hormone, but normal testosterone concentrations. Furthermore, he had incomplete epiphyseal closure with a history of continued linear growth into adulthood, as well as osteoporosis probably induced by increased bone turnover. Interestingly, both sexes of ER knock-out mice also show a 20 to 25% reduction in bone density as compared to mice expressing wild-type ERα (Korach, 1994). These knock-out mice and this patient with estrogen insensitivity syndrome reveal that ERα is crucial for proper bone maturation and mineralization.

Although ERα mutations resulting in generalized estrogen insensitivity are a rare event, acquired antiestrogen resistance and estrogen independence is frequently seen in breast cancer. Approximately two-thirds of breast tumors, at least initially, express abundant levels of ERα, and this expression is associated with a lower risk of relapse and prolonged overall survival (Osborne, 1991) of patients. ERα-expressing tumors are more likely to respond to tamoxifen therapy, an antiestrogen frequently used in the treatment of breast cancer. Unfortunately, most breast cancers eventually fail to respond to tamoxifen despite continued expression of ERα, resulting in disease recurrence and the frequent emergence of more aggressive, tamoxifen-resistant tumors. Since these tumors still express ERα, mechanisms for antiestrogen resistance other than the loss of ERα expression must exist. It has been hypothesized that alterations in the level of various ERα mRNA variants, as well as specific mutations in the ERα gene, may affect estrogen signal transduction and contribute to the development of hormone-resistant disease. However, the demonstration of this *in vivo* has been limited.

ERα variant forms fall into four major groups: (1) transcripts containing precise single or multiple exon deletions, (2) transcripts containing single nucleotide deletions and others in which several hundred nucleotides have been deleted within known exon sequences, (3) truncated transcripts, and (4) transcripts containing insertions (Table 6.3) (for review see Murphy *et al.*, 1997b; Hopp and Fuqua, 1998). This complex pattern of alternative splicing is seen in both normal tissue and tumors, and appears specific for the two ER genes since similar abundant splicing of other NR has not been found in breast cancer (Pfeffer *et al.*, 1995).

The most frequently observed and probably the most abundant of the ERα mRNA variants are those lacking either exon 4 (Pfeffer *et al.*, 1993) or exon 7 (Zhang *et al.*, 1996). The exon 4-deleted ER variant is missing parts of the DNA-binding and hinge domains, and the amino-terminal region of the ligand-binding domain, while the exon 7-deleted ER variant is lacking part of the ligand-binding domain, including the hormone-dependent transcriptional activation function AF-2. Both ER variants, when transfected into mammalian or yeast cells, act as dominant-negative inhibitors of normal ER function in certain cell types but not in others (Wang and Miksicek, 1991; Fuqua *et al.*, 1992; Koehorst *et al.*, 1994; Inoue *et al.*, 1996). These experimental observations suggest that the activity of particular ER variants might be cell-type and possibly promoter-specific. Furthermore, there is a significant positive association of the exon 4-deleted ERα mRNA variant with histological high grade and progesterone receptor-positive levels, both of which are markers for a good clinical outcome in breast cancer (Leygue *et al.*, 1996a). Even though both the exon 4-deleted ER variant and the exon 7-deleted ER variant might prove useful as biological markers for prognosis in clinical samples, it seems unlikely that they contribute significantly to the development of antiestrogen resistance since their expression is not associated with antiestrogen resistance in tamoxifen-resistant MCF-7 cells (Madsen *et al.*, 1997).

Another possible clinically relevant ERα mRNA variant is the exon 3-deleted ERα variant which is found in both normal breast epithelium and primary breast cancer (Fuqua *et al.*, 1993; Zhang *et al.*, 1996). It has been shown that the ratio in expression of the wild-type ERα and the exon 3-deleted variant is reduced about 30-fold in primary breast cancer and cell lines as compared to normal tissue (Erenburg *et al.*, 1997). Since the protein encoded by this variant is missing portions of the DNA-binding domain, it might not be surprising that it acts as a dominant-negative receptor when co-expressed with wild-type ERα in yeast, HeLa, and breast cancer cells (Wang and Miksicek, 1991; Fuqua *et al.*, 1993; Erenburg *et al.*, 1997). Stable expression of the exon 3-deleted variant in MCF-7 breast cancer cells results in a significant reduction of expression of ER-regulated genes, such as pS2 (Erenburg *et al.*, 1997). Furthermore, both *in vivo* invasiveness and estrogen-stimulated anchorage-independent growth were also noticeably reduced in the transfectants. This suppression, rather than enhancement, of the transformed phenotype by estrogen in the presence of high levels of the exon 3-deleted ER variant, and the drastic reduction of the variant compared to wild-type ERα in breast cancer cells, strongly supports the concept that the regulation of ERα mRNA splicing is altered and necessary for breast carcinogenesis.

One of the best studied of the ERα mRNA variant is the exon 5-deleted variant that encodes a truncated 40-kDa protein lacking most of the ligand-binding domain

Table 6.3 ERα mRNA variants

Variant	Sources	References
Exon 2 deleted	MCF-7	(Madsen *et al.*, 1997)
	Normal breast, breast cancer	(Leygue *et al.*, 1996a,b; Pfeffer *et al.*, 1996)
	Prolactinoma	(Chaidarun *et al.*, 1997)
Exon 3 deleted	MCF-7	(Madsen *et al.*, 1997)
	Prolactinoma	(Chaidarun *et al.*, 1997)
Exon 4 deleted	MCF-7	(Pfeffer *et al.*, 1996; Madsen *et al.*, 1997)
	Normal breast, breast cancer	(Leygue *et al.*, 1996a,b; Pfeffer *et al.*, 1996; Huang *et al.*, 1997)
	Endometrium, endometrial carcinoma	(Rice *et al.*, 1997)
	Pituitary, prolactinoma	(Chaidarun *et al.*, 1997)
Exon 5 deleted	MCF-7	(Pfeffer *et al.*, 1996; Madsen *et al.*, 1997)
	Normal breast, breast cancer	(Leygue *et al.*, 1996a,b; Pfeffer *et al.*, 1996; Gallacchi *et al.*, 1998)
	Endometrium, endometrial carcinoma	(Hu *et al.*, 1996)
	Prolactinoma	(Chaidarun *et al.*, 1997)
Exon 7 deleted	MCF-7	(Pfeffer *et al.*, 1996; Madsen *et al.*, 1997)
	Normal breast, breast cancer	(Leygue *et al.*, 1996a,b; Pfeffer *et al.*, 1996; Huang *et al.*, 1997)
	Endometrium, endometrial carcinoma	(Hu *et al.*, 1996; Rice *et al.*, 1997)
	Pituitary, prolactinoma	(Chaidarun *et al.*, 1997)
Exons 2 and 3 deleted	Normal breast	(Leygue *et al.*, 1996b)
Exons 2, 3, and 7 deleted	Breast cancer	(Leygue *et al.*, 1996a; Huang *et al.*, 1997)
Exons 2, 3, and 4 deleted	Breast cancer	(Leygue *et al.*, 1996a; Huang *et al.*, 1997)
Exons 3 and 4 deleted	Breast cancer	(Leygue *et al.*, 1996a; Huang *et al.*, 1997)
	Endometrium, endometrial carcinoma	(Hu *et al.*, 1996)
Exons 3, 4, and 5 deleted	MCF-7	(Madsen *et al.*, 1997)
	Normal breast, breast cancer	(Leygue *et al.*, 1996a,b; Pfeffer *et al.*, 1996)
Exons 3 through 7 deleted	Breast cancer	(Leygue *et al.*, 1996a; Huang *et al.*, 1997)
Parts of exon 4 through parts of exon 6	Breast cancer	(Chan and Dowsett, 1997)

Table 6.3 continued

Variant	Sources	References
Exons 4 and 7 deleted	MCF-7	(Wang and Miksicek, 1991; Madsen *et al.*, 1997)
	Normal breast, breast cancer	(Leygue *et al.*, 1996a,b; Huang *et al.*, 1997)
Exons 4 and 5 deleted	Endometrium, endometrial carcinoma	(Hu *et al.*, 1996)
Exon 7 and parts of exon 8 deleted	Endometrium, endometrial carcinoma	(Hu *et al.*, 1996)
Exon 6 duplicated	Breast cancer	(Murphy *et al.*, 1996)
Exons 3 and 4 duplicated	Breast cancer	(Murphy *et al.*, 1996)
69 bp inserted between exons 5 and 6	Breast cancer	(Murphy *et al.*, 1996)

but retaining the AF-1 transactivation function and the DNA-binding domains (Castles *et al.*, 1993; Fuqua *et al.*, 1993). This variant was first detected in ER-negative/progesterone receptor-positive tumors along with wild-type ERα, and demonstrated variable strengths of ligand-independent transcriptional activity in yeast cells (Fuqua *et al.*, 1991) as well as chicken embryo fibroblast and MCF-7 human breast cancer cells (Rea and Parker, 1996). The endogenous protein encoded by the exon 5-deleted variant is the only ERα mRNA variant which has been demonstrated to exist in breast cancer cell lines and breast tumors (Castles *et al.*, 1993; Desai *et al.*, 1997). A recent study analyzing the ratio of the exon 5-deleted variant to wild-type ERα in 32 breast cancer patients demonstrated that expression of the variant was significantly increased in relapse tissue as compared to the respective primary tumor. Thus, cells expressing higher levels of the exon 5-deleted variant may acquire resistance to tamoxifen and proliferate, leading to disease relapse as originally predicted by us (Fuqua, 1996). In contrast though, an earlier study examining the levels of the variant and wild-type ERα mRNA in 70 tamoxifen-resistant and 50 primary breast carcinomas found no significant differences in the ratio of variant to wild-type receptor (Daffada *et al.*, 1995). However, this latter study did not compare the tamoxifen-resistant tumors with their respective primary tumors as did the former study, possibly contributing to the discrepancy between the two studies. Since tamoxifen resistance is most probably multifactorial (Murphy *et al.*, 1997b), and since multiple ERα variants can occur together in any one tumor sample (Pfeffer *et al.*, 1995; Leygue *et al.*, 1996a), it seems likely that other ERα variants may also be involved in acquired tamoxifen resistance.

6.7 Conclusions

The NRs are important transcriptional activators or repressors for genes involved in many essential biological processes. Cloning and sequencing of the NRs, as well as resolution of the crystal structure of some NRs, has increased our understanding of the structure–function relationship of these transcriptional regulators with regard to disease processes, such as hormone resistance. Furthermore, recent studies examining the role of receptor-interacting proteins such as coactivators and corepressors have provided another dimension to the elucidation of hormone action. Thus, we have shown that many mutations in NRs affecting the ligand-binding or DNA-binding domains can alter transcriptional activation and have been associated with specific types of hormone resistance ranging from partial to complete hormonal insensitivity. For example, point mutations in the AR gene result in AIS with diverse clinical manifestation. However, the exact position of the point mutation and the biochemical parameters of the mutant AR do not appear to completely determine or explain the degree of virilization. Furthermore, in some cases even the same mutation can result in either CAIS or PAIS. Genetic polymorphisms within specific coactivator and corepressor gene families might conceivably contribute to the variability in clinical manifestations, but no study has yet addressed this important point. Further progress in understanding the fine details of transcriptional activation or repression should also provide new insights into the mechanisms of hormone resistance. Unlike many of the other NRs, the ERα undergoes extensive alternative splicing resulting in many different ER mRNA variants, and breast tumorigenesis or progression is thought to be influenced by alterations in the expression levels of some of the variants. Whether any of the reported NR variants or gene mutations are indeed important for the development of hormone resistance will require continued investigation.

References

Adeyemo, O., Kallio, P.J., Palvimo, J.J., Kontula, K. and Janne, O.A., 1993, A single-base substitution in exon 6 of the androgen receptor gene causing complete androgen insensitivity: the mutated receptor fails to transactivate but binds to DNA *in vitro*. *Hum. Molec. Genet.*, **2**, 1809–1812.

Aiman J., Griffin J.E., Gazak J.M., Wilson, J.D. and MacDonald, P.C., 1979, Androgen insensitivity as a cause of infertility in otherwise normal men. *N. Engl. J. Med.*, **300**, 223–227.

Allera, A., Herbst, M.A., Griffin, J.E., Wilson, J.D., Schweikert, H.U. and McPhaul, M.J., 1995, Mutations of the androgen receptor coding sequence are infrequent in patients with isolated hypospadias. *J. Clin. Endocrinol. Metab.*, **80**, 2697–2699.

Alroy, I., Towers, T.L. and Freedman, L.P., 1995, Transcriptional repression of the interleukin-2 gene by vitamin D3: direct inhibition of NFATp/AP-1 complex formation by a nuclear hormone receptor. *Molec. Cell. Biol.*, **15**, 5789–5799.

Anzick, S.L., Kononen, J., Walker, R.L., Azorsa, D.O., Tanner, M.M., Guan, X.Y., Sauter, G., Kallioniemi, O.P., Trent, J.M. and Meltzer, P.S., 1997, AIB1, a steroid receptor coactivator amplified in breast and ovarian cancer. *Science*, **277**, 965–968.

Arbizu, T., Santamaria, J., Gomez, J.M., Quilez, A. and Serra, J.P., 1983, A family with adult spinal and bulbar muscular atrophy, X-linked inheritance and associated testicular failure. *J. Neurol. Sci.*, **59**, 371–382.

Baldazzi, L., Baroncini, C., Pirazzoli, P., Balsamo, A., Capelli, M., Marchetti, G., Bernardi, F. and Cacciari, E., 1994, Two mutations causing complete androgen insensitivity: a frameshift in the steroid binding domain and a Cys→Phe substitution in the second zing finger of the androgen receptor. *Hum. Mol. Genet.*, **3**, 1169–1170.

Bamberger, C.M., Bamberger, A.M., de Castro, M. and Chrousos, G.P., 1995, Glucocorticoid receptor β, a potential endogenous inhibitor of glucocorticoid action in humans. *J. Clin. Invest.*, **95**, 2435–2441.

Baniahmad, A., Kohne, A.C. and Renkawitz, R., 1992, A transferable silencing domain is present in the thyroid hormone receptor, in the v-erbA oncogene product and in the retinoic acid receptor. *EMBO J.*, **11**, 1015–1023.

Baniahmad, C., Baniahmad, A. and O'Malley, B.W., 1994, A rapid method combining a functional test of fusion proteins *in vivo* and their purification. *Biotechniques*, **16**, 194–196.

Baniahmad, A., Leng, X., Burris, T.P., Tsai, S.Y., Tsai, M.J. and O'Malley, B.W., 1995, The tau 4 activation domain of the thyroid hormone receptor is required for release of a putative corepressor(s) necessary for transcriptional silencing. *Molec. Cell. Biol.*, **15**, 76–86.

Batch, J.A., Williams, D.M., Davies, H.R., Brown, B.D., Evans, B.A., Hughes, I.A. and Patterson, M.N., 1992, Androgen receptor gene mutations identified by SSCP in fourteen subjects with androgen insensitivity syndrome. *Hum. Molec. Genet.*, **1**, 497–503.

Beck-Peccoz, P., Chatterjee, V.K., Chin, W.W., DeGroot, L.J., Jameson, J.L., Nakamura, H., Refetoff, S., Usala, S.J. and Weintraub, B.D., 1994, Nomenclature of thyroid hormone receptor β gene mutations in resistance to thyroid hormone. First workshop on thyroid hormone resistance, 10–11 July 1993, Cambridge, UK. *J. Endocrinol. Invest.*, **17**, 283–287.

Beitel, L.K., Kazemi-Esfarjani, P., Kaufman, M., Lumbroso, R., DiGeorge, A.M., Killinger, D.W., Trifiro, M.A. and Pinsky, L., 1994, Substitution of arginine-839 by cysteine or histidine in the androgen receptor causes different receptor phenotypes in cultured cells and coordinate degrees of clinical androgen resistance. *J. Clin. Invest.*, **94**, 546–554.

Belsham, D.D., Pereira, F., Greenberg, C.R., Liao, S. and Wrogemann, K., 1995, Leu-676-Pro mutation of the androgen receptor causes complete androgen insensitivity syndrome in a large Hutterite kindred. *Hum. Mutat.*, **5**, 28–33.

Bevan, C.L., Brown, B.B., Davies, H.R., Evans, B.A., Hughes, I.A. and Patterson, M.N., 1996, Functional analysis of six androgen receptor mutations identified in patients with partial androgen insensitivity syndrome. *Hum. Molec. Genet.*, **5**, 265–273.

Bevan, C.L., Hughes, I.A. and Patterson, M.N., 1997, Wide variation in androgen receptor dysfunction in complete androgen insensitivity syndrome. *J. Steroid Biochem. Molec. Biol.*, **61**, 19–26.

Blobel, G.A., Sieff, C.A. and Orkin, S.H., 1995, Ligand-dependent repression of the erythroid transcription factor GATA-1 by the estrogen receptor. *Molec. Cell. Biol.*, **15**, 3147–3153.

Bouillon, R., Okamura, W.H. and Norman, A.W., 1995, Structure–function relationships in the vitamin D endocrine system. *Endocr. Rev.*, **16**, 200–257.

Bourguet, W., Ruff, M., Chambon, P., Gronemeyer, H. and Moras, D., 1995, Crystal structure of the ligand-binding domain of the human nuclear receptor RXR-α. *Nature*, **375**, 377–382.

Brinkmann, A.O., Jenster, G., Ris-Stalpers, C., van der Korput, J.A., Bruggenwirth, H.T., Boehmer, A.L. and Trapman, J., 1995, Androgen receptor mutations. *J. Steroid Biochem. Molec. Biol.*, **53**, 443–448.

Brinkmann, A., Jenster, G., Ris-Stalpers, C., van der Korput, H., Bruggenwirth, H., Boehmer, A. and Trapman, J., 1996, Molecular basis of androgen insensitivity. *Steroids*, **61**, 172–175.

Brown, C.J., Goss, S.J., Lubahn, D.B., Joseph, D.R., Wilson, E.M., French, F.S. and Willard, H.F., 1989, Androgen receptor locus on the human X chromosome: regional localization to Xq11–12 and description of a DNA polymorphism. *Am. J. Hum. Genet.*, **44**, 264–269.

Brown, T.A. and DeLuca, H.F., 1990, Phosphorylation of the 1,25-dihydroxyvitamin D3 receptor. A primary event in 1,25-dihydroxyvitamin D3 action. *J. Biol. Chem.*, **265**, 10025–10029.

Brown, T.R., Lubahn, D.B., Wilson, E.M., French, F.S., Migeon, C.J. and Corden, J.L., 1990, Functional characterization of naturally occurring mutant androgen receptors from subjects with complete androgen insensitivity. *Molec. Endocrinol.*, **4**, 1759–1772.

Brown, T.R., Scherer, P.A., Chang, Y.T., Migeon, C.J., Ghirri, P., Murono, K. and Zhou, Z., 1993, Molecular genetics of human androgen insensitivity. *Eur. J. Pediatr.*, **152**, S62–69.

Bruggenwirth, H.T., Boehmer, A.L., Verleun-Mooijman, M.C., Hoogenboezem, T., Kleijer, W.J., Otten, B.J., Trapman, J. and Brinkmann, A.O., 1996, Molecular basis of androgen insensitivity. *J. Steroid Biochem. Molec. Biol.*, **58**, 569–575.

Bruggenwirth, H.T., Boehmer, A.L., Lobaccaro, J.M., Chiche, L., Sultan, C., Trapman, J. and Brinkmann, A.O., 1998, Substitution of Ala564 in the first zinc cluster of the deoxyribonucleic acid (DNA)-binding domain of the androgen receptor by Asp, Asn, or Leu exerts differential effects on DNA binding. *Endocrinology*, **139**, 103–110.

Castles, C.G., Fuqua, S.A., Klotz, D.M. and Hill, S.M., 1993, Expression of a constitutively active estrogen receptor variant in the estrogen receptor-negative BT-20 human breast cancer cell line. *Cancer Res.*, **53**, 5934–5939.

Cavailles, V., Dauvois, S., Danielian, P.S. and Parker, M.G., 1994, Interaction of proteins with transcriptionally active estrogen receptors. *Proc. Natl Acad. Sci., USA*, **91**, 10009–10013.

Cavailles, V., Dauvois, S., L'Horset, F., Lopez G., Hoare, S., Kushner, P.J. and Parker, M.G., 1995, Nuclear factor RIP140 modulates transcriptional activation by the estrogen receptor. *EMBO J.*, **14**, 3741–3751.

Chaidarun, S.S., Klibanski, A. and Alexander, J.M., 1997, Tumor-specific expression of alternatively spliced estrogen receptor messenger ribonucleic acid variants in human pituitary adenomas. *J. Clin. Endocrinol. Metab.*, **82**, 1058–1065.

Chakravarti, D., LaMorte, V.J., Nelson, M.C., Nakajima, T., Schulman, I.G., Juguilon, H., Montminy, M. and Evans, R.M., 1996, Role of CBP/P300 in nuclear receptor signalling. *Nature*, **383**, 99–103.

Chan, C.M. and Dowsett, M., 1997, A novel estrogen receptor variant mRNA lacking exons 4 to 6 in breast carcinoma. *J. Steroid Biochem. Molec. Biol.*, **62**, 419–430.

Chatterjee, V.K., 1997, Resistance to thyroid hormone. *Horm. Res.*, **48**, 43–46.

Chen, J.D. and Evans, R.M., 1995, A transcriptional co-repressor that interacts with nuclear hormone receptors. *Nature*, **377**, 454–457.

Chen, J.D., Umesono, K. and Evans, R.M., 1996, SMRT isoforms mediate repression and anti-repression of nuclear receptor heterodimers. *Proc. Natl Acad. Sci., USA*, **93**, 7567–7571.

Choong, C.S., Quigley, C.A., French, F.S. and Wilson, E.M., 1996a, A novel missense mutation in the amino-terminal domain of the human androgen receptor gene in a family with partial androgen insensitivity syndrome causes reduced efficiency of protein translation. *J. Clin. Invest.*, **98**, 1423–1431.

Choong, C.S., Sturm, M.J., Strophair, J.A., McCulloch, R.K., Tilley, W.D., Leedman, P.J. and Hurley, D.M., 1996b, Partial androgen insensitivity caused by an androgen receptor mutation at amino acid 907 (Gly→Arg) that results in decreased ligand binding affinity and reduced androgen receptor messenger ribonucleic acid levels. *J. Clin. Endocrinol. Metab.*, **81**, 236–243.

Cole, T.J., Blendy, J.A., Monaghan, A.P., Krieglstein, K., Schmid, W., Aguzzi, A., Fantuzzi, G., Hummler, E., Unsicker, K. and Schutz, G., 1995, Targeted disruption of the glucocorticoid receptor gene blocks adrenergic chromaffin cell development and severely retards lung maturation. *Genes Dev.*, **9**, 1608–1621.

Collingwood, T.N., Adams, M., Tone, Y. and Chatterjee, V.K., 1994, Spectrum of transcriptional, dimerization, and dominant negative properties of twenty different mutant thyroid hormone beta-receptors in thyroid hormone resistance syndrome. *Molec. Endocrinol.*, **8**, 1262–1277.

Daffada, A.A., Johnston, S.R., Smith, I.E., Detre, S., King, N. and Dowsett, M., 1995, Exon 5 deletion variant estrogen receptor messenger RNA expression in relation to tamoxifen resistance and progesterone receptor/pS2 status in human breast cancer. *Cancer Res.*, **55**, 288–293.

Dame, M.C., Pierce, E.A. and DeLuca, H.F., 1985, Identification of the porcine intestinal 1,25-dihydroxyvitamin D3 receptor on sodium dodecyl sulfate/polyacrylamide gels by renaturation and immunoblotting. *Proc. Natl Acad. Sci., USA*, **82**, 7825–7829.

Davies, H.R., Hughes, I.A. and Patterson, M.N., 1995, Genetic counselling in complete androgen insensitivity syndrome: trinucleotide repeat polymorphisms, single-strand conformation polymorphism and direct detection of two novel mutations in the androgen receptor gene. *Clin. Endocrinol. (Oxford)*, **43**, 69–77.

De Bellis, A., Quigley, C.A., Cariello, N.F., el-Awady, M.K., Sar, M., Lane, M.V., Wilson, E.M. and French, F.S., 1992, Single base mutations in the human androgen receptor gene causing complete androgen insensitivity: rapid detection by a modified denaturing gradient gel electrophoresis technique. *Molec. Endocrinol.*, **6**, 1909–1920.

De Bellis, A., Quigley, C.A., Marschke, K.B., el-Awady, M.K., Lane, M.V., Smith, E.P., Sar, M., Wilson, E.M. and French, F.S., 1994, Characterization of mutant androgen receptors causing partial androgen insensitivity syndrome. *J. Clin. Endocrinol. Metab.*, **78**, 513–522.

Desai, A.J., Luqmani, Y.A., Walters, J.E., Coope, R.C., Dagg, B., Gomm, J.J., Pace, P.E., Rees, C.N., Thirunavukkarasu, V., Shousha, S., Groome, N.P., Coombes, R. and Ali, S., 1997, Presence of exon 5-deleted oestrogen receptor in human breast cancer: functional analysis and clinical significance. *Br. J. Cancer*, **75**, 1173–1184.

Dork, T., Schnieders, F., Jakubiczka, S., Wieacker, P., Schroeder-Kurth, T. and Schmidtke, J., 1998, A new missense substitution at a mutational hot spot of the androgen receptor in siblings with complete androgen insensitivity syndrome. *Hum. Mutat.*, **11**, 337–339.

Elgort, M.G., Zou, A., Marschke, K.B. and Allegretto, E.A., 1996, Estrogen and estrogen receptor antagonists stimulate transcription from the human retinoic acid receptor-alpha 1 promoter via a novel sequence. *Molec. Endocrinol.*, **10**, 477–487.

Ellis, H.M. and Horvitz, H.R., 1986, Genetic control of programmed cell death in the nematode C. elegans. *Cell*, **44**, 817–829.

Enmark, E., Pelto-Huikko, M., Grandien, K., Lagercrantz, S., Lagercrantz, J., Fried, G., Nordenskjold, M. and Gustafsson, J.A., 1997, Human estrogen receptor β-gene structure, chromosomal localization, and expression pattern. *J. Clin. Endocrinol. Metab.*, **82**, 4258–4265.

Erenburg, I., Schachter, B., Mira y Lopez, R. and Ossowski, L., 1997, Loss of an estrogen receptor isoform (ER alpha delta 3) in breast cancer and the consequences of its reexpression: interference with estrogen-stimulated properties of malignant transformation. *Molec. Endocrinol.*, **11**, 2004–2015.

Evans, B.A., Hughes, I.A., Bevan, C.L., Patterson, M.N. and Gregory, J.W., 1997, Phenotypic diversity in siblings with partial androgen insensitivity syndrome. *Arch. Dis. Child.*, **76**, 529–531.

Evans, R.M., 1988, The steroid and thyroid hormone receptor superfamily. *Science*, **240**, 889–895.

Fuqua, S.A., Fitzgerald, S.D., Chamness, G.C., Tandon, A.K., McDonnell, D.P., Nawaz, Z., O'Malley, B.W. and McGuire, W.L., 1991, Variant human breast tumor estrogen receptor with constitutive transcriptional activity. *Cancer Res.*, **51**, 105–109.

Fuqua, S.A., Fitzgerald, S.D., Allred, D.C., Elledge, R.M., Nawaz, Z., McDonnell, D.P., O'Malley, B.W., Greene, G.L. and McGuire, W.L., 1992, Inhibition of estrogen receptor action by a naturally occurring variant in human breast tumors. *Cancer Res.*, **52**, 483–486.

Fuqua, S.A.W., 1996, Estrogen and progesterone receptors and breast cancer, in Harris, J.R., Lippman, M.E., Morrow, M. and Hellman, S. (Eds) *Diseases of the Breast*, pp. 261–271, Philadelphia: Lippincott-Raven.

Fuqua, S.A.W., Allred, D.C., Elledge, R.M., Krieg, S.L., Benedix, M.G., Nawaz, Z., O'Malley, B.W., Greene, G.L. and McGuire, W.L., 1993, The ER-positive/PgR-negative breast cancer phenotype is not associated with mutations within the DNA binding domain. *Breast Cancer Res. Treat.*, **26**, 191–202.

Gallacchi, P., Schoumacher, F., Eppenberger-Castori, S., Von Landenberg, E.M., Kueng, W., Eppenberger, U. and Mueller, H., 1998, Increased expression of estrogen-receptor exon-5-deletion variant in relapse tissues of human breast cancer. *Int. J. Cancer*, **79**, 44–48.

Gast, A., Neuschmid-Kaspar, F., Klocker, H. and Cato, A.C., 1995, A single amino acid exchange abolishes dimerization of the androgen receptor and causes Reifenstein syndrome. *Molec. Cell. Endocrinol.*, **111**, 93–98.

Geley, S., Hartmann, B.L., Hala, M., Strasser-Wozak, E.M., Kapelari, K. and Kofler, R., 1996, Resistance to glucocorticoid-induced apoptosis in human T-cell acute lymphoblastic leukemia CEM-C1 cells is due to insufficient glucocorticoid receptor expression. *Cancer Res.*, **56**, 5033–5038.

Gilbert-Dreyfus, S., Sebaoun, C.A. and Belaisch J., 1957, Etude d'un cas familial d'androgynoidisme avec hypospadias grave gynecomastie et hyperestrogenie. *Ann. Endocr.*, **18**, 93.

Glass, C.K., Rose, D.W. and Rosenfeld, M.G., 1997, Nuclear receptor coactivators. *Curr. Opin. Cell Biol.*, **9**, 222–232.

Gottlieb, B., Lehvaslaiho, H., Beitel, L.K., Lumbroso, R., Pinsky, L. and Trifiro, M., 1998, The androgen receptor gene mutations database. *Nucleic Acids Res.*, **26**, 234–238.

Hala, M., Hartmann, B.L., Bock, G., Geley, S. and Kofler, R., 1996, Glucocorticoid-receptor-gene defects and resistance to glucocorticoid-induced apoptosis in human leukemic cell lines. *Int. J. Cancer*, **68**, 663–668.

Hanstein, B., Eckner, R., DiRenzo, J., Halachmi, S., Liu, H., Searcy, B., Kurokawa, R. and Brown, M., 1996, p300 is a component of an estrogen receptor coactivator complex. *Proc. Natl Acad. Sci., USA*, **93**, 11540–11545.

Harding, A.E., Thomas, P.K., Baraitser, M., Bradbury, P.G., Morgan-Hughes, J.A. and Ponsford, J.R., 1982, X-linked recessive bulbospinal neuronopathy: a report of ten cases. *J. Neurol. Neurosurg. Psychiatry*, **45**, 1012–1019.

Haussler, M.R., Haussler, C.A., Jurutka, P.W., Thompson, P.D., Hsieh, J.C., Remus, L.S., Selznick, S.H. and Whitfield, G.K., 1997, The vitamin D hormone and its nuclear receptor: molecular actions and disease states. *J. Endocrinol., Suppl.*, **154**, S57–73.

Hiort, O., Huang, Q., Sinnecker, G.H., Sadeghi-Nejad, A., Kruse, K., Wolfe, H.J. and Yandell, D.W., 1993, Single strand conformation polymorphism analysis of androgen receptor gene mutations in patients with androgen insensitivity syndromes: application for diagnosis, genetic counseling, and therapy. *J. Clin. Endocrinol. Metab.*, 77, 262–266.

Hiort, O., Wodtke, A., Struve, D., Zollner, A. and Sinnecker, G.H., 1994, Detection of point mutations in the androgen receptor gene using non-isotopic single strand conformation polymorphism analysis. German Collaborative Intersex Study Group. *Hum. Molec. Genet.*, **3**, 1163–1166.

Hiort, O., Sinnecker, G.H., Holterhus, P.M., Nitsche, E.M. and Kruse, K., 1996, The clinical and molecular spectrum of androgen insensitivity syndromes. *Am. J. Med. Genet.*, **63**, 218–222.

Hollenberg, S.M., Weinberger, C., Ong, E.S., Cerelli, G., Oro, A., Lebo, R., Thompson, E.B., Rosenfeld, M.G. and Evans, R.M., 1985, Primary structure and expression of a functional human glucocorticoid receptor cDNA. *Nature*, **318**, 635–641.

Holterhus, P.M., Bruggenwirth, H.T., Hiort, O., Kleinkauf-Houcken, A., Kruse, K., Sinnecker, G.H. and Brinkmann, A.O., 1997, Mosaicism due to a somatic mutation of the androgen receptor gene determines phenotype in androgen insensitivity syndrome. *J. Clin. Endocrinol. Metab.*, **82**, 3584–3589.

Hopp, T.A. and Fuqua, S.A.W., 1998, Estrogen receptor variants. *J. Mammary Gland Biology Neoplasia*, **3**, 73–83.

Horlein, A.J., Naar, A.M., Heinzel, T., Torchia, J., Gloss, B., Kurokawa, R., Ryan, A., Kamei, Y., Soderstrom, M. and Glass, C.K., 1995, Ligand-independent repression by the thyroid hormone receptor mediated by a nuclear receptor co-repressor. *Nature*, **377**, 397–404.

Horwitz, K.B., Jackson, T.A., Bain, D.L., Richer, J.K., Takimoto, G.S. and Tung, L., 1996, Nuclear receptor coactivators and corepressors. *Molec. Endocrinol.*, **10**, 1167–1177.

Hu, C., Hyder, S.M., Needleman, D.S. and Baker, V.V., 1996, Expression of estrogen receptor variants in normal and neoplastic human uterus. *Molec. Cell. Endocrinol.*, **118**, 173–179.

Huang, A., Leygue, E.R., Snell, L., Murphy, L.C. and Watson, P.H., 1997, Expression of estrogen receptor variant messenger RNAs and determination of estrogen receptor status in human breast cancer. *Am. J. Pathol.*, **150**, 1827–1833.

Hughes, M.R., Malloy, P.J., Kieback, D.G., Kesterson, R.A., Pike, J.W., Feldman, D. and O'Malley, B.W., 1988, Point mutations in the human vitamin D receptor gene associated with hypocalcemic rickets. *Science*, **242**, 1702–1705.

Huizenga, N.A., Koper, J.W., De Lange, P., Pols, H.A., Stolk, R.P., Burger, H., Grobbee, D.E., Brinkmann, A.O., De Jong, F.H. and Lamberts, S.W., 1998, A polymorphism in the glucocorticoid receptor gene may be associated with an increased sensitivity to glucocorticoids *in vivo*. *J. Clin. Endocrinol. Metab.*, **83**, 144–151.

Imai, A., Ohno, T., Iida, K., Ohsuye, K., Okano, Y. and Tamaya, T., 1995, A frame-shift mutation of the androgen receptor gene in a patient with receptor-negative complete testicular feminization: comparison with a single base substitution in a receptor-reduced incomplete form. *Ann. Clin. Biochem.*, **32**, 482–486.

Imasaki, K., Okabe, T., Murakami, H., Fujita, K., Takayanagi, R. and Nawata, H., 1995, Premature termination mutation (772Glu→stop) in the hormone-binding domain of the androgen receptor in a patient with the receptor-negative form of complete androgen insensitivity syndrome. *Endocr. J.*, **42**, 643–648.

Imasaki, K., Okabe, T., Murakami, H., Tanaka, Y., Haji, M., Takayanagi, R. and Nawata, H., 1996, Androgen insensitivity syndrome due to new mutations in the DNA-binding domain of the androgen receptor. *Molec. Cell. Endocrinol.*, **120**, 15–24.

Inoue, S., Hoshino, S., Miyoshi, H., Akishita, M., Hosoi, T., Orimo, H. and Ouchi, Y., 1996, Identification of a novel isoform of estrogen receptor, a potential inhibitor of estrogen action, in vascular smooth muscle cells. *Biochem. Biophys. Res. Commun.*, **219**, 766–772.

Jakubiczka, S., Nedel, S., Werder, E.A., Schleiermacher, E., Theile, U., Wolff, G. and Wieacker, P., 1997, Mutations of the androgen receptor gene in patients with complete androgen insensitivity. *Hum. Mutat.*, **9**, 57–61.

Jost, A., 1990, *Hormonal Control of the Masculinization of the Body*, pp. 439–442, New York: Chapman and Hall.

Kamei, Y., Xu, L., Heinzel, T., Torchia, J., Kurokawa, R., Gloss, B., Lin, S.C., Heyman, R.A., Rose, D.W., Glass, C.K. and Rosenfeld, M.G., 1996, A CBP integrator complex mediates transcriptional activation and AP-1 inhibition by nuclear receptors. *Cell*, **85**, 403–414.

Karl, M., Lamberts, S.W., Detera-Wadleigh, S.D., Encio, I.J., Stratakis, C.A., Hurley, D.M., Accili, D. and Chrousos, G.P., 1993, Familial glucocorticoid resistance caused by a splice site deletion in the human glucocorticoid receptor gene. *J. Clin. Endocrinol. Metab.*, **76**, 683–689.

Karl, M., Lamberts, S.W., Koper, J.W., Katz, D.A., Huizenga, N.E., Kino, T., Haddad, B.R., Hughes, M.R. and Chrousos, G.P., 1996a, Cushing's disease preceded by generalized glucocorticoid resistance: clinical consequences of a novel, dominant-negative glucocorticoid receptor mutation. *Proc. Assoc. Am. Physicians*, **108**, 296–307.

Karl, M., Von Wichert, G., Kempter, E., Katz, D.A., Reincke, M., Monig, H., Ali, I.U., Stratakis, C.A., Oldfield, E.H., Chrousos, G.P. and Schulte, H.M., 1996b, Nelson's syndrome associated with a somatic frame shift mutation in the glucocorticoid receptor gene. *J. Clin. Endocrinol. Metab.*, **81**, 124–129.

Kastner, P., Mark, M. and Chambon, P., 1995, Nonsteroid nuclear receptors: what are genetic studies telling us about their role in real life? *Cell*, **83**, 859–869.

Kasumi, H., Komori, S., Yamasaki, N., Shima, H. and Isojima, S., 1993, Single nucleotide substitution of the androgen receptor gene in a case with receptor-positive androgen insensitivity syndrome (complete form). *Acta Endocrinol. (Copenhagen)*, **128**, 355–360.

Kazemi-Esfarjani, P., Beitel, L.K., Trifiro, M., Kaufman, M., Rennie, P., Sheppard, P., Matusik, R. and Pinsky, L., 1993, Substitution of valine-865 by methionine or leucine in the human androgen receptor causes complete or partial androgen insensitivity, respectively with distinct androgen receptor phenotypes. *Molec. Endocrinol.*, 7, 37–46.

Kennedy, W.R., Alter, M. and Sung, J.H., 1968, Progressive proximal spinal and bulbar muscular atrophy of late onset. A sex-linked recessive trait. *Neurology*, **18**, 671–680.

Koehorst, S.G., Cox, J.J., Donker, G.H., Lopes da Silva, S., Burbach, J.P., Thijssen, J.H. and Blankenstein, M.A., 1994, Functional analysis of an alternatively spliced estrogen receptor lacking exon 4 isolated from MCF-7 breast cancer cells and meningioma tissue. *Molec. Cell. Endocrinol.*, **101**, 237–245.

Koenig, R.J., Lazar, M.A., Hodin, R.A., Brent, G.A., Larsen, P.R., Chin, W.W. and Moore, D.D., 1989, Inhibition of thyroid hormone action by a non-hormone binding c-erbA protein generated by alternative mRNA splicing. *Nature*, **337**, 659–661.

Komori, S., Sakata, K., Tanaka, H., Shima, H. and Koyama, K., 1997, DNA analysis of the androgen receptor gene in two cases with complete androgen insensitivity syndrome. *J. Obstet. Gynaecol. Res.*, **23**, 277–281.

Komori, S., Kasumi, H., Sakata, K., Tanaka, H., Hamada, K. and Koyama, K., 1998, Molecular analysis of the androgen receptor gene in 4 patients with complete androgen insensitivity. *Arch. Gynecol. Obstet.*, **261**, 95–100.

Koper, J.W., Stolk, R.P., de Lange, P., Huizenga, N.A., Molijn, G.J., Pols, H.A., Grobbee, D.E., Karl, M., de Jong, F.H., Brinkmann, A.O. and Lamberts, S.W., 1997, Lack of association between five polymorphisms in the human glucocorticoid receptor gene and glucocorticoid resistance. *Hum. Genet.*, **99**, 663–668.

Kopp, P., Kitajima, K. and Jameson, J.L., 1996, Syndrome of resistance to thyroid hormone: insights into thyroid hormone action. *Proc. Soc. Exp. Biol. Med.*, **211**, 49–61.

Korach, K.S., 1994, Insights from the study of animals lacking functional estrogen receptor. *Science*, **266**, 1524–1527.

Kristjansson, K., Rut, A.R., Hewison, M., O'Riordan, J.L. and Hughes, M.R., 1993, Two mutations in the hormone binding domain of the vitamin D receptor cause tissue resistance to 1,25 dihydroxyvitamin D3. *J. Clin. Invest.*, **92**, 12–16.

Kuiper, G.G., Faber, P.W., van Rooij, H.C., van der Korput, J.A., Ris-Stalpers, C., Klaassen, P., Trapman, J. and Brinkmann, A.O., 1989, Structural organization of the human androgen receptor gene. *J. Molec. Endocrinol.*, 2, R1–4.

Kumar, V. and Chambon, P., 1988, The estrogen receptor binds tightly to its responsive element as a ligand-induced homodimer. *Cell*, **55**, 145–156.

Kurokawa, R., Soderstrom, M., Horlein, A., Halachmi, S., Brown, M., Rosenfeld, M.G. and Glass, C.K., 1995, Polarity-specific activities of retinoic acid receptors determined by a co-repressor. *Nature*, **377**, 451–454.

La Spada, A.R., Wilson, E.M., Lubahn, D.B., Harding, A.E. and Fischbeck, K.H., 1991, Androgen receptor gene mutations in X-linked spinal and bulbar muscular atrophy. *Nature*, **352**, 77–79.

La Spada, A.R., Roling, D.B., Harding, A.E., Warner, C.L., Spiegel, R., Hausmanowa-Petrusewicz, I., Yee, W.C. and Fischbeck, K.H., 1992, Meiotic stability and genotype–phenotype correlation of the trinucleotide repeat in X-linked spinal and bulbar muscular atrophy. *Nat. Genet.*, **2**, 301–304.

Labuda, M., Fujiwara, T.M., Ross, M.V., Morgan, K., Garcia-Heras, J., Ledbetter, D.H., Hughes, M.R. and Glorieux, F.H., 1992, Two hereditary defects related to vitamin D metabolism map to the same region of human chromosome 12q13–14. *J. Bone Miner. Res.*, **7**, 1447–1453.

Lamberts, S.W., Huizenga, A.T., de Lange, P., de Jong, F.H. and Koper, J.W., 1996, Clinical aspects of glucocorticoid sensitivity. *Steroids*, **61**, 157–160.

Lazar, M.A., 1993, Thyroid hormone receptors: multiple forms, multiple possibilities. *Endocr. Rev.*, **14**, 184–193.

Lazar, M.A., Hodin, R.A. and Chin, W.W., 1989, Human carboxyl-terminal variant of alpha-type c-erbA inhibits trans-activation by thyroid hormone receptors without binding thyroid hormone. *Proc. Natl Acad. Sci., USA*, **86**, 7771–7774.

Lazennec, G., Ediger, T.R., Petz, L.N., Nardulli, A.M. and Katzenellenbogen, B.S., 1997, Mechanistic aspects of estrogen receptor activation probed with constitutively active estrogen receptors: correlations with DNA and coregulator interactions and receptor conformational changes. *Molec. Endocrinol.*, **11**, 1375–1386.

LeDouarin, B., Zechel, C., Garnier, J.M., Lutz, Y., Tora, L., Pierrat, P., Heery, D., Gronemeyer, H., Chambon, P. and Losson, R., 1995, The N-terminal part of TIF1, a putative mediator of the ligand-dependent activation function (AF-2) of nuclear receptors, is fused to B-raf in the oncogenic protein T18. *EMBO J.*, **14**, 2020–2033.

Lee, J.W., Ryan, F., Swaffield, J.C., Johnston, S.A. and Moore, D.D., 1995, Interaction of thyroid-hormone receptor with a conserved transcriptional mediator. *Nature*, **374**, 91–94.

Leygue, E., Huang, A., Murphy, L.C. and Watson, P.H., 1996a, Prevalence of estrogen receptor variant messenger RNAs in human breast cancer. *Cancer Res.*, **56**, 4324–4327.

Leygue, E.R., Watson, P.H. and Murphy, L.C., 1996b, Estrogen receptor variants in normal human mammary tissue. *J. Natl Cancer Inst.*, **88**, 284–290.

Lobaccaro, J.M., Lumbroso, S., Belon, C., Galtier-Dereure, F., Bringer, J., Lesimple, T., Heron, J.F., Pujol, H. and Sultan, C., 1993a, Male breast cancer and the androgen receptor gene. *Nat. Genet.*, **5**, 109–110.

Lobaccaro, J.M., Lumbroso, S., Belon, C., Galtier-Dereure, F., Bringer, J., Lesimple, T., Namer, M., Cutuli, B.F., Pujol, H. and Sultan, C., 1993b, Androgen receptor gene mutation in male breast cancer. *Hum. Molec. Genet.*, **2**, 1799–1802.

Lobaccaro, J.M., Lumbroso, S., Berta, P., Chaussain, J.L. and Sultan, C., 1993c, Complete androgen insensitivity syndrome associated with a de novo mutation of the androgen receptor gene detected by single strand conformation polymorphism, *J. Steroid Biochem. Mol. Biol.*, **44**, 211–216.

Lobaccaro, J.M., Lumbroso, S., Poujol, N., Belon, C. and Sultan, C., 1994, Molecular genetics of androgen insensitivity syndromes. *Cell. Molec. Biol. (Noisy-le-grand)*, **40**, 301–308.

Lobaccaro, J.M., Poujol, N., Chiche, L., Lumbroso, S., Brown, T.R., and Sultan, C., 1996, Molecular modelling and *in vitro* investigations of the human androgen receptor DNA-binding domain: application for the study of two mutations. *Molec. Cell. Endocrinol.*, **116**, 137–147.

Lu, B., Leygue, E., Dotzlaw, H., Murphy, L.J., Murphy, L.C. and Watson, P.H., 1998, Estrogen receptor-beta mRNA variants in human and murine tissues. *Molec. Cell. Endocrinol.*, **138**, 199–203.

Lubs, H.A., Jr., Vilar, O. and Bergenstal, D.M., 1959, Familial male pseudohermaphrodism with labial testes and partial feminization: endocrine studies and genetic aspects. *J. Clin. Endocr. Metab.*, **19**, 1110.

Luisi, B.F., Xu, W.X., Otwinowski, Z., Freedman, L.P., Yamamoto, K.R. and Sigler, P.B., 1991, Crystallographic analysis of the interaction of the glucocorticoid receptor with DNA. *Nature*, **352**, 497–505.

Lumbroso, S., Lobaccaro, J.M., Belon, C., Amram, S., Bachelard, B., Garandeau, P. and Sultan, C., 1994, Molecular prenatal exclusion of familial partial androgen insensitivity (Reifenstein syndrome). *Eur. J. Endocrinol.*, **130**, 327–332.

Lumbroso, S., Lobaccaro, J.M., Georget, V., Leger, J., Poujol, N., Terouanne, B., Evain-Brion, D., Czernichow, P. and Sultan, C., 1996, A novel substitution (Leu707Arg) in exon 4 of the androgen receptor gene causes complete androgen resistance. *J. Clin. Endocrinol. Metab.*, **81**, 1984–1988.

MacDonald, P.N., Dowd, D.R., Nakajima, S., Galligan, M.A., Reeder, M.C., Haussler, C.A., Ozato, K. and Haussler, M.R., 1993, Retinoid X receptors stimulate and 9-cis retinoic acid inhibits 1,25-dihydroxyvitamin D3-activated expression of the rat osteocalcin gene. *Molec. Cell. Biol.*, **13**, 5907–5917.

Madsen, M.W., Reiter, B.E., Larsen, S.S., Briand, P. and Lykkesfeldt, A.E., 1997, Estrogen receptor messenger RNA splice variants are not involved in antiestrogen resistance in sublines of MCF-7 human breast cancer cells. *Cancer Res.*, 57, 585–589.

Malloy, P.J., Hochberg, Z., Tiosano, D., Pike, J.W., Hughes, M.R. and Feldman, D., 1990, The molecular basis of hereditary 1,25-dihydroxyvitamin D3 resistant rickets in seven related families. *J. Clin. Invest.*, **86**, 2071–2079.

Malloy, P.J., Weisman, Y. and Feldman, D., 1994, Hereditary 1α,25-dihydroxyvitamin D-resistant rickets resulting from a mutation in the vitamin D receptor deoxyribonucleic acid-binding domain. *J. Clin. Endocrinol. Metab.*, **78**, 313–316.

Malloy, P.J., Eccleshall, T.R., Gross, C., Van Maldergem, L., Bouillon, R. and Feldman, D., 1997, Hereditary vitamin D resistant rickets caused by a novel mutation in the vitamin D receptor that results in decreased affinity for hormone and cellular hyporesponsiveness. *J. Clin. Invest.*, **99**, 297–304.

Malmgren, H., Gustavsson, J., Tuvemo, T. and Dahl, N., 1996, Rapid detection of a mutation hot-spot in the human androgen receptor. *Clin. Genet.*, **50**, 202–205.

Mangelsdorf, D.J., Thummel, C., Beato, M., Herrlich, P., Schutz, G., Umesono, K., Blumberg, B., Kastner, P., Mark, M. and Chambon, P., 1995, The nuclear receptor superfamily: the second decade. *Cell*, **83**, 835–839.

Marcelli, M., Tilley, W.D., Wilson, C.M., Wilson, J.D., Griffin, J.E. and McPhaul, M.J., 1990, A single nucleotide substitution introduces a premature termination codon into the androgen receptor gene of a patient with receptor-negative androgen resistance. *J. Clin. Invest.*, **85**, 1522–1528.

Marcelli, M., Tilley, W.D., Zoppi, S., Griffin, J.E., Wilson, J.D. and McPhaul, M.J., 1991, Androgen resistance associated with a mutation of the androgen receptor at amino acid 772 (Arg→Cys) results from a combination of decreased messenger ribonucleic acid levels and impairment of receptor function. *J. Clin. Endocrinol. Metab.*, **73**, 318–325.

Marcelli, M., Zoppi, S., Wilson, C.M., Griffin, J.E. and McPhaul, M.J., 1994, Amino acid substitutions in the hormone-binding domain of the human androgen receptor alter the stability of the hormone receptor complex. *J. Clin. Invest.*, **94**, 1642–1650.

McDermott, M.T. and Ridgway, E.C., 1993, Thyroid hormone resistance syndromes. *Am. J. Med.*, **94**, 424–432.

McEwan, I.J., Wright, A.P. and Gustafsson, J.A., 1997, Mechanism of gene expression by the glucocorticoid receptor: role of protein–protein interactions. *Bioessays*, **19**, 153–160.

McPhaul, M.J., Marcelli, M., Tilley, W.D., Griffin, J.E., Isidro-Gutierrez, R.F. and Wilson, J.D., 1991, Molecular basis of androgen resistance in a family with a qualitative abnormality of the androgen receptor and responsive to high-dose androgen therapy. *J. Clin. Invest.*, **87**, 1413–1421.

McPhaul, M.J., Marcelli, M., Zoppi, S., Wilson, C.M., Griffin, J.E. and Wilson, J.D., 1992, Mutations in the ligand-binding domain of the androgen receptor gene cluster in two regions of the gene. *J. Clin. Invest.*, **90**, 2097–2101.

Menasce, L.P., White, G.R., Harrison, C.J. and Boyle, J.M., 1993, Localization of the estrogen receptor locus (ESR) to chromosome 6q25.1 by FISH and a simple post-FISH banding technique. *Genomics*, **17**, 263–265.

Moalli, P.A. and Rosen, S.T., 1994, Glucocorticoid receptors and resistance to glucocorticoids in hematologic malignancies. *Leuk. Lymphoma*, **15**, 363–374.

Morris, J.M., 1953, Syndrome of testicular feminization in male pseudohermaphrodites. *Amer. J. Obst. Gynec.*, **65**, 1192.

Mowszowicz, I., Lee, H.J., Chen, H.T., Mestayer, C., Portois, M.C., Cabrol, S., Mauvais-Jarvis, P. and Chang, C., 1993, A point mutation in the second zinc finger of the DNA-binding domain of the androgen receptor gene causes complete androgen insensitivity in two siblings with receptor-positive androgen resistance. *Molec. Endocrinol.*, **7**, 861–869.

Murphy, L.C., Wang, M., Coutt, A. and Dotzlaw, H., 1996, Novel mutations in the estrogen receptor messenger RNA in human breast cancers. *J. Clin. Endocrinol. Metab.*, **81**, 1420–1427.

Murphy, L.C., Dotzlaw, H., Leygue, E., Douglas, D., Coutts, A. and Watson, P.H., 1997a, Estrogen receptor variants and mutations. *J. Steroid Biochem. Molec. Biol.*, **62**, 363–372.

Murphy, L.C., Leygue, E., Dotzlaw, H., Douglas, D., Coutts, A. and Watson, P.H., 1997b, Oestrogen receptor variants and mutations in human breast cancer. *Ann. Med.*, **29**, 221–234.

Nagashima, T., Seko, K., Hirose, K., Mannen, T., Yoshimura, S., Arima, R., Nagashima, K. and Morimatsu, Y., 1988, Familial bulbo-spinal muscular atrophy associated with testicular atrophy and sensory neuropathy (Kennedy–Alter–Sung syndrome). Autopsy case report of two brothers. *J. Neurol. Sci.*, **87**, 141–152.

Nakai, A., Sakurai, A., Macchia, E., Fang, V. and DeGroot, L.J., 1990, The roles of three forms of human thyroid hormone receptor in gene regulation. *Molec. Cell. Endocrinol.*, **72**, 143–148.

Nakao, R., Yanase, T., Sakai, Y., Haji, M. and Nawata, H., 1993, A single amino acid substitution (gly743→val) in the steroid-binding domain of the human androgen receptor leads to Reifenstein syndrome. *J. Clin. Endocrinol. Metab.*, **77**, 103–107.

Ogawa, S., Inoue, S., Orimo, A., Hosoi, T., Ouchi, Y. and Muramatsu, M., 1998, Cross-inhibition of both estrogen receptor alpha and beta pathways by each dominant negative mutant. *FEBS Lett.*, **423**, 129–132.

Onate, S.A., Tsai, S.Y., Tsai, M.J. and O'Malley, B.W., 1995, Sequence and characterization of a coactivator for the steroid hormone receptor superfamily. *Science*, **270**, 1354–1357.

Osborne, C.K., 1991, Receptors, in Harris J.R., Hellman, S., Henderson, I.C. and Kinne, D.W. (Eds) *Breast Diseases*, pp. 301–325, Philadelphia: Lippincott.

Pettersson, K., Grandien, K., Kuiper, G.G. and Gustafsson, J.A., 1997, Mouse estrogen receptor β forms estrogen response element-binding heterodimers with estrogen receptor α. *Molec. Endocrinol.*, **11**, 1486–1496.

Pfeffer, U., Fecarotta, E., Castagnetta, L. and Vidali, G., 1993, Estrogen receptor variant messenger RNA lacking exon 4 in estrogen-responsive human breast cancer cell lines. *Cancer Res.*, **53**, 741–743.

Pfeffer, U., Fecarotta, E. and Vidali, G., 1995, Coexpression of multiple estrogen receptor variant messenger RNAs in normal and neoplastic breast tissues and in MCF-7 cells. *Cancer Res.*, **55**, 2158–2165.

Pfeffer, U., Fecarotta, E., Arena, G., Forlani, A. and Vidali, G., 1996, Alternative splicing of the estrogen receptor primary transcript normally occurs in estrogen receptor positive tissues and cell lines. *J. Steroid Biochem. Molec. Biol.*, **56**, 99–105.

Pinsky, L., Trifiro, M., Kaufman, M., Beitel, L.K., Mhatre, A., Kazemi-Esfarjani, P., Sabbaghian, N., Lumbroso, R., Alvarado, C. and Vasiliou, M., 1992, Androgen resistance due to mutation of the androgen receptor. *Clin. Invest. Med.*, **15**, 456–472.

Poujol, N., Lobaccaro, J.M., Chiche, L., Lumbroso, S. and Sultan, C., 1997, Functional and structural analysis of R607Q and R608K androgen receptor substitutions associated with male breast cancer. *Molec. Cell. Endocrinol.*, **130**, 43–51.

Prior, L., Bordet, S., Trifiro, M.A., Mhatre, A., Kaufman, M., Pinsky, L., Wrogeman, K., Belsham, D.D., Pereira, F. and Greenberg, C., 1992, Replacement of arginine 773 by cysteine or histidine in the human androgen receptor causes complete androgen insensitivity with different receptor phenotypes. *Am. J. Hum. Genet.*, **51**, 143–155.

Quigley, C.A., De Bellis, A., Marschke, K.B., el-Awady, M.K., Wilson, E.M. and French, F.S., 1995, Androgen receptor defects: historical, clinical, and molecular perspectives. *Endocr. Rev.*, **16**, 271–321.

Radmayr, C., Culig, Z., Glatzl, J., Neuschmid-Kaspar, F., Bartsch, G. and Klocker, H., 1997, Androgen receptor point mutations as the underlying molecular defect in 2 patients with androgen insensitivity syndrome. *J. Urol.*, **158**, 1553–1556.

Rastinejad, F., Evilia, C. and Lu, P., 1995, Studies of nucleic acids and their protein interactions by 19F NMR. *Methods Enzymol.*, **261**, 560–575.

Rea, D. and Parker, M.G., 1996, Effects of an exon 5 variant of the estrogen receptor in MCF-7 breast cancer cells. *Cancer Res.*, **56**, 1556–1563.

Refetoff, S., 1994, Resistance to thyroid hormone: an historical overview. *Thyroid*, **4**, 345–349.

Refetoff, S., DeWind, L.T. and DeGroot, L.J., 1967, Familial syndrome combining deaf-mutism, stuppled epiphyses, goiter and abnormally high PBI: possible target organ refractoriness to thyroid hormone. *J. Clin. Endocrinol. Metab.*, **27**, 279–294.

Refetoff, S., Weiss, R.E. and Usala, S.J., 1993, The syndromes of resistance to thyroid hormone. *Endocr. Rev.*, **14**, 348–399.

Reifenstein, E.C., Jr., 1947, Hereditary familial hypogonadism. *Proc. Amer. Fed. Clin. Res.*, **3**, 86.

Renaud, J.P., Rochel, N., Ruff, M., Vivat, V., Chambon, P., Gronemeyer, H. and Moras, D., 1995, Crystal structure of the RAR-γ ligand-binding domain bound to all-trans retinoic acid. *Nature*, **378**, 681–689.

Rice, L.W., Jazaeri, A.A. and Shupnik, M.A., 1997, Estrogen receptor mRNA splice variants in pre- and postmenopausal human endometrium and endometrial carcinoma. *Gynecol. Oncol.*, **65**, 149–157.

Ris-Stalpers, C., Trifiro, M.A., Kuiper, G.G., Jenster, G., Romalo, G., Sai, T., van Rooij, H.C., Kaufman, M., Rosenfield, R.L. and Liao, S., 1991, Substitution of aspartic acid-686 by histidine or asparagine in the human androgen receptor leads to a functionally inactive protein with altered hormone-binding characteristics. *Molec. Endocrinol.*, **5**, 1562–1569.

Ris-Stalpers, C., Hoogenboezem, T., Sleddens, H.F., Verleun-Mooijman, M.C., Degenhart, H.J., Drop, S.L., Halley, D.J., Oosterwijk, J.C., Hodgins, M.B. and Trapman, J., 1994, A practical approach to the detection of androgen receptor gene mutations and pedigree analysis in families with x-linked androgen insensitivity. *Pediatr. Res.*, **36**, 227–234.

Ritchie, H.H., Hughes, M.R., Thompson, E.T., Malloy, P.J., Hochberg, Z., Feldman, D., Pike, J.W. and O'Malley, B.W., 1989, An ochre mutation in the vitamin D receptor gene causes hereditary 1,25-dihydroxyvitamin D3-resistant rickets in three families. *Proc. Natl Acad. Sci., USA*, **86**, 9783–9787.

Rodien, P., Mebarki, F., Mowszowicz, I., Chaussain, J.L., Young, J., Morel, Y. and Schaison, G., 1996, Different phenotypes in a family with androgen insensitivity caused by the same M780I point mutation in the androgen receptor gene. *J. Clin. Endocrinol. Metab.*, **81**, 2994–2998.

Ross, C.A., Becher, M.W., Colomer, V., Engelender, S., Wood, J.D. and Sharp, A.H., 1997, Huntington's disease and dentatorubral-pallidoluysian atrophy: proteins, pathogenesis and pathology. *Brain Pathol.*, **7**, 1003–1016.

Ruizeveld de Winter, J.A., Trapman, J., Vermey, M., Mulder, E., Zegers, N.D. and van der Kwast, T.H., 1991, Androgen receptor expression in human tissues: an immunohisto-chemical study. *J. Histochem. Cytochem.*, **39**, 927–936.

Rut, A.R., Hewison, M., Kristjansson, K., Luisi, B., Hughes, M.R. and O'Riordan, J.L., 1994, Two mutations causing vitamin D resistant rickets: modelling on the basis of steroid hormone receptor DNA-binding domain crystal structures. *Clin. Endocrinol.*, **41**, 581–590.

Saijo, T., Ito, M., Takeda, E., Huq, A.H., Naito, E., Yokota, I., Sone, T., Pike, J.W. and Kuroda, Y., 1991, A unique mutation in the vitamin D receptor gene in three Japanese patients with vitamin D-dependent rickets type II: utility of single-strand conformation polymorphism analysis for heterozygous carrier detection. *Am. J. Hum. Genet.*, **49**, 668–673.

Sakurai, A., Nakai, A. and DeGroot, L.J., 1989, Expression of three forms of thyroid hormone receptor in human tissues. *Molec. Endocrinol.*, **3**, 392–399.

Sande, S. and Privalsky, M.L., 1996, Identification of TRACs (T3 receptor-associating cofactors), a family of cofactors that associate with, and modulate the activity of, nuclear hormone receptors. *Molec. Endocrinol.*, **10**, 813–825.

Saunders, P.T., Padayachi, T., Tincello, D.G., Shalet, S.M. and Wu, F.C., 1992, Point mutations detected in the androgen receptor gene of three men with partial androgen insensitivity syndrome. *Clin. Endocrinol. (Oxford)*, **37**, 214–220.

Schwabe, J.W., Chapman, L., Finch, J.T. and Rhodes, D., 1993, The crystal structure of the estrogen receptor DNA-binding domain bound to DNA: how receptors discriminate between their response elements. *Cell*, **75**, 567–578.

Sinnecker, G.H., Hiort, O., Nitsche, E.M., Holterhus, P.M. and Kruse, K., 1997, Functional assessment and clinical classification of androgen sensitivity in patients with mutations of the androgen receptor gene. German Collaborative Intersex Study Group. *Eur. J. Pediatr.*, **156**, 7–14.

Sleddens, H.F., Oostra, B.A., Brinkmann, A.O. and Trapman, J., 1992, Trinucleotide repeat polymorphism in the androgen receptor gene (AR). *Nucleic Acids Res.*, **20**, 1427.

Sleddens, H.F., Oostra, B.A., Brinkmann, A.O. and Trapman, J., 1993, Trinucleotide (GGN) repeat polymorphism in the human androgen receptor (AR) gene. *Hum. Molec. Genet.*, **2**, 493.

Smith, E.P., Boyd, J., Frank, G.R., Takahashi, H., Cohen, R.M., Specker, B., Williams, T.C., Lubahn, D.B. and Korach, K.S., 1994, Estrogen resistance caused by a mutation in the estrogen-receptor gene in a man. *N. Engl. J. Med.*, **331**, 1056–1061.

Sobue, G., Hashizume, Y., Mukai, E., Hirayama, M., Mitsuma, T. and Takahashi, A., 1989, X-linked recessive bulbospinal neuronopathy. A clinicopathological study. *Brain*, **112**, 209–232.

Sone, T., Marx, S.J., Liberman, U.A. and Pike, J.W., 1990, A unique point mutation in the human vitamin D receptor chromosomal gene confers hereditary resistance to 1, 25–dihydroxyvitamin D3. *Molec. Endocrinol.*, **4**, 623–631.

Stein, B. and Yang, M.X., 1995, Repression of the interleukin-6 promoter by estrogen receptor is mediated by NF-κB and C/EBPβ. *Molec. Cell. Biol.*, **15**, 4971–4979.

Studzinski, G.P., McLane, J.A. and Uskokovic, M.R., 1993, Signaling pathways for vitamin D-induced differentiation: implications for therapy of proliferative and neoplastic diseases. *Crit. Rev. Eukaryot. Gene Expr.*, **3**, 279–312.

Sultan, C., Lumbroso, S., Poujol, N., Belon, C., Boudon, C. and Lobaccaro, J.M., 1993, Mutations of androgen receptor gene in androgen insensitivity syndromes. *J. Steroid Biochem. Molec. Biol.*, **46**, 519–530.

Tanenbaum, D.M., Wang, Y., Williams, S.P. and Sigler, P.B., 1998, Crystallographic comparison of the estrogen and progesterone receptor's ligand binding domains. *Proc. Natl Acad. Sci., USA*, **95**, 5998–6003.

Thompson, E.B., 1994, Apoptosis and steroid hormones. *Molec. Endocrinol.*, **8**, 665–673.

Tincello, D.G., Saunders, P.T., Hodgins, M.B., Simpson, N.B., Edwards, C.R., Hargreaves, T.B. and Wu, F.C., 1997, Correlation of clinical, endocrine and molecular abnormalities with *in vivo* responses to high-dose testosterone in patients with partial androgen insensitivity syndrome. *Clin. Endocrinol. (Oxford)*, **46**, 497–506.

Trifiro, M., Prior, R.L., Sabbaghian, N., Pinsky, L., Kaufman, M., Nylen, E.G., Belsham, D.D., Greenberg, C.R. and Wrogemann, K., 1991, Amber mutation creates a diagnostic *MaeI* site in the androgen receptor gene of a family with complete androgen insensitivity. *Am. J. Med. Genet.*, **40**, 493–499.

Usala, S.J., Bale, A.E., Gesundheit, N., Weinberger, C., Lash, R.W., Wondisford, F.E., McBride, O.W. and Weintraub, B.D., 1988, Tight linkage between the syndrome of generalized thyroid hormone resistance and the human c-erbA β gene. *Molec. Endocrinol.*, **2**, 1217–1220.

Voegel, J.J., Heine, M.J., Zechel, C., Chambon, P. and Gronemeyer, H., 1996, TIF2, a 160 kDa transcriptional mediator for the ligand-dependent activation function AF-2 of nuclear receptors. *EMBO J.*, **15**, 3667–3675.

vom Baur, E., Zechel, C., Heery, D., Heine, M.J., Garnier, J.M., Vivat, V., Le Douarin, B., Gronemeyer, H., Chambon, P. and Losson, R., 1996, Differential ligand-dependent interactions between the AF-2 activating domain of nuclear receptors and the putative transcriptional intermediary factors mSUG1 and TIF1. *EMBO J.*, **15**, 110–124.

Wagner, R.L., Apriletti, J.W., McGrath, M.E., West, B.L., Baxter, J.D. and Fletterick, R.J., 1995, A structural role for hormone in the thyroid hormone receptor. *Nature*, **378**, 690–697.

Walters, M.R., 1992, Newly identified actions of the vitamin D endocrine system. *Endocr. Rev.*, **13**, 719–764.

Wang, Y. and Miksicek, R.J., 1991, Identification of a dominant negative form of the human estrogen receptor. *Molec. Endocrinol.*, **5**, 1707–1715.

Warrick, J.M., Paulson, H.L., Gray-Board, G.L., Bui, Q.T., Fischbeck, K.H., Pittman, R.N. and Bonini, N.M., 1998, Expanded polyglutamine protein forms nuclear inclusions and causes neural degeneration in *Drosophila*. *Cell*, **93**, 939–949.

Webb, P., Lopez, G.N., Uht, R.M. and Kushner, P.J., 1995, Tamoxifen activation of the estrogen receptor/AP-1 pathway: potential origin for the cell-specific estrogen-like effects of antiestrogens. *Molec. Endocrinol.*, **9**, 443–456.

Weidemann, W., Linck, B., Haupt, H., Mentrup, B., Romalo, G., Stockklauser, K., Brinkmann, A.O., Schweikert, H.U. and Spindler, K.D., 1996, Clinical and biochemical investigations and molecular analysis of subjects with mutations in the androgen receptor gene. *Clin Endocrinol (Oxford)*, **45**, 733–739.

Weinberger, C., Thompson, C.C., Ong, E.S., Lebo, R., Gruol, D.J. and Evans, R.M., 1986, The c-erb-A gene encodes a thyroid hormone receptor. *Nature*, **324**, 641–646.

Weis, K.E., Ekena, K., Thomas, J.A., Lazennec, G. and Katzenellenbogen, B.S., 1996, Constitutively active human estrogen receptors containing amino acid substitutions for tyrosine 537 in the receptor protein. *Molec. Endocrinol.*, **10**, 1388–1398.

Wellington, C.L., Ellerby, L.M., Hackam, A.S., Margolis, R.L., Trifiro, M.A., Singaraja, R., McCutcheon, K., Salvesen, G.S., Propp, S.S., Bromm, M., Rowland, K.J., Zhang, T., Rasper, D., Roy, S., Thornberry, N., Pinsky, L., Kakizuka, A., Ross, C.A., Nicholson, D.W., Bredesen, D.E. and Hayden, M.R., 1998, Caspase cleavage of gene products associated with triplet expansion disorders generates truncated fragments containing the polyglutamine tract. *J. Biol. Chem.*, **273**, 9158–9167.

Werner, S. and Bronnegard, M., 1996, Molecular basis of glucocorticoid-resistant syndromes. *Steroids*, **61**, 216–221.

Wiese, R.J., Goto, H., Prahl, J.M., Marx, S.J., Thomas, M., al-Aqeel, A. and DeLuca, H.F., 1993, Vitamin D-dependency rickets type II: truncated vitamin D receptor in three kindreds. *Molec. Cell. Endocrinol.*, **90**, 197–201.

Wilde, J., Moss, T. and Thrush, D., 1987, X-linked bulbo-spinal neuronopathy: a family study of three patients. *J. Neurol. Neurosurg. Psychiatry*, **50**, 279–284.

Wilson, J.D., 1992, Syndromes of androgen resistance. *Biol. Reprod.*, **46**, 168–173.

Wilson, J.D., Griffin, J.E., George, F.W. and Leshin, M., 1981, The role of gonadal steroids in sexual differentiation. *Recent Prog. Horm. Res.*, **37**, 1–39.

Wooster, R., Mangion, J., Eeles, R., Smith, S., Dowsett, M., Averill, D., Barrett-Lee, P., Easton, D.F., Ponder, B.A. and Stratton, M.R., 1992, A germline mutation in the androgen receptor gene in two brothers with breast cancer and Reifenstein syndrome. *Nat. Genet.*, **2**, 132–134.

Wurtz, J.M., Bourguet, W., Renaud, J.P., Vivat, V., Chambon, P., Moras, D. and Gronemeyer, H., 1996, A canonical structure for the ligand-binding domain of nuclear receptors. *Nat. Struct. Biol.*, **3**, 206.

Yagi, H., Ozono, K., Miyake, H., Nagashima, K., Kuroume, T. and Pike, J.W., 1993, A new point mutation in the deoxyribonucleic acid-binding domain of the vitamin D receptor in a kindred with hereditary 1,25-dihydroxyvitamin D-resistant rickets. *J. Clin. Endocrinol. Metab.*, **76**, 509–512.

Yeh, S. and Chang, C., 1996, Cloning and characterization of a specific coactivator, ARA70, for the androgen receptor in human prostate cells. *Proc. Natl Acad. Sci., USA*, **93**, 5517–5521.

Yoh, S.M., Chatterjee, V.K. and Privalsky, M.L., 1997, Thyroid hormone resistance syndrome manifests as an aberrant interaction between mutant T3 receptors and transcriptional corepressors. *Molec. Endocrinol.*, **11**, 470–480.

Yoshizawa, T., Handa, Y., Uematsu, Y., Takeda, S., Sekine, K., Yoshihara, Y., Kawakami, T., Arioka, K., Sato, H., Uchiyama, Y., Masushige, S., Fukamizu, A., Matsumoto, T. and Kato, S., 1997, Mice lacking the vitamin D receptor exhibit impaired bone formation, uterine hypoplasia and growth retardation after weaning. *Nat. Genet.*, **16**, 391–396.

Zhang, Q.X., Hilsenbeck, S.G., Fuqua, S.A. and Borg, A., 1996, Multiple splicing variants of the estrogen receptor are present in individual human breast tumors. *J. Steroid Biochem. Molec. Biol.*, **59**, 251–260.

Zoppi, S., Marcelli, M., Deslypere, J.P., Griffin, J.E., Wilson, J.D. and McPhaul, M.J., 1992, Amino acid substitutions in the DNA-binding domain of the human androgen receptor are a frequent cause of receptor-binding positive androgen resistance. *Molec. Endocrinol.*, **6**, 409–415.

Zoppi, S., Wilson, C.M., Harbison, M.D., Griffin, J.E., Wilson, J.D., McPhaul, M.J. and Marcelli, M., 1993, Complete testicular feminization caused by an amino-terminal truncation of the androgen receptor with downstream initiation. *J. Clin. Invest.*, **91**, 1105–1112.

7

VARIATION IN GENES REGULATING THE IMMUNE SYSTEM AND RELATIONSHIP TO DISEASE

Ann B. Begovich and Jorge R. Oksenberg

7.1 Introduction

In order to maintain immune homeostasis, the immune system must be capable of responding specifically and appropriately to a wide array of antigenic challenges from foreign agents, such as viruses and bacteria, which have the ability to mutate rapidly. Consequently, molecular polymorphism of the molecules involved in the adaptive immune process [specifically molecules encoded by the major histocompatibility complex (MHC), T-cell receptors (TCRs), and immunoglobulins (Igs)] is an important characteristic of the vertebrate immune system. The mechanisms by which polymorphisms are generated and maintained, however, differ depending on the molecule involved. For TCRs (the antigen receptor on T cells) and Igs (the antigen receptor on B cells), diversity is generated through a process of somatic rearrangement, in which different gene segments are assembled to create rearranged receptor gene transcripts. A single functional antigen receptor molecule is expressed by each T and B cell and its progeny. This rearrangement process is capable of generating as many as 10^9 different populations (clones) of immune cells, each with unique antigen-binding characteristics. In addition, Igs have the ability to mutate somatically, increasing the diversity even more. Diversity within the genes of the MHC, however, is achieved by allelic variation between individuals at the population level and each individual expresses a maximum of two alleles at any given MHC locus.

The key initial step for most immune responses is T-cell recognition of an antigen not in its intact form but rather as peptide fragments derived from the foreign protein and bound to MHC molecules. This antigen presentation step is necessary for triggering of effector T cells, for provision of "help" for B-cell activation and differentiation into immunoglobulin (antibody)-producing cells, and for activation of macrophages. Multiple cell-surface molecules interact during this antigen presentation step but the trimolecular complex – MHC/antigenic peptide/TCR – is the focal point for initiation and propagation of most immune responses to protein antigens.

139

Clearly, this immune process is essential for the recognition, response, and clearing of foreign pathogens. However, the intracellular machinery involved in antigen processing is not restricted to foreign antigens; "self" antigens are also processed and presented generating T cells whose receptors are specific for "self" antigens. In normal healthy individuals these autoreactive T cells are kept in check by mechanisms leading to self-tolerance. Central tolerance occurs in the developing thymus and leads to the clonal deletion of lymphocytes bearing receptors for ubiquitous self-antigens (antigens expressed by most cell types). Autoreactive T cells that make it to the periphery are kept in check by T-cell inactivation (anergy) or other regulatory T cells. Neither of these processes, however, is full proof and the potential for autoimmunity exists.

The following is a review of the genetic variation in the two genetically encoded entities that make up the trimolecular complex, the loci that encode the molecules of the major histocompatibility complex and those that encode the T-cell antigen receptor, and their role in disease. In addition, given the recent and exciting advances in the understanding of host genetic variation and susceptibility to HIV infection as well as progression to AIDS we also discuss genetic variation in another class of molecules that help to regulate the immune system, the chemokine receptors.

7.2 Human major histocompatibility complex, HLA

7.2.1 Overview

The human MHC, referred to as the human leukocyte antigen (HLA) complex, encompasses approximately 4 Mb and is located on the short arm of chromosome six as shown in Figure 7.1. It contains a number of genes that play a central role in the regulation of the immune response including the genes for two distinct classes (class I and class II) of highly polymorphic cell surface molecules that bind and present processed antigens in the form of peptides to T lymphocytes. As discussed above, this presentation step is crucial in initiating both cellular and humoral immune responses.

The class I molecules, HLA-A, HLA-B, and HLA-C, are found on most nucleated cells. They are cell-surface glycoproteins that bind and present processed peptides derived primarily from endogenously synthesized proteins (e.g. viral and tumor peptides) to CD8$^+$ T cells. These heterodimers consist of an HLA-encoded alpha chain associated with the non-MHC-encoded polypeptide, β_2-microglobulin. While β_2-microglobulin is monomorphic, the alpha-chain genes are extremely polymorphic, as shown in Table 7.1, with the variability localized primarily to exons two and three (Mason and Parham, 1998), which encode the amino-terminal extracellular domains that function as the peptide-binding site. Within these two exons, the polymorphism is concentrated into discrete clusters that lie within a relatively conserved framework region. Analysis of the HLA class I crystal structure has shown that these polymorphic residues line the peptide-binding cleft and interact directly with peptide and/or the T-cell receptor (Bjorkman *et al.*, 1987a,b; Garrett *et al.*, 1989; Madden *et al.*, 1992).

The class II molecules are encoded in the HLA-D region seen in Figure 7.1. These cell-surface glycoproteins consist of an HLA-encoded alpha and beta chain

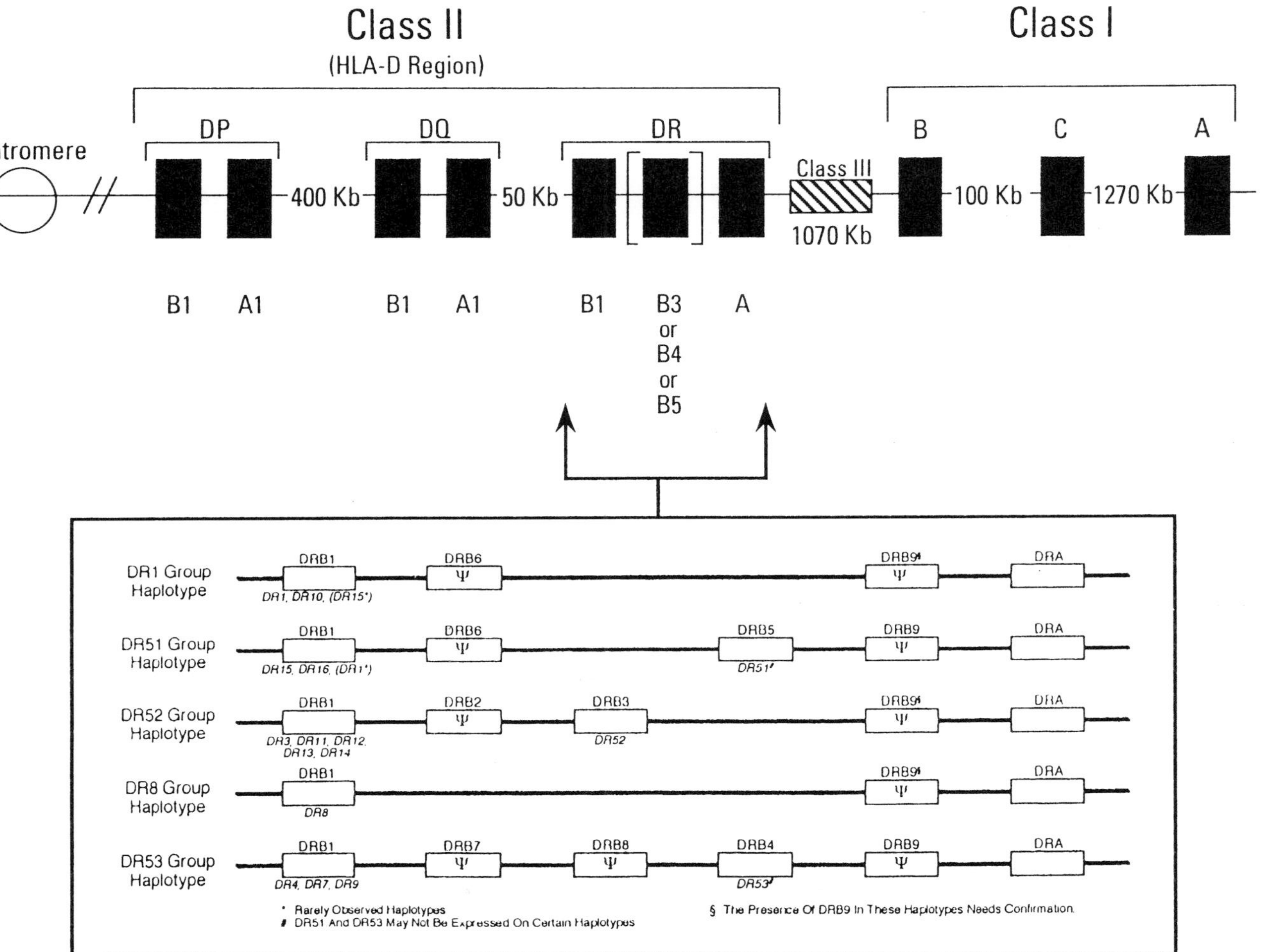

Figure 7.1 Map of the expressed genes of the HLA class I and class II regions. The insert below shows the genomic organization of the HLA-DR region. The serological specificities encoded by a gene appear underneath in boldface type. Ψ Indicates non-expressed genes. (See text for details.)

Table 7.1 Allelic diversity at the HLA class I and II loci[a]

Locus	No. of alleles
HLA-A	122
HLA-B	254
HLA-C	72
HLA-DRA	2
HLA-DRB1	215
HLA-DRB3	19
HLA-DRB4	8
HLA-DRB5	13
HLA-DQA1	19
HLA-DQB1	38
HLA-DPA1	15
HLA-DPB1	84

[a]The number of alleles listed here includes silent variants and is the allele count as of November 1998 (S.G.E. Marsh, personal communication).

associated as heterodimers on the cell surface of antigen-presenting cells such as B cells and macrophages. Class II molecules also serve as receptors for processed peptides; however, these peptides are derived predominantly from membrane and extracellular proteins (e.g. bacterial peptides), and they are presented to CD4[+] T-lymphocytes. The HLA-D region contains several class II genes and has three subregions: HLA-DR, HLA-DQ, and HLA-DP. Both the DQ and DP regions contain one functional gene for each of their alpha and beta chains. The HLA-DR subregion contains one functional gene for the alpha chain; the number of functional genes for the beta chain varies from one to two as seen in Figure 7.1. All individuals express a DRB1-encoded polymorphic polypeptide, which is found on the cell surface in association with the DRA-encoded polypeptide. The other functional DRB genes, DRB3, DRB4, and DRB5, encode polypeptides that are also found on the cell surface in association with the DRA-encoded polypeptide but at a lower level and only in certain class II haplotypes. With the exception of the DRA gene, the genes encoding the functional class II molecules are also polymorphic as seen in Table 7.1, with virtually all of the polymorphism localized to a single exon, exon two (Marsh, 1998). Figure 7.2 shows that as with the class I molecules, class II molecules also display clustering of polymorphic residues in the peptide-binding cleft (Brown *et al.*, 1993). This extensive polymorphism in both the class I and II loci and its localization to the peptide-binding groove have led many to propose that HLA polymorphism is maintained in the population because different allelic products have the capacity to bind and present different peptides. Consequently, the more alleles a population has, the better it is able to cope with a large array of infectious pathogens.

Until recently, HLA typing was performed using a combination of serological (Terasaki *et al.*, 1964) and cellular assays (Bach and Voynow, 1966). Serology, the most commonly used technique, relies on the microcytotoxicity assay in which an antiserum (or monoclonal antibody) is mixed with live lymphocytes and allowed to bind to the expressed class I or II molecules. Specific binding is then detected

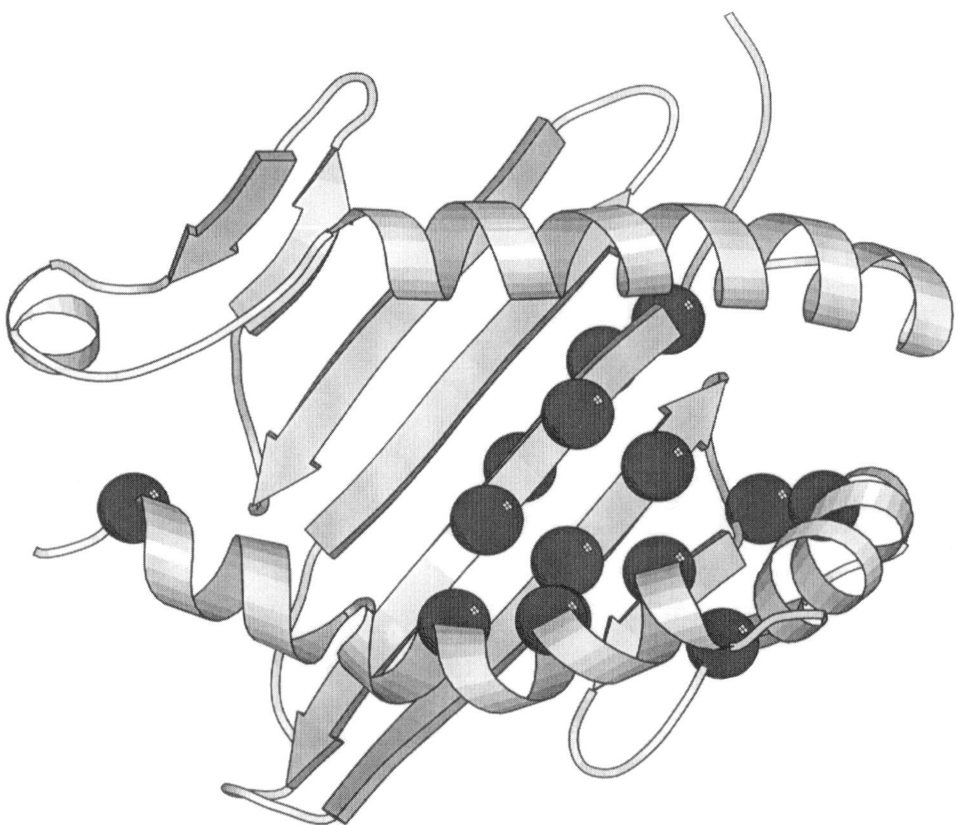

Figure 7.2 Localization of the polymorphic residues to the peptide-binding cleft of the HLA class II heterodimer, HLA-DR. Circles indicate the polymorphic residues, which line the cleft either on the floor of the groove or directed inwards from the walls where they can interact directly with the peptide. All the polymorphism is located in the β chain of the DRB molecule; the HLA-DRA gene, which encodes the α helix along the top of the molecule and the four β-pleated sheets on the left side of the molecule, is monomorphic.

with the addition of complement, which lyses the cells and allows the uptake of a dye. Over the years this technique has been refined; it is quite reliable, easily reproduced, does not require sophisticated instrumentation and provides results in a few hours. There are, however, limitations, the most important being the level of resolution. For example, the DR4 serologic specificity is found on 32 distinct allelic products (Bodmer *et al.*, 1999). The majority, if not all, of these alleles type identically with serologic reagents; however, T cells can distinguish many, if not all, of these, suggesting that these allelic differences are immunologically important.

The development of the polymerase chain reaction (PCR) (Saiki *et al.*, 1985) has facilitated the development of highly efficient and powerful molecular methods for genetic typing, including sequencing-based typing (SBT), sequence-specific priming (SSP), and sequence-specific oligonucleotide (SSO) probing. Some of these methods have been reviewed elsewhere (Begovich and Erlich, 1995). Figure 7.3 is a schematic of one such technique, the reverse line assay, which is based on the hybridization of PCR product, labeled with biotinylated primers during the amplification, to an array of immobilized probes on a nylon membrane. The presence of PCR product bound to a specific probe is detected using streptavidin-HRP (horseradish peroxidase) and the genetic type is determined by the pattern of positive probes that can be seen in Figure 7.4. The application of these powerful techniques to the HLA loci has proved instrumental in the careful and complete characterization of allelic diversity of these loci and is absolutely necessary before the role of these molecules in disease can be ascertained.

The genes within the 4-Mb region that encode the HLA antigens are inherited in a Mendelian, co-dominant fashion and the linear array of these loci on a particular chromosome is referred to as a haplotype. One unique characteristic of the MHC region is the observation that certain HLA antigens occur on the same haplotype more often then expected by chance. This phenomenon is termed linkage disequilibrium. There is extremely strong disequilibrium between the DR and DQ regions and weak, but statistically significant, disequilibrium between DPB1 and the DR/DQ region which is explained by a few DR–DQ–DP haplotypes with high positive disequilibrium (Begovich *et al.*, 1992). Other population and family studies have confirmed the existence of these extended haplotypes and have shown

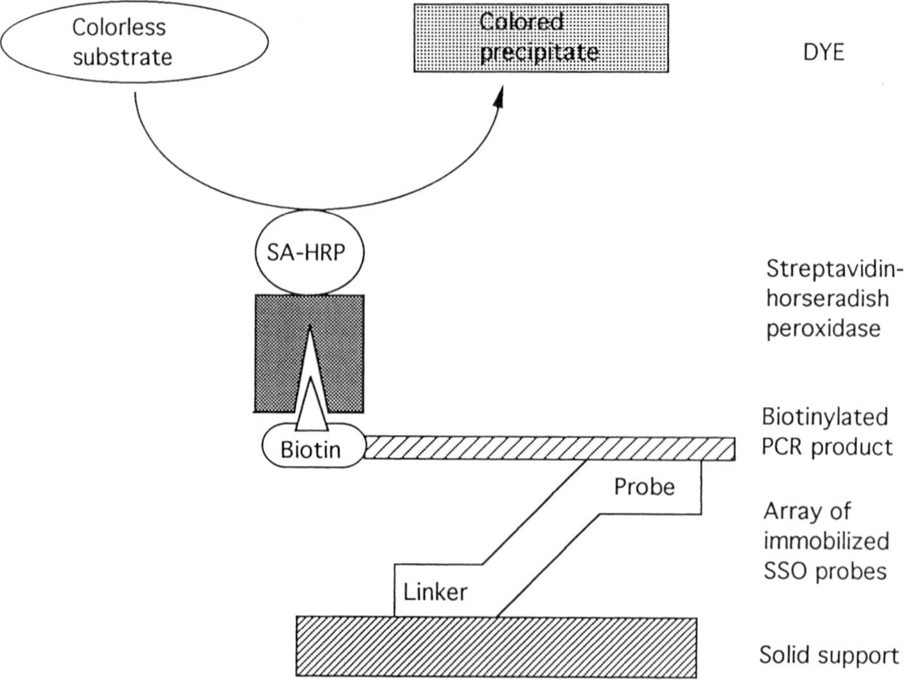

Figure 7.3 Schematic of the reverse line assay, which uses an array of immobilized probes. SA-HRP, streptavidin-horseradish peroxidase; PCR, polymerase chain reaction; SSO, sequence-specific oligonucleotide.

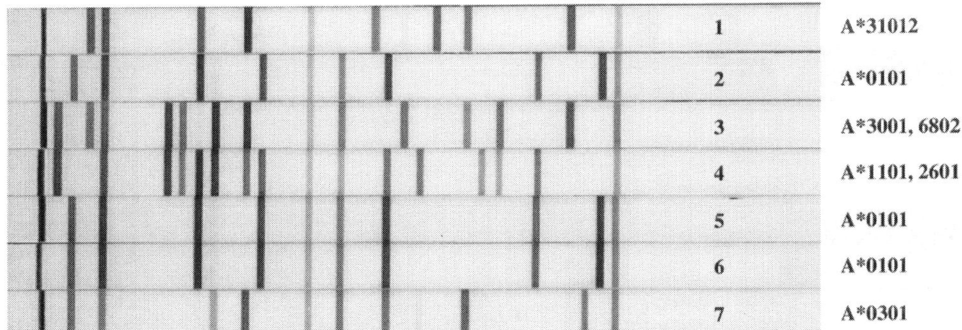

Figure 7.4 Application of the 35-probe HLA-A reverse line assay to panel of DNAs from seven different lymphoblastoid cell lines. The HLA-A type, based on the interpretation of the probe hybridization patterns, is given to the right of each strip.

that some of them extend from A through DP (Alper *et al.*, 1992). Examples of these extended haplotypes include the common Caucasian A*0101–B*0801–DRB1*0301–DQB1*0201–DPB1*0101 haplotype (Alper *et al.*, 1992; Thomson, 1995) and the type I diabetes-susceptible haplotype A*3002–B*1801–DRB1*0301–DQB1*0201–DPB1*0202 (Noble *et al.*, 1998). The presence of linkage disequilibrium within the HLA region creates difficulties in establishing which genes actually confer susceptibility and which are secondarily associated because of linkage disequilibrium. Clearly, both the identification of these extended haplotypes and the knowledge of their frequency in various populations are necessary to understand the role that the individual MHC loci play in disease.

Although multiple alleles have been described for all but the DRA1 locus seen in Table 7.1, most populations contain only a subset of these alleles (Charron, 1997; Teresaki and Gjertson, 1997). For example, although over 200 DRB1 alleles have been defined, an analysis of 266 independent chromosomes from the Caucasian CEPH families revealed only 27 of these DRB1 alleles (Begovich *et al.*, 1992). Some alleles appear to be unique to particular populations. For example, DRB1*1105 has only been found in Filipinos (Bugawan *et al.*, 1994) while DRB1*0807 appears to be unique to South American Amerindians (Mack and Erlich, 1998).

There is also substantial variation in allele and haplotype frequencies among different human populations (Imanishi *et al.*, 1992; Charron, 1997; Teresaki and Gjertson, 1997) which is thought to reflect the past history of selection, admixture, drift and the population bottlenecks that each population has been subject to. As discussed above, there are 32 distinct polypeptides (DRB1*0401→DRB1*0432) that make up the DR4 serogroup (Bodmer *et al.*, 1999); however, the distribution of these alleles is quite variable in different populations (Teresaki and Gjertson, 1997). Table 7.2 shows the DR4 allele frequencies found in four ethnically distinct populations typed by a single laboratory using the same PCR-based technology. The frequency of DR4 alleles ranges from over 37% in the Amerindian population to less than 1% in the African population. DRB1*0401, the most common DR4 allele in the Caucasian population, is absent from the Amerindian and African

Table 7.2 Frequency of DR4 alleles in four ethnically diverse populations

	Caucasian[a] 2N = 266	*Asian*[b] 2N = 210	*Amerindian*[c] 2N = 98	*African*[d] 2N = 344
All DR4 alleles	0.211	0.1477	0.377	0.0058
DRB1*0401	0.094	0.0048		
DRB1*0402	0.004			
DRB1*0403/6	0.015	0.0524	0.041	
DRB1*0404	0.079		0.020	
DRB1*0405	0.004	0.0905		0.0058
DRB1*0407	0.015			
DRB1*0411			0.316	

[a]Data from the Caucasian CEPH families (Begovich *et al.*, 1992).
[b]Data from a Filipino population (Bugawan *et al.*, 1994).
[c]Data from the Ticuna, a South American Amerindian population (Mack and Erlich, 1998).
[d]Data from a population from Central Cameroon (Begovich *et al.*, in preparation).

populations and relatively infrequent in the Asian population; DRB1*0411, on the other hand, is only found in the Amerindian population where it is extremely common.

Analysis of these same four populations also illustrates the extensive haplotypic variability among populations. For example, the frequency of DR8 haplotypes ranges from over 26% in the Amerindian population to less than 2% in the Asian population, as shown in Table 7.3. In addition to frequency differences, there are also differences in the composition of these haplotypes; DRB1*0801 is associated with DQB1*0402 in the Caucasian and Amerindian population but with an entirely different DQB1 allele, *0601, in the Asian population, also shown in Table 7.3. These population differences have an impact on HLA and disease studies. First, they underscore the need to use ethnically matched control and patient populations in any disease study. Second, they can be quite useful in understanding the role that an individual MHC allele might play in disease. For example, narcolepsy, which is a neurological disorder characterized by excessive daytime sleepiness, is strongly associated with the DRB1*1501–DQA1*0102–DQB1*0602 haplotype in Japanese and Caucasian patients (Juji *et al.*, 1984; Matsuki *et al.*, 1985; Billiard *et al.*, 1986; Rogers *et al.*, 1997). In these two ethnic groups, linkage disequilibrium between the DRB1*1501 and DQB1*0602 alleles is almost 100% making it impossible to determine which of these two genes actually confers susceptibility. By studying African Americans, where linkage disequilibrium between DRB1*1501 and DQB1*0602 is less tight and these alleles are associated with other DRB1 and DQB1 alleles, Mignot and colleagues (1994) were able to show that it is the DQB1*0602 allele, not DRB1*1501, that predisposes individuals to narcolepsy.

Although there is significant variation in the class I and II allele frequencies among human populations, almost all populations studied to date have a common feature, namely a relatively even distribution of allele frequencies at the HLA-A, -B, -DR, and -DQ loci. (HLA-C has not been extensively analyzed and, for reasons that remain unclear, the DP loci have a more skewed allele distribution.)

Table 7.3 Frequency of DR8 haplotypes in four ethnically diverse populations

DRB1–DQA1–DQB1 haplotypes	Caucasian[a] 2N = 266	Asian[b] 2N = 210	Amerindian[c] 2N = 98	African[d] 2N = 344
All DR8 haplotypes	0.035	0.015	0.266	0.071
0801–0401–0402	0.031		0.010	
0801–NT–0601[e]		0.005		
0802–0401–0402	0.004		0.031	
0803–NT–0601[e]		0.010		
0804–0401–0301				0.045
0804–0501–0301				0.011
0804–0401–0402				0.009
0804–0102–0602				0.006
0807–0401–0402			0.225	

[a]Data from the Caucasian CEPH families (Begovich *et al.*, 1992).
[b]Data from a Filipino population (Bugawan *et al.*, 1994).
[c]Data from the Ticuna, a South American Amerindian population (Mack and Erlich, 1998).
[d]Data from a population from Central Cameroon (Begovich *et al.*, in preparation).
[e]Samples in the Filipino data set were not typed (NT) for the DQA1 locus.

This distribution, as measured by the Ewens–Watterson F-Statistic (Ewens, 1972; Watterson, 1978), differs significantly from that expected for a locus where mutation and random genetic drift are the only forces affecting allele frequencies (conditions of neutrality) and suggests these loci have an evolutionary history of balancing selection. Balancing selection is expected to maintain multiple alleles in a population at appreciable frequencies due to heterozygote advantage, frequency dependent selection, and/or other selective forces. There are, however, exceptions to this observation. One is the distribution of DRB1 alleles in a population from the Philippines where the DRB1*1502 allele and the DRB1*1502–DQB1*0502–DPB1*0101 haplotype are significantly more frequent than expected (Bugawan *et al.*, 1994). Another example is the extremely high frequency of the A*2402 allele in Papua New Guinea Highlanders (Bugawan *et al.*, 1999). In both instances, the evidence suggests that these unusual distributions are due to strong directional selection.

In summary, although the HLA molecules are extremely polymorphic, these polymorphisms do not occur randomly throughout the molecule. They are localized almost exclusively to the peptide-binding groove where they directly impact the specificity of the bound peptide and T-cell receptor interactions and consequently the immune response. As discussed in more detail below, these molecules are also associated with susceptibility and resistance to a number of diseases. Although these associations are most likely to be the result of the function that these molecules play in the normal immune response, the exact mechanisms remain unknown. In order to understand the mechanisms of susceptibility and resistance to a disease (whether the disease is cancer, infectious or autoimmune) one must first be able to unequivocally identify the individual MHC molecules implicated. Consequently, it is important that highly reliable typing methods for characterizing the allelic polymorphism at these loci be used to elucidate the patterns of linkage disequilibrium across this region and the inter-population allelic and haplotypic variation.

7.2.2 HLA and autoimmune disease

The role of the class I and II molecules in the immune response is twofold. As discussed above, class I and II molecules present on the cell surface of antigen-presenting cells bind processed antigen and present it to T cells initiating an immune response (Unanue and Allen, 1987; Brodsky *et al.*, 1996). In addition, class I and class II molecules present on the stromal cells of the thymus help to mold the specificity of the mature T-cell repertoire (Kappler *et al.*, 1987; MacDonald *et al.*, 1988; Teh *et al.*, 1988). Consequently, an individual's ability to respond to an antigen, whether the antigen is foreign or self, is partly determined by the amino acid sequences of its class I and II molecules. Therefore, it is not surprising that polymorphisms within the genes of the MHC define specific haplotypes associated with resistance and susceptibility to diseases including autoimmune disorders, such as type-I diabetes, rheumatoid arthritis, and multiple sclerosis (Nepom and Erlich, 1991; Thorsby, 1997), certain cancers such as Hodgkin's disease (Bodmer *et al.*, 1989; Klitz *et al.*, 1994; Oza *et al.*, 1994; Taylor *et al.*, 1996) and cervical carcinoma (Apple *et al.*, 1994), and infectious diseases such as malaria (Hill *et al.*, 1991a). In some diseases, single genes within these haplotypes are themselves implicated in disease susceptibility (e.g. B27 and ankylosing spondylitis) while in other diseases a specific heterodimer, whether encoded in *cis* (on the same chromosome) or *trans* (on homologous chromosomes), is implicated (e.g. DQA1*0501:DQB1*02 and celiac disease). Finally, in certain diseases, such as type-I diabetes and cervical carcinoma, there seems to be a complex interaction between alleles as multiple loci within the MHC.

These disorders have a complex etiology that involves both genetic and environmental factors. First, it is important to note that the MHC loci function as susceptibility loci. This means that a particular class I or II molecule is found more frequently in patients with a particular disease than in normal healthy individuals; however, most individuals with a disease-associated class I or II allele never develop disease. Second, although the HLA molecules play a role in susceptibility and resistance to these disorders, it is clear that other genes and unknown environmental factors are necessary to precipitate disease in those having a genetic predisposition. The following is a review of three autoimmune diseases where genetic variation within the genes of the HLA complex plays a role in disease susceptibility and/or resistance. It is not intended to be a thorough review of all MHC-associated autoimmune diseases but rather presents examples of different types of associations.

7.2.3 Single locus effects – ankylosing spondylitis

The strongest of all HLA and disease associations is the class I serotype HLA-B27 and ankylosing spondylitis (AS). AS is a chronic disease of young adults, especially men, characterized by inflammation, fibrosis and ossification of articulations of the axial skeleton. HLA-B27 is the major genetic determinant; more than 95% of AS patients are HLA-B27 positive compared with 7% of unaffected individuals

(Brewerton *et al.*, 1973) resulting in a relative risk (RR) of more than 100. (Relative risk is a measure of disease risk in an individual with a particular HLA-type relative to one without this genetic marker.) The high incidence of HLA-B27 in ethnically diverse patients (Gonzalez-Roces *et al.*, 1997) and the observation that HLA-B27 transgenic rats develop a spontaneous disease resembling AS (Hammer *et al.*, 1990) argue strongly for a direct role of B27 molecules in the disease. However, only 4% of HLA-B27-positive individuals develop the disease, which suggests that other genes and environmental factors play a role in disease pathogenesis. The role of an environmental trigger is supported by the observation that B27-transgenic rats raised in a germ-free environment do not develop inflammatory disease (Taurog *et al.*, 1994; Rath *et al.*, 1996).

At the amino acid sequence level, the B27 serotype has been subdivided into 14 distinct alleles (Bodmer *et al.*, 1999). The frequency of these alleles varies between the different ethnic groups. B*2705 is found in all populations and may be the progenitor from which all other B27 alleles arose (Lopez de Castro, 1998). B*2702 is unique to Caucasians; B*2704, *2706, and *2707 are found predominantly in Asians; and B*2703 is frequent in West Africans (Khan, 1996; Gonzalez-Roces *et al.*, 1997). The remaining B27 alleles are relatively rare and population data are not available. There is general agreement that four of these alleles (2702, 2704, 2705, and 2707) are positively associated with AS in a number of populations while two (2706 and 2709) are negatively associated (Gonzalez-Roces *et al.*, 1997; Lopez de Castro, 1998). Whether B*2703 is positively or negatively associated with disease remains unclear (Hill *et al.*, 1991b; Brown *et al.*, 1997; Gonzalez-Roces *et al.*, 1997) and the remaining seven alleles are too infrequent to draw any significant conclusions about their role in susceptibility to AS.

The sequences of 12 of the 14 B27 alleles are available (Marsh, 1998) and amino acid sequence comparisons show that there are 17 polymorphic residues within this group of 12 alleles. Individual alleles differ from one another by as few as one to as many as 12 amino acids (Lopez de Castro, 1998). Not surprisingly, the majority of these amino acid differences lie in the peptide-binding groove. Examination of the amino acid differences between disease-associated and non-associated B27 subtypes has shown that the two resistant subtypes (B*2706 and B*2709) differ from the disease-associated subtypes at amino acid 116. This residue lines the floor of one of the two major pockets of the B27 peptide-binding groove (Madden *et al.*, 1992) where it is thought to play a critical role in stabilization of the class I-peptide complex (Parker *et al.*, 1994).

Given that the function of class I molecules is to present peptide to CD8$^+$ T cells, it has been proposed that the cytotoxic T lymphocyte (CTL) response to an arthritogenic peptide(s) of endogenous or microbial origin, presented and bound by B27 molecules, may trigger disease pathogenesis (Benjamin and Parham, 1990). Recent work in the transgenic-rat model supports this hypothesis (Zhou *et al.*, 1998). Male rats expressing both the B*2705 allele and a minigene encoding a high-affinity B27 ligand had a significantly reduced prevalence of arthritis compared with male rats transgenic for B*2705 only. It is thought that the high-affinity minigene-encoded peptide was able to displace the usual B27-bound endogenous peptide population, which includes the arthritogenic peptide(s), significantly reducing disease incidence and supporting the hypothesis that B27-related arthritis requires binding of a specific peptide or set of peptides to the AS-susceptible B27 subtypes. Work is currently underway to identify these peptides and to elucidate the mechanism by which they induce arthritis.

7.2.4 Single heterodimer effects – celiac disease

Celiac disease (CD) is an autoimmune disorder of the small intestinal mucosa that is precipitated by the ingestion of gluten, the major protein component of wheat, barley, rye, and oats. The prevalence of this disease in the Western world is estimated at between 0.2% and 0.5% (Catassi *et al.*, 1994); however, this is probably an underestimate as not all cases of CD are symptomatic. In fact, the presentation of this disease is quite variable, ranging from absence of significant symptoms to life-threatening health problems (Troncone *et al.*, 1996). CD is most common in Europe where the frequency increases in a gradient towards the north and west of the continent. The highest incidence is in the west of Ireland where one person in 300 may be affected (Mylotte *et al.*, 1973). Malabsorption with chronic diarrhea, iron deficiency, and growth retardation are often the main symptoms. At the histological level the small bowel presents various degrees of mucosal lesions and intraeptitheal lymphocyte invasion indicative of an autoimmune process (Marsh, 1992). Treatment of patients with a gluten-free diet is quite effective and usually results in morphologic normalization of the mucosal lesions and disease remission.

CD has a multigenic and multifactoral origin. Evidence of a genetic predisposition is supported by a 10% prevalence in first-degree relatives and an 80–100% concordance in monozygotic twins compared to 20% in dizygotic twins (Stokes *et al.*, 1976; Sollid and Thorsby, 1993). A significant portion of CD disease predisposition maps to the HLA region and studies initially demonstrated an association with the class I serotype B8 (Falchuk, 1972). This association was later shown to be due to linkage disequilibrium between B8 and susceptible alleles in the class II region (Ek *et al.*, 1978; Tosi *et al.*, 1983; Roep *et al.*, 1988). In northern Europeans, DR3 seems to be a susceptibility allele, together with any other DR allele, while in southern Europeans DR7 is associated with CD almost only in combination with DR5 (Mearin *et al.*, 1983; Morellini *et al.*, 1988; Hall *et al.*, 1992). Molecular analysis of the DQA1 and DQB1 alleles on these haplotypes showed that these two patient populations shared the same DQ heterodimer, DQA1*0501–DQB1*0201 also referred to as DQ2 (Sollid *et al.*, 1989). The DQA1 and DQB1 loci on the same chromosome (in *cis*) in DR3 individuals encode this heterodimer, while in DR5/DR7 heterozygotes this DQ heterodimer is encoded by the same two genes but in the *trans* position. The DR7 haplotype supplies DQB1*0201 while the DR5 haplotype supplies DQA1*0501. The association of CD with DR3 in the north and DR5/DR7 in the south can be explained by the normal frequency of these alleles in these areas. DR3 is very common in the north and relatively infrequent in southern Europe while the opposite is true for the DR5 and DR7 haplotypes. Greater than 90% of CD patients carry DQ genes that encode this heterodimer while the majority of the remaining 10% express the DR4-associated DQ heterodimer, DQB1*0301–DQB1*0302 (DQ8). Thus CD appears to be primarily associated with DQ2 and to a lesser extent DQ8.

CD is a model disease for the study HLA disease associations. The susceptible HLA molecules have been identified, the disease-inducing agent, gluten, is known and it is relatively easy to biopsy and isolate T cells from the intestinal mucosa of CD patients for functional studies. Characterization of gluten-specific T cells from the intestinal mucosa of CD patients has resulted in a series of CD4$^+$ T-cell lines that recognize proteins from the gliadin fraction of gluten (Lundin *et al.*, 1993). The majority of these T cells, which use a large array of different T-cell receptors, recognized gliadin when presented by DQ2 if the patient was DQ2$^+$ or DQ8 if the patient carried the genes for this molecule (Lundin *et al.*, 1993, 1994a,b). The DQ2 peptide-binding motif has been described by two independent groups (Johansen *et al.*, 1996; van de Wal *et al.*, 1996; Vartdal *et al.*, 1996) and the search is on for peptides within gliadin that contain this motif. Together these data suggest that preferential binding and presentation to T cells in the intestinal mucosa of an as yet unknown gluten-derived peptide(s) by the disease-associated DQ heterodimers (DQ2 and DQ8) may be the mechanism behind the HLA associations with CD. The next step will be to define the other genetic, non-HLA factors contributing to CD susceptibility and to explain what triggers disease pathogenesis in some genetically susceptible individuals and not in others. Two published genome scans have failed to identify any regions that overlap (Zhong *et al.*, 1996; Greco *et al.*, 1998).

7.2.5 Multi-locus effects – type I diabetes

Type I diabetes is an autoimmune disorder characterized by T-cell-mediated destruction of the insulin-producing islet cells of the pancreas. The incidence of this disease is highest in Scandinavians and Sardinians, lowest in Japanese, and affects about 0.2–0.4% of Caucasians. The concordance rate is 30–70% for monozygotic twins but only 10–20% in HLA-identical siblings (Rotter and Landlaw, 1984). In addition, the risk to siblings sharing no HLA haplotypes with the affected sibling is greater than the prevalence of this disease in the general population (Thomson, 1995). Thus, although linkage studies (Davies *et al.*, 1994) have shown that the HLA region is the major genetic determinant, contributing 44% of the genetic risk (Risch, 1987), it is clear that other genes and unknown environmental triggers are necessary for pathogenesis.

Initial association studies in Caucasian type I diabetics using serological reagents showed increased frequencies of the HLA B8 and B15 serotypes and later DR3 and DR4, the class II molecules in linkage disequilibrium with these HLA-B markers (Svejgaard *et al.*, 1980; Tiwari and Terasaki, 1985). Both the DR3 and DR4 serotypes are strongly associated with disease; over 93% of patients carry at least one of these DR serotypes compared to 43% of controls and there is a significant increase in risk for DR3/DR4 heterozygotes (Thomson, 1988; Caillat-Zucman *et al.*, 1997). DR2, on the other hand, is decreased in patients and appears to confer protection (Tiwari and Terasaki, 1985; Caillat-Zucman *et al.*, 1997). Molecular characterization of the susceptible DR3 and DR4 haplotypes has shown that it is the DQA1 and DQB1 genes on these susceptible haplotypes that are most strongly associated with disease (DQA1*0501–DQB1*0201 with DR3 and DQA1*0301–DQB1*0302 with DR4). The increased risk of DR3/DR4 heterozygotes is thought to be due to the DQ heterodimer encoded in *trans* by this genotype,

DQA1*0301–DQB1*0201. This is supported by the finding that DR3/DR9 heterozygotes (a rare genotype in Caucasians found more frequently in Asians and Blacks) encode the same *trans*-associated heterodimer and are also associated with a significant increase in risk (Hu *et al.*, 1993).

Characterization of the amino acid sequences of these susceptible DQ alleles has led to the somewhat simplistic model that the charge of amino acid 57 in the DQB1 molecule mediates susceptibility. Alleles with a neutral amino acid residue (Ala, Ser, or Val) at position 57 are associated with disease susceptibility while those containing the charged residue, aspartic acid (Asp57), are associated with dominant protection (Todd *et al.*, 1987; Horn *et al.*, 1988; Morel *et al.*, 1988). Further studies, however, have suggested that these disease associations are much more complex. For instance, Caucasian DR7 haplotypes contain a DQB1 allele with Ala at position 57 but are not susceptible and in Japan, Asp57-containing DQB1 alleles are associated with susceptibility to type I diabetes (Lundin *et al.*, 1989; Yamagata *et al.*, 1989; Awata *et al.*, 1990; Khalil *et al.*, 1990).

The DQ heterodimer clearly plays an important role in predisposition but extensive family studies and studies of type I diabetes in non-Caucasian populations indicate that other loci within the MHC also influence susceptibility to this disease. For example, although there are a number of different DR4 alleles associated with the susceptible DQA1*0301–DQB1*0302 haplotype, only a subset of these DR–DQ haplotypes actually confer susceptibility (Sheehy *et al.*, 1989; Tait and Harrison, 1991; Cucca *et al.*, 1993; Erlich *et al.*, 1993). Table 7.4 summarizes the results of such a study from the Twelth International Histocompatibility Workshop (Caillet-Zucman *et al.*, 1997). DQB1*0302 in association with DRB1*0401 or *0402 predisposes to disease. However, a strong protective effect was observed when DQB1*0302 was associated with *0403, *0404, or *0407. This indicates that these three DRB1 alleles can overcome predisposition by the highly susceptible DQB1*0302 allele. Now that molecular techniques are available to fully characterize the DP and class I loci, there is increasing evidence that alleles at these loci may also play a role in susceptibility to type I diabetes (Noble *et al.*, 1996, 1998).

Table 7.4 Common DRB1*04 subtype frequencies among DQB1*0302 individuals

DRB1*04	Patients (n = 747)	%	Controls (n = 234)	%	OR[a]
0401	346	46.3	82	35.0	1.6
0402	144	19.3	25	10.7	2.0
0403	19	2.5	26	11.1	0.21
0404	106	14.2	60	35.6	0.48
0405	112	15	24	10.3	
0406	1	0.1	1	0.4	
0407	3	0.4	6	2.6	0.15
0408	8	1.1	5	2.1	
All other DR4s	8	1.0	5	2.1	

[a]OR is the odds ratio and is defined as the disease risk of an individual with a particular genetic marker relative to someone without the marker.

The search for other (non-HLA) genetic factors involved in type I diabetes suscep-tibility has uncovered numerous candidates including a marker within the 5′ flanking region of the insulin gene (termed IDDM2) and other markers on chromosomes 2, 8, 11, and 15. Therefore, while our understanding of the complexity of the genetics of susceptibility to type I diabetes is relatively far along, very little is known about the mechanism. Although numerous models based on peptide binding have been proposed (Nepom, 1990; Sheehy, 1992), the target antigen has not been identified, making the necessary functional studies difficult.

7.2.6 HLA and cancer

Susceptibility and resistance to certain types of cancers including nasopharyngeal carcinoma (Lu *et al.*, 1990), Hodgkin's disease (Amiel, 1967; Bodmer *et al.*, 1989; Klitz *et al.*, 1994; Oza *et al.*, 1994; Taylor *et al.*, 1996), and acute lymphoblastic leukemia (Taylor *et al.*, 1995) have been associated with genes within the MHC. Perhaps the most interesting HLA and cancer association is with the human papil-lomavirus (HPV)-induced cervical carcinoma. HPVs are thought to be the causative agents of most cervical intraepithelial (CIN) lesions that can spontaneously regress or progress to cervical cancer. There are several types of HPVs that differ in their ability to transform the cervical epithelium. Some, such as HPV16 and 18, are considered high risk and are often associated with high-grade lesions (CIN3), carci-noma *in situ*, or invasive carcinoma. By contrast, other HPV types are associated with low-grade lesions that do not progress. This, in addition to the fact that the rate of HPV infection is relatively high among cytologically normal women (6–53% depending on the population), suggests that additional environmental factors and the host immunological response are probably important for the devel-opment of disease.

Analysis of a Hispanic population from New Mexico, where the incidence of cervical cancer is nearly twice as high as for non-Hispanic white women, showed that specific DR–DQ haplotypes were significantly associated with both susceptibility and resistance to cervical cancer (Apple *et al.*, 1994). Specifically, the DRB1*1501–DQB1*0602 haplotype was increased among women with HPV16-associated invasive carcinoma; this haplotype was not increased among cancer cases that were associated with HPV types other than 16. A similar trend was seen in a British patient population (Duggan-Keen *et al.*, 1996). In addition, the DRB1*1302–DQB1*0604 haplotype was significantly decreased in the Hispanic cancer patients suggesting it confers protection (Apple *et al.*, 1994).

The susceptible DRB1*1501–DQB1*0602 haplotype is known to be in linkage disequilibrium with the B7 serotype (Begovich *et al.*, 1992); consequently it is not surprising that the B7 serotype was found to be increased in a British patient population (Duggan-Keen *et al.*, 1996). Apple and colleagues (1997) examined the distribution of B7 in their Hispanic population and found it to be significantly increased among both HPV16-associated CIN3 and HPV16-associated cancers relative to controls. However, the relationship between these two HLA markers differed among the two HPV16-associated cancer groups. Although both B7 and DRB1*1501 are independent risk factors for HPV16-associated CIN3, neither of these alleles are independently increased in HPV16-associated cancer. Rather, only individuals with both B7 and DRB1*1501 are increased in the HPV16-associated cancers suggesting that HLA-B7 and DRB1*1501, although in linkage disequilibrium, interact to confer an increased risk for the development of invasive cervical cancer following infection with HPV16. Together these data suggest that specific HLA molecules may interact and influence the immune response to specific HPV-encoded epitopes and affect the risk of cervical disease.

7.2.7 HLA and infectious diseases

As discussed above, the extensive allelic variation within the genes of the MHC has led many to propose that HLA polymorphism is maintained in the population because different allelic products have the capacity to bind and present different peptides and therefore defend against different foreign pathogens (Snell, 1968; Doherty and Zinkernagel, 1975). Until recently, there have been very few data in the human to support this model of parasite-driven MHC diversity. However, a recent case-control study of malaria in West African children identified both a class I antigen, HLA-Bw53, and an HLA class II haplotype (DRB1*1302–DQB1*0501) that were independently associated with protection from severe *Plasmodium falciparum* malaria (Hill *et al.*, 1991a). The observation that both Bw53 and the protective class II haplotype are common in West Africa but rare in other racial groups is regarded as additional evidence that this infectious agent is directing the evolution of these genes in this geographical area.

The simplest explanation for these associations is that T cells respond specifically to peptides from *P. falciparum* presented in the context of these protective class I and II molecules and provide protective immunity to malaria in man (Hill *et al.*, 1991a). To better understand the Bw53-protective effect, the sequence motif for Bw53-bound peptides was determined and used to identify peptides from the four pre-erythrocytic-stage *P. falciparum* antigens that have been cloned and sequenced (Hill *et al.*, 1992). Several peptides bearing this motif were identified and tested for their ability to elicit CTL responses from lymphocytes from Africans naturally exposed to malaria. A single nonamer peptide from liver-stage specific antigen (LSA-1) was able to elicit a secondary cytotoxic T-cell response in Bw53-positive individuals. Sequence analysis showed that this epitope is conserved in the local *P. falciparum* population.

The above examples clearly show the pivotal role that variation in one component of the trimolecular complex, the MHC molecules, plays in disease predisposition as well as resistance. Next we examine the genes that make up the T-cell antigen receptors.

7.3 The T-cell antigen receptor

7.3.1 Overview

The antigen receptor of circulating T lymphocytes is a heterodimer comprising two variable, but clonally distributed, glycoproteins of molecular weight about 40–60 kDa, either α/β, present in over 95% of peripheral blood lymphocytes, or γ/δ, non-covalently associated with five other invariant molecules collectively known as the

CD3 complex. Overall, the structure of the TCR proteins is similar to that of immunoglobulins with variable (V), diversity (D), joining (J), and constant (C) regions (Davis, 1990). The exons encoding the different component domains are assembled by a site-specific recombination reaction, known as V(D)J recombination, to create unique functional TCR transcripts (Schatz *et al.*, 1992). The antigen-binding domains of the TCR have a β-barrel structure, in which a conserved framework of β-strands support three hypervariable loops termed complementary-determining regions (CDRs). The putative CDR1 and CDR2 loops are encoded within the germline sequences of the V gene segments, while the CDR3 loops are encoded by the V(D)J junction sequences, plus non-germline additions or deletions that generate the structural diversity necessary to recognize antigens. The CDR3 loops provide the primary contact with the antigen peptide lying in the MHC grove, while anchor stretches in the flanking CDR1 and two loops contact the alpha helices of the MHC (Jorgensen *et al.*, 1992; Garboczi *et al.*, 1996).

The genes for the four different receptor chains are located in three gene complexes as shown in Figure 7.5 (Moss *et al.*, 1992; Concannon, 1997). The germline organization of the TCR alpha (A) gene region has yet to be determined, but approximately 45 AV (alpha-variable) and 60 to 70 AJ (alpha-joining) genes are spread over 100 kb on chromosome 14q11. About 4 kb downstream lies a single AC (alpha-constant) gene. Interestingly, the TCR delta (D) gene complex lies between AV and AJ. The number of different DV genes is limited, but an enormous potential exists for δ chain junctional diversity. There are three DD and three DJ segments, and up to three DD genes may be utilized in one transcript.

The TCR beta (B) and TCR gamma (G) genes are found on the long and short arms of chromosome 7 respectively. The TCRB gene complex spans around

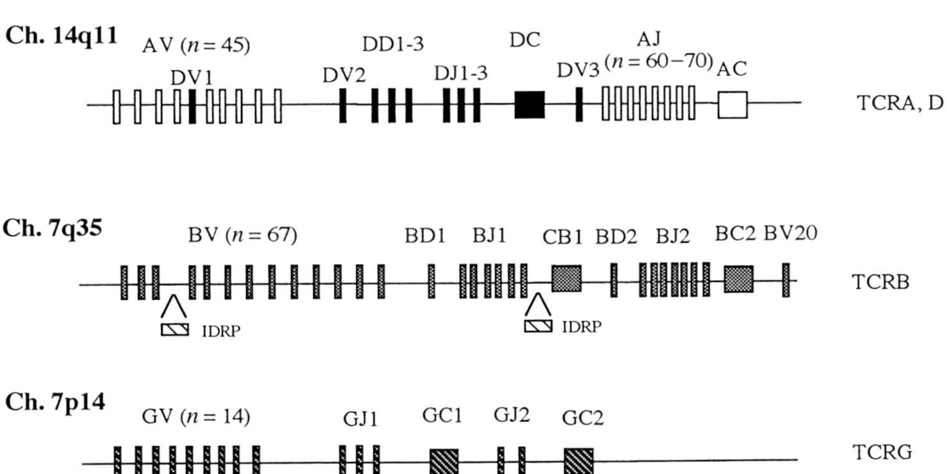

Figure 7.5 Germline organization of human TCR gene complexes. The locus names A, B, G, and D correspond to the α, β, γ, and δ chains, and the genetic elements are described as V (variable), D (diversity), J (joining), and C (constant). The number of functional genes for multi-member gene complexes (AV, AJ, BV, and GV), is given in parentheses. (See text for details.)

575 kb at 7q35. Rowen and colleagues (1996) have determined the entire sequence of a contiguous 685 kb stretch covering the entire TCRB locus. There are two BC genes, 13 BJ genes and two BD genes. In addition, the locus includes 67 BV genes (some haplotypes contain two identical copies of BV13S2); of them 47 are predicted to be functional. The TCRG complex includes 14 GV genes (eight of which appear to be functional), located upstream of two sets of GJ segments, each with a constant region complex. In addition, a cluster of six orphan BV genes was recently localized on chromosome 9p21 (Robinson *et al.*, 1993). Although several of these genes maintain correct open reading frames, their role remains unclear since no evidence of their transcription has been observed.

7.3.2   Germline TCR gene polymorphism

Early studies used restriction fragment length polymorphisms (RFLPs) with full-length cDNA or C-region probes together with Southern blot analysis to identify polymorphisms in the TCR regions (Concannon, 1997). Frequently, these polymorphisms lay outside coding regions. Although the biological relevance of RFLP markers is difficult to assess, they can be useful as genetic markers to follow segregation of disease traits in family studies and to construct haplotypes (Wei *et al.*, 1994). Seboun (1997) provides a comprehensive summary of published TCRA and TCRB RFLPs. For genetic studies, however, microsatellites and single-nucleotide polymorphisms (SNPs) are more informative and efficient. More than 20 microsatellites have now been described in the TCR regions (Charmley *et al.*, 1995; Buchmayer *et al.*, 1996).

Productive sequence variations have also been identified in the coding segment of AV and BV genes (Concannon, 1997). Some polymorphisms result in structural changes, such as the creation of a stop codon in the BV20 gene (Charmley *et al.*, 1993). Other polymorphisms lie within CDRs where they may affect the recognition of the HLA–peptide complex. One polymorphism in the recombination signal sequence 3′ of TCRBV3S1 was reported to affect rearrangement efficiency and, consequently, the level of circulating T cells carrying BV3S1 (Posnett *et al.*, 1994). PCR-based assays for several of the detected genomic variations have been developed, and allele and genotype frequencies in populations of interest can be determined (Wei *et al.*, 1995). The comparison of diversity in TCRAV germline polymorphisms in multiple populations revealed allele conservation in different populations, suggesting that TCR polymorphism predates the divergence of human subpopulations (Ibberson *et al.*, 1998). This is consistent with the high degree of similarity between human and other primate TCRB gene sequences (Jaeger *et al.*, 1994; Uccelli *et al.*, 1997).

Contrary to the immunoglobulin genes, somatic hypermutation is generally thought not to contribute to diversity of the V region of the TCR. Nevertheless, somatic hypermutation was identified in some TCR sequences derived from antigen-specific T cells homing to the spleen following immunization with antigen. Hypermutation seems to be restricted to the TCRA chain genes (Ikuta *et al.*, 1985; Zheng *et al.*, 1994). In addition, structural analysis of the TCR loci by pulse field electrophoresis (PFGE) has revealed the presence of size polymorphisms as well (Seboun *et al.*, 1989a; Zhao *et al.*, 1994). Two families of insertion/deletion-related polymorphisms (IDRP) were detected in the TCRB germline as shown in Figure 7.5. They appear as two independent allelic *Sfi* I polymorphisms of 330/300 kb and 145/139 kb and were mapped to the variable and constant regions of the TCR complex respectively. Other, less-frequent IDRPs were also identified. IDRPs can be used to trace the parental origin of clonal T-cell populations (Seboun *et al.*, 1992).

7.3.3   *TCR gene polymorphism and immune disorders*

TCR-RFLP markers were used extensively in population-based case-control studies to investigate genetic association with autoimmune disorders such as type I diabetes (IDDM), systemic lupus erythematosus (SLE), myasthenia gravis (MG), rheumatoid arthritis (RA), and multiple sclerosis (MS), and are reviewed in Concannon (1992). Allergic (Moffatt *et al.*, 1997) and infectious (Sachdeva *et al.*, 1997) disorders have also been studied. While some correlations were reported, results have been difficult to replicate or extend beyond particular populations. The analysis of candidate genes (receptors, interleukins, growth factors, myelin components, etc.) in case-control studies, although straightforward, is a problematic approach in diseases of partially known etiology, or when the gene has a moderate effect. An additional limitation is imposed by the difficulty in defining a perfectly matched control group, and biologically irrelevant associations can arise from population migration and admixture. Family-based studies, on the other hand, can identify unambiguous TCR haplotypes by segregation analysis, and confirm a true genetic effect through positive linkage, association, or both. Two independent linkage studies of the TCRB genes were carried out in MS-affected sib pairs collected in the US (Seboun *et al.*, 1989b) and the UK (Wood *et al.*, 1995), and appeared to confirm the presence of a genetic effect of this locus in disease susceptibility. Co-inheritance of TCRB haplotypes identical by descent was increased in MS sibling pairs compared with expected values, whereas the distribution of haplotype sharing was random when patients were compared with their unaffected siblings. The UK study also indicated complementation with the HLA locus. The role of TCRB haplotypes in the context of particular HLA-DR types was confirmed for a cohort of Canadian relapsing-progressive but not relapsing-remitting MS patients (Hockertz *et al.*, 1998). A similar approach in IDDM-affected sib pairs yielded no evidence for linkage with TCRB genes (Concannon *et al.*, 1990). All these studies used primarily RFLP-defined markers.

7.3.4   *The expressed repertoire of TCR genes*

TCR chain genes are encoded as discontinuous gene segments that rearrange somatically during T-cell differentiation in the thymus to express a unique TCR molecule. The V(D)J recombination reaction generates much of the required diversity in the antigen-binding site of the TCR. Hence, when considering the diversity of TCR gene products and their influence on the normal immune response or in disease, the physical rearrangements of TCR genes add a layer of complexity to the inherent differences in the germline configuration. During the rearrangement process in the embryonic thymus, T cell–stromal cell interactions result in competitive processes of positive and negative selection to ensure, on the one hand, self-tolerance, and on the other hand immune competence by the T-cell repertoire (Goodnow, 1996). It was recently estimated that transgenic mice expressing a single peptide–MHC class II complex positively selects at least 10^5 different BV rearrangements, indicating the great plasticity and significant size of the circulating repertoire (Gapin et al., 1998). In addition to germline availability and V(D)J-recombination, the actual repertoire expressed in the periphery depends upon additional factors including: (a) non-random V–J rearrangements; (b) multiple rearrangements of the alpha chain genes; (c) exposure to antigens and superantigens; and (d) peripheral idiotypic interactions such as suppression and anergy.

Overall, the pattern of TCR expression is unique to each individual, but certain TCRV genes are consistently expressed to a greater degree in either CD4 or CD8 positive cells (Grunewald et al., 1991). Twin and family studies clearly demonstrate the role of the HLA genotype in the shaping of the repertoire, i.e. quantitative TCR usage of V-region genes correlates with HLA-haplo-identity (Malhotra et al., 1992). These studies also indicate that other polymorphic loci elsewhere in the genome (Clarke et al., 1995), and non-genomic factors like recent antigenic challenges or aging, have a marked effect on TCR usage.

7.3.5 TCR usage in human immune disorders

Conserved amino acids at the VJ junction of the α chain or the VDJ junction of the β chain are indicative of clonal expansion among a population of T cells. Furthermore, when T-cell populations carrying particular V gene products rearranged to common CDR3 motifs are over-represented among a population of infiltrating lymphocytes, an antigen-specific T-cell response may be inferred. Not surprisingly, a large number of investigators have analyzed the composition of the TCR repertoire in several autoimmune disorders, both in the periphery and in the affected tissues (Oksenberg and Steinman, 1995). Most studies focus on V and CDR3 segment utilization by the TCRB transcripts, using PCR-based methods to analyze sequence and length, combined with quantitative amplification or antibody staining. What is the likelihood of detecting clonal TCR rearrangements associated with pathogenic T cells in a cellular infiltrate in inflamed tissue? The homing of T cells to the central nervous system in experimental demyelinating disease has an antigen-specific and a non-specific phase (Yednock et al., 1992). Within a few hours after the intravenous injection of a pathogenic (able to mediate encephalitogenicity) myelin-reactive T cell line expressing Vβ 8.2 receptor until day 4, the brain lesions are comprised almost entirely of Vβ 8.2 T cells. By contrast, during active clinical disease with paralysis, the TCR genes rearranged in the brain are quite diverse, including not only BV8.2, but also nearly the entire repertoire of BV genes. After the animals recover from the acute attack, around day 15 post-infusion, the TCR repertoire in the lesions is again quite restricted. In the human demyelinating disease, MS, TCRBV gene transcripts in acute active lesions are quite diverse, whereas in chronic lesions they are more restricted. The detection of TCRV genes and CDR3 sequences, both in the MS brain and peripheral blood, similar to those expressed by myelin-reactive encephalitogenic T cells in rodents, suggested that a pathogenic T-cell-mediated immune response to myelin basic protein might be occurring in the central nervous system of MS patients (Oksenberg *et al.*, 1993; Karpuj *et al.*, 1997).

TCR studies in rheumatoid arthritis have demonstrated the existence of highly diverse TCR repertoires in both synovial fluid and synovial tissue. In most studies, however, it was noted that T-cell populations present at the site of inflammation were found to express a selective pattern of TCR V-genes which differed from the pattern found in peripheral blood-derived T cells (Struyk *et al.*, 1995).

In addition to autoimmunity, the composition of the expressed TCR repertoire has been extensively studied in infectious disorders, allotransplantation, immunodeficiency, atherosclerosis, and cancer (Oksenberg *et al.*, 1992). Of special interest are recent studies, which use tetrameric MHC–peptide complexes to isolate antigen-specific T cells. McMichael and colleagues have made use of these tetrameric complexes in conjunction with anti-BV chain-specific monoclonal antibodies to show that chronically HIV-infected $CD8^+$ T cells are clonally expanded and display cytotoxic activity (Wilson *et al.*, 1998). It is obvious that the precise definition of the TCR repertoire involved in a particular disease may have therapeutic implications given the success of reversing experimental autoimmune disorders by directing specific forms of immunotherapy against TCR-V region gene products (Steinman *et al.*, 1993). TCR molecules have been a therapeutic target for T-cell leukemia and lymphomas (Janson *et al.*, 1989), as well as for human autoimmunity (Bourdette *et al.*, 1994; Steinman, 1995; Wilson *et al.*, 1997). Conversely, T-cell clones useful to the host can be expanded *ex vivo* to boost a specific immune response. With clinical protocols already in place, the application of this type of specific immunotherapy will depend on our understanding of the genetic and epigenetic contributions to the development of the TCR repertoire.

7.4 Chemokines and chemokine receptors

7.4.1 Overview

Chemokines are members of an expanding family of small serum proteins of between 7 and 16 kDa, primarily involved in selective trafficking and homing of leukocytes to sites of infection and inflammation, leukocyte maturation in the bone marrow, tissue repair and vascularization, and hematopoiesis and renewal of circulating leukocytes (Premack and Schall, 1996; Baggiolini, 1998; Tachibana *et al.*, 1998; Zou *et al.*, 1998). Their amino acid sequences contain four highly conserved cysteine (Cys) amino acid residues forming essential disulfide bonds. Based on this conserved structural property they are divided into four major subgroups (Ward *et al.*, 1998). The α-subfamily of chemokines (CXC), which in humans are all encoded in chromosome 4q12–q21, has the first two NH_2-terminal Cys separated by one non-conserved amino acid residue. The β-subfamily (CC), found in a cluster on chromosome 17q11–21, has these Cys in juxtaposition. The C (or γ) family has a lone NH_2-terminal Cys residue, while the CX3C (or δ) family has the Cys separated by three intervening amino acids. Nearly 50 chemokines have been described to date. Most fall into the CXC and CC groups, since only one C and one CX3C have been identified in humans. The aberrant secretion of chemokines has been detected in a wide variety of inflammatory and infectious diseases (Luster, 1998).

The distribution of allelic differences, in both coding and regulatory regions of multiple cytokine genes, and their role in disease susceptibility and pathogenesis are under intense investigation.

The specific effects of chemokines are mediated by single-chain, hepta-helical receptors coupled to GTP-binding proteins. The N-terminal region and three extra-cellular loops act in concert to bind the chemokine ligand (Luster, 1998). The C-terminal domain and three intracellular domains participate in signal transduction. Chemokines often share receptors, and a number of target cells express multiple receptor types, resulting in functional overlapping (Zack-Howard *et al.*, 1996). The chemokine/ligand promiscuity does not usually cross family boundaries, i.e. CXC-receptors (CXCR) and CC-receptors (CCR) only recognize chemokines of the corresponding subfamilies. Interestingly, virally encoded sequences represent shared chemokine receptors that have been transduced into the viral genomes during evolution. For example, human herpes virus 6, Karposi's sarcoma-associated herpes virus 8, an alternative open reading frame for cytomegalovirus, and the molluscum contagiosum virus encode CC chemokine analogs (Premack and Schall, 1996). The role of viral chemokine receptor homologues in the infectious process is unknown. Perhaps the first chemokine receptor linked with human disease is DARC (Duffy antigen receptor for cytokines), a receptor originally identified in erythrocytes (Power and Wells, 1996). It is a promiscuous receptor because it binds a number of CXC and CC chemokines. In addition, DARC functions as a receptor for the malarial parasite *Plasmodium vivax*. In certain ethnic groups, the gene is repressed in some individuals due to a point mutation in the erythroid promoter. Absence of DARC expression confers resistance to *P. vivax*.

7.4.2 *Chemokine receptors and HIV infection*

CCR5, a CC-type receptor that binds the chemokines RANTES, MIP-1a and MIP-1b, has been recently shown to mediate the entry into target cells of M-tropic HIV-1 strains (Alkhatib *et al.*, 1996; Deng *et al.*, 1996; Dragic *et al.*, 1996). The CMKBR5 gene located on chromosome 3p21.3 encodes CCR5. Individuals homozygous for a mutant allele, which carries a 32 bp deletion (labeled as Δccr5 or Δ32), are highly resistant to HIV infection (Dean *et al.*, 1996; Liu *et al.*, 1996) as seen in Table 7.5. In heterozygous individuals there is infection, but the rate of disease progression is lowered, with a typical delay of 2–4 years (Dean *et al.*, 1996). The deletion causes a frameshift at amino acid 185, which results in a stop codon and premature truncation within the third extracellular domain, preventing expression of the receptor in those homozygous individuals. There is no obvious phenotype associated with the δ32 homozygous genotype. A gene frequency of approximately 10% and homozygosity frequency of 1% was found for Δccr5 in Caucasian populations of Northern Europe (Martinson *et al.*, 1997). A North to South gradient was also observed, with the highest allele frequencies in Finnish and Mordvinian populations (16%), and the lowest in Sardinia (4%) (Libert *et al.*, 1998). Outside of Europe, Δccr5 is seen at frequencies above 5% in the Caucasus and at 2–5% in populations from Saudi Arabia, India, and Pakistan. The general absence of the Δccr5 allele in sub-Saharan Africa, Oceania, the Americas, and in non-human primates, suggests that this mutation has a recent and single origin in northeastern

Table 7.5 Chemokine and chemokine receptor polymorphisms associated with HIV infection and disease progression

Receptor	Ligand specificity	Cellular expression of receptor	HIV-1 associated polymorphisms
CCR2B	MCP-1, 2, 3, 4	Monocytes, basophils, activated T cells, myeloid precursor cells	CCR2B-641V/I (disease progression)
CCR5	RANTES, MIP-1a, MIP-1b	Activated monocytes, T cells, macrophages, dendritic cells, microglia, neurons, astrocytes, epithelium, endothelium, fibroblasts, vascular smooth muscle	CCR5Δ32 (infection and disease progression); CCR5-m303 (infection and disease progression); CCR5-59029-G (disease progression)
CXCR4	SDF-1	Activated and naive T cells, B cells, monocytes, macrophages, dendritic cells, granulocytes, neurons, microglia	SDF1-3'A (disease progression)

Europe. Random genetic drift or, more likely, selective advantage can explain this geographic pattern. On the bases of this distribution and intrahaplotypic variation determined by flanking microsatellites, it was estimated that the $\Delta 32$ mutation occurred around 4000 years ago (Carrington *et al.*, 1997; Libert *et al.*, 1998). On the other hand, Stephens *et al.* (1998) propose a more recent estimate of the origin of this mutation, about 700 years ago. It has been estimated that CCR5-$\Delta 32$ heterozygotes represent 20% of all long-term non-progressors (Stewart *et al.*, 1997). In a recent study, a less dramatic protective effect of the $\Delta 32$ mutation was observed in intravenously HIV-infected hemophilic patients ($n = 84$) compared to HIV-positive individuals infected by sexual contact ($n = 160$) (Malo *et al.*, 1998).

Additional CCR5 polymorphisms were also reported. For example, a rarer single-point mutation at position 303 (T→A, CCR5-m303), that introduces a premature stop codon, was recently identified in an individual carrying the $\Delta 32$ mutation in the second allele (Quillent *et al.*, 1998). The m303 was inherited as a single Mendelian trait, and a small population-based genotyping analysis showed that three out of 209 healthy blood donors were heterozygous for the mutant allele. Ansari-Lari and colleagues (1997) described seven additional CCR5 variants detected in African-American, Hispanics, Chinese, and/or Japanese, but not in a group of 50 American Caucasians. Carrington and colleagues (1997) reported 16 mutations in the coding region of the CCR5 gene (four overlap with the Ansari-Lari study), all but three of which are non-synonymous. Five were detected exclusively in African-Americans, an observation consistent with an adaptive accumulation of function-altering alleles as a consequence of historic selective environmental pressures. The effects of these polymorphisms on the HIV-1 co-receptor function of CCR5 remain to be determined.

A biallelic polymorphism (A/G) was identified in the CCR5 promoter region (position 59 029) (McDermott *et al.*, 1998). Although both alleles are common in the population, 59 029-G had 45% lower *in vitro* promoter activity. Interestingly, in a cohort of HIV-1 seroconvertors lacking both CCR5 $\Delta 32$ and CCR2-641 (see below), 59 029 G/G individuals progressed to AIDS on average 3.8 years more slowly than 59 029 A/A individuals, further supporting the role of CCR5 in HIV-1 pathogenesis. This effect may be the result of reduced CCR5 mRNA production. Higher cellular expression of CCR5 on CD4[+] T cells correlates with activation and advance disease progression (Ostrowski *et al.*, 1998).

The CCR2B gene also maps to the 3p21.3 region, and its product is the receptor for the monocyte attractants MCP-1, 2, 3, and 4. A recent study reported that a conservative substitution of valine by isoleucine in the coding first *trans*-membrane region of CCR2B (CCR2-64I) influences disease progression similar to δccr5, but not infection or transmission of HIV (Smith *et al.*, 1997). Analysis of a Swiss HIV cohort revealed that 33% of 52 long-term non-progressors have the CCR2 mutation compared to 19% in 241 typical progressors (Rizzardi *et al.*, 1998). The mechanism by which the CCR2-64I mutation influences protection remains unclear. It has been speculated that the influence of the CCR2-64I allele on disease progression may be related to CCR5 heterozygosity because of the strong linkage disequilibrium observed between CCR2-64I and a mutation (CCR5-59 653T) in the regulatory region of the closely mapped CCR5 gene (Kostrikis *et al.*, 1998). CCR2 and CCR5 proteins have over 80% homology, and although they have different chemokine-binding profiles, both receptors respond to RANTES. Nevertheless, the effects of CCR2-64I and Δccr5 seem to be independent (Smith *et al.*, 1997). In total, it has been estimated that CCR5 or CCR2B mutations account for about 30% of patients who remain AIDS-free for more than 16 years after infection (Smith *et al.*, 1997; Stewart, 1998).

Delay in progression to AIDS and death has also been observed in individuals homozygous for a common polymorphism in the 3′ untranslated region of the chemokine stromal-derived factor 1 (SDF-1) gene. SDF-1 is the principal ligand for CXCR4, the major coreceptor of T-tropic HIV-1 strains (Winkler, 1998). The SDF-1 gene is located in 10q11.2, and the recessive clinical benefit of the mutation appears to be stronger than that conferred by the CCR5-Δ32 or CCR2B-64I mutations. SDF1-3′A has similar frequencies in Asians (0.275) and Caucasians (0.211) but it is less common in African-Americans (0.057) (Stewart, 1998). The polymorphism is in a highly conserved region, suggesting that it may serve as a target for *cis*-acting factors influencing transcript abundance, synthesis, transport, stability, or splice product abundance (Winkler *et al.*, 1998).

A major impact of the discovery of the HIV-coreceptors is the potential availability of new therapeutic and preventive opportunities (Dimitrov and Broder, 1997). For example a non-chemotactic derivative of RANTES lacking the first eight N-terminal amino acids is effective in competing with the binding site of the receptor, then inhibiting HIV infection. Truncated chemokines could be potentially useful as inhibitors of HIV infection because they are unlikely to elicit inflammation, affect cell trafficking, or enhance virus replication. Complexes of coreceptors with viral envelope glycoprotein, associated with CD4, may also provide neutralizing conformational epitopes necessary for an effective vaccination strategy. Knowledge of receptor and coreceptor functions, as well as their germline polymorphisms, will certainly have implications for the design and development of therapeutic protocols and clinical trials. It should be noted, however, that the mutations conferring resistance to HIV infection and progression are not infallible, and during the course of infection HIV variants emerge that can also use an expanded range of coreceptors for infection (Connor *et al.*, 1997).

Acknowledgments

We thank P. Moonsamy, R. Saiki, and J. Wodehouse for help in the production of Figure 7.4 and P. Travers for Figure 7.2. J.R. Oksenberg is a fellow of the Esther A. and Joseph Klingenstein Fund for the Neurosciences.

References

Alkhatib, G., Combadiere, C., Broder, C.C., Feng, Y., Kennedy, P.E., Murphy, P.M., and Berger, E.A., 1996, CC CKR5: a RANTES, MIP-1alpha, MIP-1beta receptor as a fusion cofactor for macrophages-tropic HIV-1. *Science*, 272, 1955–1958.
Alper, C.A., Awdeh, Z. and Yunis, E.J., 1992, Conserved, extended MHC haplotypes. *Experimental Clinical Immunogenetics*, 9, 58–71.
Amiel, J., 1967, Study of the leukocyte phenotypes in Hodgkin's disease, in Terasaki, P.I. (Ed.) *Histocompatibility Testing*, pp. 79–81, Copenhagen: Munksgaard.

Ansari-Lari, M.A., Liu, X.-M., Metzker, M.L., Rut, A.R., and Gibbs, R.A., 1997, The extent of genetic variation in the CCR5 gene. *Nature Genetics*, 16, 221–222.

Apple, R.J., Erlich, H.A., Klitz, W., Manos, M.M., Becker, T.M. and Wheeler, C.M., 1994, HLA DR–DQ associations with cervical carcinoma show papillomavirus-type specificity. *Nature Genetics*, 6, 157–162.

Apple, R.J., Lin, P., Becker, T.M., Erlich, H.A. and Wheeler, C.M., 1997, Both HLA-B7 and DRB1*1501 appear to be required to confer risk to HPV16-associated invasive cancer, but function as independent risk factors for HPV16-associated severe dysplasia, in Charron, D. (Ed.) *HLA: Genetic Diversity of HLA – Functional and Medical Implication*, pp. 758–760, Sevres: EDK.

Awata, T., Kuzuya, T., Matsuda, A., Iwamoto, Y., Kanazawa, Y., Okuyama, M. and Juji, T., 1990, High frequency of aspartic acid at position 57 of HLA-DQ beta-chain in Japanese IDDM patients and nondiabetic subjects, *Diabetes*, 39, 266–269.

Bach, F.H. and Voynow, N.K., 1966, One-way stimulation in mixed leukocyte cultures. *Science*, 153, 545–547.

Baggiolini, M., 1998, Chemokines and leukocyte traffic. *Nature*, 392, 565–568.

Begovich, A.B. and Erlich, H.A., 1995, HLA typing for bone marrow transplantation – new polymerase chain reaction-based methods. *Journal of the American Medical Association*, 273, 586–591.

Begovich, A.B., McClure, G.R., Suraj, V.C., Helmuth, R.C., Fildes, N., Bugawan T.L., Erlich, H.A. and Klitz, W., 1992, Polymorphism, recombination, and linkage disequilibrium within the HLA class II region. *Journal of Immunology*, 148, 249–258.

Benjamin, R. and Parham, P., 1990, Guilt by association: HLA-B27 and ankylosing spondylitis. *Immunology Today*, 11, 137–142.

Billiard, M., Seignalet, J., Besset, A. and Cadilhac, J., 1986. HLA-DR2 and narcolepsy. *Sleep*, 9, 149–152.

Bjorkman, P.J., Saper, M.A., Samraoui, B., Bennett, W.S., Strominger, J.L. and Wiley, D.C., 1987a, Structure of the human class I histocompatibility antigen, HLA-A2. *Nature*, 329, 506–512.

Bjorkman, P.J., Saper, M.A., Samraoui, B., Bennett, W.S., Strominger, J.L. and Wiley, D.C., 1987b, The foreign antigen binding site and T cell recognition regions of class I histocompability antigens. *Nature*, 329, 512–518.

Bodmer, J.G., Tonks, S., Oza, A. M., Lister, T.A. and Bodmer, W.F., 1989, HLA-DP based resistance to Hodgkin's disease. *Lancet*, i, 1455–1456.

Bodmer, J.G., Marsh, S.G.E., Albert, E.D., Bodmer W. F., Bontrop, R.E., Dupont, B., Erlich, H.A., Hansen, J.A., Mach, B., Mayr, W.R., Parham, P., Petersdorf, E.W., Sasazuki, T., Schreuder, G.M. Th., Strominger, J.L., Svejgaard, A. and Terasaki, P.I., 1999, Nomenclature for factors of the HLA system, 1998. *Tissue Antigens*, 53, 407–446.

Bourdette, D.N., Whitham, R.H., Chou, Y.K., Morrison, W.J., Atherton, J., Kenny, C., Liefeld, D., Hashim, G.A., Offner, H. and Vandenbark, A.A., 1994, Immunity of TCR peptides in multiple sclerosis. *Journal of Immunology*, 152, 2510–2519.

Brewerton, D.A., Hart, F.D., Nicholls, A., Caffrey, M., James, D.C. and Sturrock, R.D., 1973, Ankylosing spondylitis and HLA-B27. *Lancet*, i, 904–907.

Brodsky, F.M., Lem, L. and Bresnahan, P.A., 1996, Antigen processing and presentation. *Tissue Antigens*, 47, 464–471.

Brown, J.H., Jardetzky, T.S., Gorga, J.C., Stern, L.J., Urban, R.G., Strominger J.L. and Wiley, D.C., 1993, Three-dimensional structure of the human class II histocompatibility antigen HLA-DR1. *Nature*, 364, 33–39.

Brown, M.A., Jepson, A., Young, A., Whittle, H.C., Greenwood, B.M. and Wordsworth, B.P., 1997, Ankylosing spondylitis in West Africa – evidence for a non-HLA-B27 protective effect. *Annals of Rheumatoid Disease*, 56, 68–70.

Buchmayer, H., Rumpold, H. and Mannhalter, C., 1996, Identification of a variable number tandem repeat region in the human T cell receptor alpha-delta (TCRAD) locus. *Human Genetics*, **3**, 333–335.

Bugawan, T.L., Chang, J.D., Klitz, W. and Erlich, H.A., 1994, PCR/oligonucleotide probe typing of HLA class II alleles in a Filipino population reveals an unusual distribution of HLA haplotypes. *American Journal of Human Genetics*, **54**, 331–340.

Bugawan, T.L., Mack, S.J., Stone King, M., Saha, M., Beck, H.P. and Erlich, H.A., 1997, HLA class I allele distributions in six Pacific/Asian populations: evidence of selection at the HLA-A locus. *Tissue Antigens*, **53**, 311–319.

Caillat-Zucman, S., Djilali-Saiah, I., Timsit, J., Bonifacio, E., Sepe, V., Collins, P., Bottazzo, G.F., Maclaren, N., Delamaire, M., Martin, S., Yamamoto, A.M., McWeeney, S., Valdes, A.M., Babron, M.C., Thorsby, E., Clerget-Darpoux, F., Thomson, G., Bach, J.F. and the participating centers, 1997, Insulin dependent diabetes mellitus (IDDM): 12th International Histocompatibility Workshop study, in Charron D. (Ed.) *HLA: Genetic Diversity of HLA – Functional and Medical Implication*, pp. 389–398, Sevres: EDK.

Carrington, M., Kissner, T., Gerrard, B., Ivanov, S., O'Brien, S.J. and Dean, M., 1997, Novel alleles of the chemokine-receptor gene CCR5. *American Journal of Human Genetics*, **61**, 1261–1267.

Catassi, C., Ratsch, I.M., Fabiani, E., Rossini, M., Bordicchia, F., Candela, F., Coppa, G.V. and Giorgi, P.L., 1994, Coeliac disease in the year 2000: exploring the iceberg. *Lancet*, **343**, 200–203.

Charmley, P., Wang, K., Hood, L. and Nickerson, D.A., 1993, Identification and physical mapping of polymorphic human T cell receptor V beta gene with a frequent null allele. *Journal of Experimental Medicine*, **177**, 135–145.

Charmley, P., Concannon, P., Hood, L. and Rowen, L., 1995, Frequency and polymorphism of simple sequence repeats in a contiguous 685 kilobase DNA sequence containing the human T-cell receptor beta chain gene complex. *Genomics*, **56**, 963–969.

Charron, D., 1997, *HLA: Genetic Diversity of HLA – Functional and Medical Implication*, Sevres: EDK.

Clarke, G.R., Reyburn, H., Lancaster, F.C. and Boylston, A.W., 1995, Bimodal distribution of V beta 2+ CD4+ T cells in human peripheral blood. *European Journal of Immunology*, **151**, 837–842.

Concannon, P., 1992, T-cell receptor gene polymorphisms: application in the study of autoimmune disease, in Rose, N.R. (Ed.) *Manual of Clinical Laboratory Immunology*, 4th edn, pp. 885–889, Washington DC: American Society for Microbiology.

Concannon, P., 1997, Introduction to the biology and genetics of human T cell receptor alpha and beta chain genes, in Oksenberg, J.R. (Ed.) *The Antigen Human T cell Receptor: Selected Protocols and Applications*, pp. 1–7, Austin: Landes Bioscience.

Concannon, P., Wright, J.A., Wright, L.G., Sylvester, D.R. and Spielman, R.S., 1990, T cell receptor genes and insulin dependent diabetes mellitus (IDDM): no evidence for linkage from affected sib pairs. *American Journal of Human Genetics*, **47**, 45–52.

Connor, R.I., Sheridan, K.E., Ceradini, D., Choe, S. and Landau, N. R., 1997, Change in coreceptor use correlates with disease progression in HIV-1 infected individuals. *Journal of Experimental Medicine*, **185**, 621–628.

Cucca, F., Muntoni, F., Lampis, R., Frau, F., Argiolas, L., Silvetti, M., Angius, E., Cao, A., De Virgilis, S. and Congia, M., 1993, Combinations of specific DRB1, DQA1, DQB1 haplotypes are associated with insulin-dependent diabetes mellitus in Sardinia. *Human Immunology*, **37**, 85–94.

Davies, J.L., Kawaguchi, Y., Bennett, S.T., Copeman, J.B., Cordell, H.J., Pritchard, L.E., Reed, P.W., Gough, S.C., Jenkins, S.C., Palmer, S.M., Balfour, K.M., Rowe, B.R., Farrall, M., Barnett, A.H., Bain, S.C. and Todd, J.A., 1994, A genome-wide search for human type 1 diabetes susceptibility genes. *Nature*, **371**, 130–136.

Davis, M.M., 1990, T cell receptor gene diversity and selection. *Annual Review of Biochemistry*, **59**, 475–496.

Dean, M., Carrington, M., Winkler, C., Huttley, G.A., Smith, M.W., Allikmets, R., Goedert, J.J., Buchbinder, S.P., Vittinghoff, E., Gomperts, E., Donfield, S., Vlahov, D., Kaslow, R., Saah, A., Rinaldo, C., Detels, R. and O'Brien, S.J., 1996, Genetic restriction of HIV-1 infection and progression to AIDS by a deletion of the CKR5 structural allele. *Science*, **273**, 1856–1862.

Deng, H., Liu, R., Ellmeier, W., Choe, S., Unutmaz, D., Burkhart, M., Di Marzio, P., Marmon, S., Sutton, R.E., Hill, C.M., Davis, C.B., Peiper, S.C., Schall, T.J., Littman, D.R. and Landau N.R., 1996, Identification of a major co-receptor for primary isolates of HIV-1. *Nature*, **381**, 661–666.

Dimitrov, D.S. and Broder, C.C., 1997, HIV and membrane receptors, in *Medical Intelligence Unit*, pp. 67–68, Austin: Landes Bioscience.

Doherty, P.C. and Zinkernagel, R.M., 1975, A biological role for the major histocompatibility antigens. *Lancet*, **i**, 1406–1409.

Dragic, T., Litwin, V., Allaway, G.P., Martin, S.R., Huang, Y., Nagashima, K.A., Cayanan, C., Maddon, P.J., Koup, R.A., Moore, J.P.and Paxton, W.A., 1996, HIV-1 entry into CD4+ cells is mediated by the chemokine receptor CC-CKR-5. *Nature*, **381**, 667–673.

Duggan-Keen, M.F., Keating, P.J., Stevens, F.R., Sinnott, P., Snijders, P.J.F., Walboomers, J.M.M., Davidson, S., Hunter, R.D., Dyer, P.A. and Stern, P.L., 1996, Immunogenetic factors in HPV-associated cervical cancer: influence on disease progression. *European Journal of Immunogenetics*, **23**, 275–284.

Ek, J., Albrechtsen, D., Solheim B.G. and Thorsby, E., 1978, Strong association between the HLA-Dw3-related B cell alloantigen -DRw3 and coeliac disease. *Scandanavian Journal of Gastroenterology*, **13**, 229–233.

Erlich, H.A., Zeidler, A., Chang, J., Shaw, S., Raffel, L.J., Klitz, W., Beshkov, Y., Costin, G., Pressman, S., Bugawan T. and Rotter, J.I., 1993, HLA class II alleles and susceptibility and resistance to insulin dependent diabetes mellitus in Mexican-American families. *Nature Genetics*, **3**, 358–364.

Ewens, W.J., 1972, The sampling theory of selectively neutral alleles. *Theoretical Population Biology*, **3**, 87–112.

Falchuk, Z.M., Rogentine, G.N. and Strober, W., 1972, Predominance of histocompatibility antigen HLA8 in patients with gluten-sensitive enteropathy. *Journal of Clinical Investigation*, **51**, 1602–1605.

Gapin, L., Fukui, Y., Kannellopoulos, J., Sano, T., Casrouge, A., Malier, V., Beaudoing, E., Gautheret, D., Claverie, J.-M., Sazazuki, Y. and Kourilsky, P., 1998, Quantitative analysis of the T cell repertoire selected by a single peptide–major histocompatibility complex. *Journal of Experimental Medicine*, **187**, 1871–1883.

Garboczi, D.N., Ghosh, P., Utz, U., Fan, Q.R., Biddison, W.E. and Wiley, D.C., 1996, Structure of the complex between human T-cell receptor, viral peptide and HLA-A2. *Nature*, **384**, 134–141.

Garrett, T.P.J., Saper, M.A., Bjorkman, P.J., Strominger, J.L. and Wiley, D.C., 1989, Specificity pockets for the side chains of peptide antigens in HLA-Aw68. *Nature*, **342**, 692–696.

Goodnow, C.C., 1996, Balancing immunity and tolerance: deleting and tuning lymphocyte repertoires. *Proceedings of the National Academy of Sciences, United States of America*, **93**, 2264–2271.

Gonzalez-Roces, S., Alvarez, M.V., Gonzalez, S., Dieye, A., Makni, H., Woodfield, D.G., Housan, L., Konenkov, V., Abbadi, M.C., Grunnet, N., Coto, E. and Lopez-Larrea, C., 1997, HLA-B27 polymorphism and worldwide susceptibility to ankylosing spondylitis. *Tissue Antigens*, **49**, 116–123.

Greco, L., Corazza, G., Babron, M.-C., Clot, F., Fulchignoni-Lataud, M.-C., Percopo, S., Zavattari, P., Bouguerra, F., Dib, C., Tosi, R., Troncone, R., Ventura, A., Mantavoni, W., Magazzu, G., Gatti, R., Lazzari, R., Giunta, A., Perri, F., Iacono, G., Cardi, E., de Virgiliis, S., Cataldo, F., De Angelis, G., Musumeci, S., Ferrari, R., Balli, F., Bardella, M.-T., Volta, U., Catassi, C., Torre, G., Eliaou J.-F., Serre, J.-L. and Clerget-Darpoux, F., 1998, Genome search in celiac disease. *American Journal of Human Genetics*, **62**, 669–675.

Grunewald, J., Janson, C.H. and Wigzell, H., 1991, Biased expression of individual T cell receptor V gene segments in CD4+ and CD8+ human peripheral blood T lymphocyte. *European Journal of Immunology*, **21**, 819–822.

Hall, M.A., Mazzilli, M.C., Satz, M.L., Barboni, F., Bartova, A., Brunnler, G., Ciclitira, P.J., Corazza, G.R., Ferrante, P., Gerok, W., Herrara, M., Kollek, A., Keller, E., Lanchbury, J.S., Mantovani, V., Muser, K., Petronzelli, F., Roschmann, E., Theiler, G., Volk, B.A., Welsh, K.I., Wienker, T. and Albert, E., 1992, Coeliac disease study, in Tsuji, K., Aizawa, M. and Sasazuki, T. (Eds), *HLA 1991*, vol. 1, pp. 722–729, Oxford: Oxford University Press.

Hammer, R.E., Maika, S.D., Richardson, J.A., Tang, J.-P. and Taurog, J.D., 1990, Spontaneous inflammatory disease in transgenic rats expressing HLA-B27 and human β2-m: an animal model of HLA-B27–associated human disorders. *Cell*, **63**, 1099–1112.

Hill, A.V.S., Allsopp, C.E.M., Kwatkowski, D., Anstey, N.M., Twumasi, P., Rowe, P.A., Bennett, S., Brewster, D., McMichael, A.J. and Greenwood, B.M., 1991a, Common West African antigens are associated with protection from severe malaria. *Nature*, **352**, 595–600.

Hill, A.V.S., Allsopp, C.E.M., Kwiatkowski, D., Anstey, N.M., Greenwood, B.M. and McMichael, A.J., 1991b, HLA class I typing by PCR: HLA-B27 and an African B27 subtype. *Lancet*, **337**, 640–642.

Hill, A.V.S., Elvin, J., Willis, A.C., Aidoo, M., Allsopp, C.E.M., Gotch, F.M., Gao, X.M., Takiguchi, M., Greenwood, B.M., Townsend, A.R.M., McMichael, A.J. and Whittle, H.C., 1992, Molecular analysis of the association of HLA-B53 and resistance to severe malaria. *Nature*, **360**, 434–439.

Hockertz, M.K., Paty, D.W. and Beall, S.S., 1998, Susceptibility to relapsing–progressive multiple sclerosis is associated with inheritance of genes linked to the variable region of the TcR beta locus: use of affected family-based controls. *American Journal of Human Genetics*, **62**, 373–385.

Horn, G.T., Bugawan, T.L., Long, C.M. and Erlich, H.A., 1988, Allelic sequence variation of the HLA-DQ loci: relationship to serology and to insulin-dependent diabetes suscepti-bility. *Proceedings of the National Academy of Sciences, United States of America*, **85**, 6012–6016.

Hu, C.-Y., Allen, M., Chuang, L.-M., Lin, B.J. and Gyllensten, U., 1993, Association of insulin-dependent diabetes mellitus in Taiwan with HLA class II DQB1 and DRB1 alleles. *Human Immunology*, **38**, 105–114.

Ibberson, M.R., Copier, J.P., Llop, E., Navarrete, C., Hill, A.V.S., Cruickshank, J.K. and So, A.K.L., 1998, T-cell receptor variable alpha polymorphism in European, Chinese, South American, Afro-Caribbean, and Gambian populations. *Immunogenetics*, **47**, 124–130.

Ikuta, K., Ogura, T., Shimizu, A. and Honjo, T., 1985, Low frequency of somatic muta-tion in beta chain variable region genes of human T-cell receptors. *Proceedings of the National Academy of Sciences, United States of America*, **82**, 7701–7705.

Imanishi, T., Akaza, T., Kimura, A., Tokunaga, K. and Gojobori, T., 1992, W15.1: allele and haplotype frequencies for HLA and complement loci in various ethnic groups, in Tsuji, M., Aizawa, M. and Saszuki, T. (Eds), *HLA 1991, Proceedings of the Eleventh International Histocompatibility Workshop and Conference*, vol. 1, pp. 83–108, Oxford: Oxford University Press.

Jaeger, E., Bontrop, R. and Lanchbury, J., 1994, Structure, diversity and evolution of the T-cell receptor VB gene repertoire in primates. *Immunogenetics*, **40**(3), 184–191.

Janson, C.H., Tehrani, M., Mellstedt, H. and Wigsell, H., 1989, Anti-idiotypic monoclonal antibody to a T cell chronic lymphatic leukemia. *Cancer Immunology and Immunotherapy*, 28, 225–232.

Johansen, B.H., Vartdal, F., Eriksen, J.A., Thorsby, E. and Sollid, L.M., 1996, Identification of a putative motif for binding of peptides to HLA-DQ2. *International Immunology*, 8, 177–182.

Jorgensen, J.L., Esser, U., Fazekas de St. Groth, B., Reay, P. and Davis, M.M., 1992, Mapping TCR-peptide contacts by variant peptide immunization of single-chain transgenics. *Nature*, 355, 224–230.

Juji, T., Satake, M., Honda, Y. and Doi, Y., 1984, HLA antigens in Japanese patients with narcolepsy. All the patients were DR2 positive. *Tissue Antigens*, 24, 316–319.

Kappler, J.W., Roehm, N. and Marrack, P., 1987, T cell tolerance by clonal elimination in the thymus. *Cell*, 49, 273–280.

Karpuj, M.A., Steinamn, L. and Oksenberg, J.R., 1997, Multiple sclerosis: a polygenic disease involving epistatic interactions, germline rearrangements and environmental effects. *Neurogenetics*, 1, 21–28.

Khalil, I., d'Auriol, L., Gobet, M., Morin, L., Lepage, V., Deschamps, I., Park, M.S., Degos, L., Galibert, F. and Hors, J., 1990, A combination of HLA-DQ beta ASP57-negative and HLA DQ alpha Arg52 confers susceptibility to insulin-dependent diabetes mellitus. *Journal of Clinical Investigation*, 85, 1315–1319.

Khan, M.A., 1996, Spondyloarthropathies – spondyloarthropathies. *Current Opinion in Rheumatology*, 8, 267–268.

Klitz, W., Aldrich, C.L., Fildes, N., Horning, S.J. and Begovich, A.B., 1994, Localization of predisposition to Hodgkin disease in the HLA class II region. *American Journal of Human Genetics*, 54, 497–505.

Kostrikis, L.G., Huang, Y., Moore, J.P., Wolinsky, S.M., Zhang, L., Guo, Y., Deutsch, L., Phair, J., Neumann, A.U. and Ho, D.D., 1998, A chemokine receptor CCR2 allele delays HIV-1 disease progression and is associated with a CCR5 promoter mutation. *Nature Medicine*, 4, 350–353.

Libert, F., Cochaux, P., Beckman, G., Samson, M., Aksenova, M., Cao, A., Czeizel, A., Claustres, M, de la Rua, C., Ferrari, M., Ferrec, C., Glover, G., Grinde, B., Guran, S., Kucinskas, V., Lavinha, J., Mercier, B., Ogur, G., Peltonen, L., Rosatelli, C., Schwartz, M., Spitsyn, V., Timar, L., Beckman, L., Vassart, G., et al., 1998, The delta CCR5 mutation conferring protection against HIV-1 in Caucasian populations has a single and recent origin in northeastern Europe. *Human Molecular Genetics*, 7, 399–406.

Liu, R., Paxton, W.A., Choe, S., Ceradini, D., Martin, S.R., Horuk, R., MacDonald, M.E., Stuhlmann, H., Koup, R.A. and Landau, N.R., 1996, Homozygous defect in HIV-1 coreceptor accounts for resistance of some multiply-exposed individuals to HIV-infection. *Cell*, 86, 367–378.

Lopez de Castro, J., 1998, The pathogenetic role of HLA-B27 in chronic arthritis. *Current Opinions in Immunology*, 10, 59–66.

Lu, S.J., Day, N.E., Degos, L., Lepage, V., Wange, P.C., Chan, S.H., Simons, M., McKnight, B., Easton, D., Zeng, Y. and de-The, G., 1990, Linkage of a nasopharyngeal carcinoma susceptibility locus to the HLA regions. *Nature*, 346, 470–471.

Lundin, K.E.A., Ronningen, K.S., Aono, S., Spurkland, A., Gaudernack, G., Isshiki, G. and Thorsby, E., 1989, HLA-DQ antigens and DQ beta amino acid 57 of Japanese patients with insulin-dependent diabetes mellitus: detection of a DRw8DQw8 haplotype. *Tissue Antigens*, 34, 233–241.

Lundin, K.E.A., Scott, H., Hansen, T., Paulsen, G., Halstensen, T.S., Fausa, O., Thorsby, E. and Sollid, L.M., 1993, Gliadin-specific, HLA-DQ(α1*0501, β1*0201) restricted T cells isolated from the small intestinal mucosa of celiac disease patients. *Journal of Experimental Medicine*, 178, 187–196.

Lundin, K.E.A., Gjertsen H.A., Scott, H., Sollid, L.M. and Thorsby, E, 1994a, Function of DQ2 and DQ8 as HLA susceptibility molecules in celiac disease. *Human Immunology*, 41, 24–27.

Lundin, K.E.A., Scott, H., Fausa, O., Thorsby, E. and Sollid, L.M., 1994b, T cells from the small intestinal mucosa of a DR4, DQ7/DR4, DQ8 celiac disease patient preferentially recognize gliadin when presented by DQ8. *Human Immunology*, 41, 285–291.

Luster, A.D., 1998, Chemokines – chemotactic cytokines that mediate inflammation, *The New England Journal of Medicine*, 338, 436–445.

Mack, S.J. and Erlich, H.A., 1998, HLA class II polymorphism in the Ticuna of Brazil: evolutionary implications of the DRB1*0807 allele, *Tissue Antigens*, 51, 41–50.

Madden, D.R., Gorga, J.C., Strominger, J.L. and Wiley, D.C., 1992, The three-dimensional structure of HLA-B27 at 2.1 Å resolution suggests a general mechanism for tight peptide binding to MHC. *Cell*, 70, 1035–1048.

Malhotra, U., Spielman, R. and Concanon, P., 1992, Variability in the T cell receptor V beta gene usage in human peripheral blood lymphocytes. Studies of identical twins, siblings, and IDDM patients. *Journal of Immunology*, 149, 1802–1808.

Malo, A., Rommel, F., Bogner, J., Gruber, R., Schramm, W., Goebel, F.D., Riethmuller, G. and Wank, R., 1998, Lack of protection from HIV infection by the mutant HIV coreceptor CCR5 intravenously HIV infected hemophilia patients. *Immunobiology*, 198, 485–488.

Marsh, M.N., 1992, Gluten, major histocompatibility complex, and the small intestine. A molecular and immunobiologic approach to the spectrum of gluten sensitivity ('celiac sprue'). *Gastroenterology*, 102, 330–354.

Marsh, S.G.E., 1998, HLA class II region sequences, 1998. *Tissue Antigens*, 51, 467–507.

Martinson, J.J., Chapman, N.H., Rees, D.C., Liu, Y.-T. and Clegg, J.B., 1997, Global distribution of the CCR5 gene 32-basepair deletion. *Nature Genetics*, 16, 100–103.

Mason, P.M. and Parham, P., 1998, HLA class I region sequences, 1998. *Tissue Antigens*, 51, 417–466.

Matsuki, K., Juji, T., Tokunaga, K., Naohara, T., Satake, M. and Honda, Y., 1985, Human histocompatibility leukocyte antigen (HLA) haplotype frequencies estimated from the data on HLA class I, II, and III antigens in 111 Japanese narcoleptics. *Journal of Clinical Investigation*, 76, 2078–2083.

McDermott, D.H., Zimmerman, P.A., Guignard, F., Kleeberger, C.A., Leitman, S.F. and Murphy, P.M., 1998, CCR5 promoter polymorphism and HIV disease progression. Multicenter AIDS cohort study. *Lancet*, 12, 866–870.

MacDonald, H.R., Schneider, R., Lees, R.K., Howe, R.C., Acha-Orbea, H., Festenstein, H., Zinkernagel, R.M. and Hengartner, H., 1988, T-cell receptor V beta use predicts reactivity and tolerance to Mls[a]-encoded antigens. *Nature*, 332, 40–45.

Mearin, M.L., Biemond, I., Pena, A.S., Polanco, I., Vazquez, C., Schreuder, G. Th. M., de Vries, R.R.P. and van Rood, J.J., 1983, HLA-DR phenotypes in Spanish coeliac children: their contribution to the understanding of the genetics of the disease. *Gut*, 24, 532–537.

Mignot, E., Lin, X., Arrigoni, J., Macaubas, C., Olive, F., Hallmayer, J., Underhill, P., Guilleminault, C., Dement, W.C. and Grumet, F.C., 1994, DQB1*0602 and DQA1*0102 (DQ1) are better markers than DR2 for narcolepsy in Caucasian and Black Americans. *Sleep*, 17, S60–S67.

Moffatt, M.F., Schou, C., Faux, J.A. and Cookson, W.O., 1997, Germline TCR-A restriction of immunoglobulin E responses to allergen. *Immunogenetics*, 46, 226–230.

Morel, P.A., Dorman, J.S., Todd, J.A., McDevitt, H.O. and Trucco, M., 1988, Aspartic acid at position 57 of the HLA-DQ beta chain protects against type I diabetes: a family study. *Proceedings of the National Academy of Sciences, United States of America*, 85, 8111–8115.

Morellini, M., Trabace, S., Mazzilli, M.C., Lulli, P., Cappellacci, R., Bonamico, M., Margarit, I. and Gandini, E., 1988, A study of HLA class II antigens in an Italian paediatric population with coeliac disease. *Disease Markers*, 6, 23–28.

Moss, P.A.H., Rosenberg, W.M.C. and Bell, J.I., 1992, The human T cell receptor in health and disease. *Annual Review of Immunology*, **10**, 71–96.

Mylotte, M., Egan-Mitchell, B., McCarthy, C.F. and McNicholl, B., 1973, Incidence of coeliac disease in the West of Ireland. *British Medical Journal*, i, 703–705.

Nepom, G. and Erlich, H.A., 1991, MHC-class II molecules and autoimmunity. *Annual Review of Immunology*, **9**, 493–525.

Nepom, G.T., 1990, A unified hypothesis for the complex genetics of HLA associations with IDDM. *Diabetes*, **39**, 1153–1157.

Noble, J.A., Valdes, A.M., Cook, M., Klitz, W., Thomson, G. and Erlich, H.A., 1996, The role of HLA class II genes in insulin-dependent diabetes mellitus: molecular analysis of 180 Caucasian, multiplex families. *American Journal of Human Genetics*, **59**, 1134–1148.

Noble, J.A., Valdes, A.M., Apple, R., Bugawan T.L., Thomson, G. and Erlich, H.A., 1998, Multiple HLA loci influence susceptibility to type I diabetes. *Human Immunology*, **59**(supplement 1), 76 (abs).

Oksenberg, J.R. and Steinman, L., 1995, Detection an analysis of T cell receptor gene products by RT-PCR, in Siebert P.D. and Larrick J.W. (Eds). *Reverse Transcriptase PCR*, pp. 61–85, London, Ellis Horwood.

Oksenberg, J.R., Panzara, M.A. and Steinman, L., 1992, *The Polymerase Chain Reaction and the Analysis of the T cell Receptor Repertoire*, Austin: R.G. Landes.

Oksenberg, J.R., Panzara, M.A., Begovich, A.B., Mitchell, D., Erlich, H.A., Murray, R.S., Shimonkevitz, R., Sherrit, M., Rothbard, J., Bernard, C.C.A. and Steinman, L., 1993, Selection for T-cell receptor V-D-J gene rearrangements with specificity for a myelin basic protein peptide in brain lesions of multiple sclerosis. *Nature*, **362**, 68–70.

Ostrowski, M.A., Justement, S.J., Catanzaro, A., Hallahan, C.A., Ehler, L.A., Mizell, S.B., Kumar, P.N., Mican, J.A., Chun, T.W. and Fauci, A.S., 1998, Expression of chemokine receptors CXCR4 and CCR5 in HIV-infected and uninfected individuals. *Journal of Immunology*, **161**, 3195–3201.

Oza, A.M., Tonks, S., Lim, J., Fleetwood, M.A., Lister, T.A., Bodmer, J. and Collaborating Centers, 1994, A clinical and epidemiological study of human leukocyte antigen-DPB alleles in Hodgkin's disease. *Cancer Research*, **54**, 5101–5105.

Parker, K.C., Biddison, W.E. and Coligan, J.E., 1994, Pocket mutations of HLA-B27 show that anchor residues act cumulatively to stabilize peptide binding. *Biochemistry*, **33**, 7736–7743.

Posnett, D.N., Vissinga, C.S., Pambuccian, C., Wei, S., Robinson, M.A., Kostyu, D. and Concannon, P., 1994, Level of human TCRBV3S1 expression correlates with allelic polymorphism in the spacer region of the recombination signal sequence. *Journal of Experimental Medicine*, **179**, 1707–1711.

Power, C.A. and Wells, T.N.C., 1996, Cloning and characterization of human chemokine receptors. *Trends in Pharmaceutical Sciences*, **17**, 209–213.

Premack, B.A. and Schall, T.J., 1996, Chemokine receptors: gateways to inflammation and infection. *Nature Medicine*, **2**, 1174–1178.

Quillent, C., Oberlin, E., Braun, J., Rousset, D., Gonzalez-Canali, G., Metais, P., Montagnier, L., Virelizier, J.-L., Arenzana-Seisdedos, F. and Beretta, A., 1998, HIV-resistant phenotype conferred by combination of two separate inherited mutations of the CCR5 gene. *Lancet*, **351**, 14–18.

Rath, H.C., Herfarth, H.H., Ikeda, J.S., Grenther, W.B., Hamm, T.E., Balish, E., Taurog, J.D., Hammer, R.E., Wilson, K.H. and Sartor, R.B., 1996, Normal luminal bacteria, especially Bacteroides species, mediate chronic colitis, gastritis, and arthritis in HLA-B27/human β2 miroglobulin transgenic rats. *Journal of Clinical Investigation*, **98**, 945–953.

Risch, N., 1987, Assessing the role of HLA-linked and unlinked determinants of disease. *American Journal of Human Genetics*, **40**, 1–14.

Rizzardi, G.P., Morawetz, R.A., Vicenzi, E., Ghezzi, S., Poli, G., Lazzarini, A., Pantaleo, G. and Cohort., S.H., 1998, CCR2 polymorphism and HIV disease. *Nature Medicine*, **4**, 252–253.

Robinson, M.A., Mitchell, M.P., Wei, S., Day, C.E., Zhao, T.M. and Concannon, P., 1993, Organization of human T-cell receptor beta-chain genes: clusters of V beta genes are present on chromosome 7 and 9. *Proceedings of the National Academy of Sciences, United States of America*, **90**, 2433–2437.

Roep, B.O., Bontrop, R.E., Pena, A.S., van Eggermond, M.C., van Rood, J.J. and Giphart, M.J., 1988, An HLA-DQ alpha allele identified at DNA and protein level is strongly associated with celiac disease. *Human Immunology*, **23**, 271–279.

Rogers, A.E., Meehan, J., Guilleminault, C., Grumet, F.C. and Mignot, E., 1997, HLA DR15 (DR2) and DQB1*0602 typing studies in 188 narcoleptic patients with cataplexy. *Neurology*, **48**, 1550–1556.

Rotter, J.I. and Landlaw, E.M., 1984, Measuring the genetic contribution of a single locus to a multilocus disease. *Clinical Genetics*, **26**, 529–542.

Rowen, L., Koop, B.F. and Hood, L., 1996, The complete 685-kilobase DNA sequence of the human beta T cell receptor locus. *Science*, **272**, 1755–1762.

Sachdeva, G., Kaur, G., Bhutani, L.K. and Bamezai, R., 1997, Genetic variations at the TCR gamma locus in circulating peripheral blood mononuclear cells of clinically categorized leprosy patients. *Human Genetics*, **100**, 30–34.

Saiki, R.K., Scharf, S., Faloona, F., Mullis, K.B., Horn, G.T., Erlich, H.A. and Arnheim, N., 1985, Enzymatic amplification of β-globin genomic sequences and restriction site analysis for diagnosis of sickle cell anemia. *Science*, **230**, 1350–1354.

Schatz, D.G., Oettinger, M.A. and Schlissel, M.S., 1992, V(D)J recombination: molecular biology and regulation. *Annual Review of Immunology*, **10**, 359–383.

Seboun, E., 1997, Analysis of TCR genes at the genomic level, in Oksenberg, J.R. (Ed.) *The Antigen T Cell Receptor: Selected Protocols and Applications*, pp. 461–488, Austin, Landes Bioscience.

Seboun, E., Robinson, M.A., Kindt, T.J. and Hauser, S.L., 1989a, Insertion/deletion-related polymorphisms in human T cell receptor B gene complex. *Journal of Experimental Medicine*, **170**, 1263–1270.

Seboun, E., Robinson, M.A., Doolittle, T.H., Ciulla, T.A., Kindt, T.J. and Hauser, S.L., 1989b, A susceptibility locus for multiple sclerosis is linked to the T cell receptor beta chain complex. *Cell*, **57**, 1095–1100.

Seboun, E., Joshi, N. and Hauser, S.L., 1992, Haplotypic origin of beta-chain genes expressed by human T-cell clones. *Immunogenetics*, **36**, 363–368.

Sheehy, M.J., 1992, HLA and insulin-dependent diabetes: a protective perspective. *Diabetes*, **41**, 123–129.

Sheehy, M.J., Scharf, S.J., Rowe, J.R., Neme de Gimenez, M.H., Meske, L.M., Erlich, H.A. and Nepom, B.S., 1989, A diabetes-susceptible HLA haplotype is best defined by a combination of HLA-DR and -DQ alleles. *Journal of Clinical Investigation*, **83**, 830–835.

Smith, M.W., Dean, M., Carrington, M., Winkler, C., Huttley, G.A., Lomb, D.A., Goedert, J.J., O'Brien, T.R., Jacobson, L.P., Kaslow, R., Buchbinder, S., Vittinghoff, E., Vlahov, D., Hoots, K., Hilgartner, M.W. and O'Brien, S.J., 1997, Contrasting genetic influence of CCR2 and CCR5 variants on HIV-1 infection and disease progression. *Science*, **277**, 959–965.

Snell, G.D., 1968, The H-2 locus of the mouse: observations and speculations concerning its comparative genetics and its polymorphism. *Folia Biologia*, **14**, 335–358.

Sollid, L.M. and Thorsby, E., 1993, HLA susceptibility genes in celiac disease: genetic mapping and role in pathogenesis. *Gastroenterology*, **105**, 910–922.

Sollid, L.M., Markussen, G., Ek, J., Gjerde, H., Vartdal, F. and Thorsby, E., 1989, Evidence for a primary association of celiac disease to a particular HLA-DQ α/β heterodimer. *Journal of Experimental Medicine*, **169**, 345–350.

Steinman, L., 1995, Escape from "horror autotoxicuss": pathogenesis and treatment of autoimmune disease. *Cell*, **80**, 7–10.

Steinman, L., Lindsey, J.W., Alters, S. and Hodgkinson, S., 1993, From treatment of experimental allergic encephalomyelitis to clinical trials in multiple sclerosis, in Bach, J.F. (Ed.) *Monoclonal Antibodies and Peptide Therapy in Autoimmune Diseases*, pp. 253–260, New York, Marcel Dekker, Inc.

Stephens, J.C., Reich, D.E., Goldstein, D.B., Shin, H.D., Smith, M.W., Carrington, M., Winkler, C., Huttley, G.A., Allikmets, R., Schriml, L., Gerrard, B., Malasky, M., Ramos, M.D., Morlot, S., Tzetis, M., Oddoux, C., di Giovine, F.S., Nasioulas, G., Chandler, D., Aseev, M., Hanson, M., Kalaydjieva, L., Glavac, D., Gasparini, P., Dean, M., *et al.*, 1998, Dating the origin of the CCR5-D32 AIDS-resistant allele by the coalescence of haplotypes. *American Journal of Human Genetics*, **62**, 1507–1515.

Stewart, G., 1998, Chemokine genes – beating the odds. *Nature Medicine*, **4**, 275–277.

Stewart, G.J., Ashton, L.J., Biti, R.A., French, R.A., Bennetts, B.H., Newcombe, N.R., Benson, E.M., Carr, A., Cooper, D.A. and Kaldor, J.M., 1997, Increased frequency of CCR5-Delta 32 heterozygotes among long-term non-progressors with HIV-1 infection. *AIDS*, **11**, 1833–1838.

Stokes, P.L., Ferguson, R., Holmes, G.K. and Cooke, W.T., 1976, Familial aspects of coeliac disease. *Quarterly Journal of Medicine*, **45**, 567–582.

Struyk, L., Hawes, G.E., Chatila, M.K., Breedveld, F.C., Kurnick, J.T. and van den Elsen, P.J., 1995, T cell receptors in rehumatoid arthritis. *Arthritis and Rheumatism*, **38**, 577–589.

Svejgaard, A., Platz, P. and Ryder, L.P., 1980, Insulin dependent diabetes mellitus, in Terasaki, P. (Ed.) *Histocompatibility Testing*, pp. 638–656, Los Angeles: University of California Press.

Tachibana, K., Hirota, S., Lizasa, H., Yoshida, H., Kawabata, K., Kataoka, Y., Kitamura, Y., Matsushima, K., Yoshida, N., Nishikawa, S., Kishimoto, T. and Nagasawa, T., 1998, The chemokine receptor CXCR4 is essential for vascularization of the gastrointestinal tract. *Nature*, **393**, 591–594.

Tait, B.D. and Harrison, L.C., 1991, Overview: the major histocompatibility complex and insulin dependent diabetes mellitus. *Baillieres Clinical Endocrinology and Metabolism*, **5**, 211–228.

Taurog, J.D., Richardson, J.A., Croft, J.T., Simmons, W.A., Zhou, M, Fernandez-Sueiro, J.L., Balish, E. and Hammer, R.E., 1994, The germfree state prevents development of gut and joint inflammatory disease in HLA-B27 transgenic rats. *Journal of Experimental Medicine*, **180**, 2359–2364.

Taylor, G.M., Robinson, M.D., Binchy, A., Birch, J.M., Stevens, R.F., Jones, P.M., Carr, T., Dearden, S. and Gokhale, D.A., 1995, Preliminary evidence of an association between HLA-DPB1*0201 and childhood common acute lymphoblastic leukaemia supports an infectious aetiology. *Leukemia*, **9**, 440–443.

Taylor, G.M., Gokhale, D.A., Crowther, D., Woll, P., Harris, M., Alexander, F., Jarrett, R. and Cartwright, R.A., 1996, Increased frequency of HLA-DPB1*0301 in Hodgkin's disease suggests that susceptibility is HVR-sequence and subtype-associated. *Leukemia*, **10**, 854–859.

Teh, H.S., Kisielow, P., Scott, B., Kishi, H., Uematsu, Y., Bluthmann, H. and von Boehmer, H., 1988, Thymic major histocompatibility complex antigens and the alpha beta T-cell receptor determine the CD4/CD8 phenotype of T cells. *Nature*, **335**, 229–233.

Teresaki, P.I. and Gjertson, D.W., 1997, *HLA 1997*, Los Angeles: UCLA Tissue Typing Laboratory.

Terasaki, P.I., Mandell, M., Van De Water, J. and Edgington T.E., 1964, Human blood lymphocyte cytotoxicity reactions with allogeneic antisera. *Annals of the New York Academy of Sciences*, **120**, 322–334.

Thomson, G., 1988, HLA and disease associations: models for insulin dependent diabetes mellitus and the study of complex human genetic disorders. *Annual Review of Genetics*, **22**, 31–50.

Thomson, G., 1995, HLA disease associations: models for the study of complex human genetic disorders. *Critical Reviews in Clinical Laboratory Science*, **32**, 183–219.

Thorsby, E., 1997, Invited anniversary review: HLA associated diseases. *Human Immunology*, **53**, 1–11.

Tiwari, J.L. and Terasaki, P.I., 1985, *HLA and Disease*, New York: Springer.

Todd, J.A., Bell, J.I. and McDevitt, H.O., 1987, HLA-DQ beta gene contributes to susceptibility and resistance to insulin-dependent diabetes mellitus. *Nature*, **329**, 599–604.

Tosi, R., Vosmara, D., Tanigaki, N., Ferrara, G.B., Cicimarra, F., Buffolano, W., Follo, D. and Auricchio, S., 1983, Evidence that coeliac disease is primarily associated with a DC locus allelic specificity. *Clinical Immunology and Immunopathology*, **28**, 395–404.

Troncone, R., Greco, L. and Auricchio, S., 1996, Gluten-sensitive enteropathy. *Pediatric Clinics of North America*, **43**, 355–373.

Uccelli, A., Oksenberg, J.R., Jeong, M.C., Genain, C.P., Rombos, T., Jaegert, E.M., Giunti, D., Lanchbury, J.S. and Hauser, S.L., 1997, Characterization of the TCRB chain repertoire in the New World monkey *Callithrix jacchus*. *Journal of Immunology*, **158**, 1201–1207.

Unanue, E.R. and Allen, P.M., 1987, The basis for the immunoregulatory role of macrophages and other accessory cells. *Science*, **236**, 551–557.

van de Wal, Y., Kooy, Y.M.C., Drijfhout, J.W., Amons, R. and Koning, F., 1996, Peptide binding characteristics of the coeliac disease-associated DQ(α1*0501, β1*0201) molecule. *Immunogenetics*, **44**, 246–253.

Vartdal, F., Johansen, B.H., Friede, T., Thorpe, C.J., Stevanovic, S., Ericksen, J.E., Sletten, K., Thorsby, E., Rammensee, H.-G. and Sollid, L.M., 1996, The peptide binding motif of the disease associated HLA-DQ (α1*0501, β1*0201) molecule. *European Journal of Immunology*, **26**, 2764–2772.

Ward, S.G., Bacon, K. and Westwick, J., 1998, Chemokines and T lymphocytes: more than an attraction. *Immunity*, **9**, 1–11.

Watterson, G.A., 1978, The homozygosity test of neutrality. *Genetics*, **88**, 405–417.

Wei, S., Charmley, P., Robinson, M.A. and Concannon, P., 1994, The extent of the human germline T-cell receptor V beta gene segment repertoire. *Immunogenetics*, **40**, 27–36.

Wei, S., Charmley, P., Birchfield, R.I. and Concannon, P., 1995, Human T-cell receptor BV gene polymorphism and multiple sclerosis. *American Journal of Human Genetics*, **56**, 963–969.

Wilson, D.A., Golding, A.B., Smith, R.A., Dafashy, T., Nelson, J., Smith, L., Carlo, D.J., Brostoff, S.W. and Gold, D.P., 1997, Results of a phase I clinical trial of a T-cell receptor peptide vaccine in patients with multiple sclerosis. I. Analysis of T-cell receptor utilization in CSF cell populations. *Journal of Neuroimmunology*, **76**, 15–28.

Wilson, J.D.K., Ogg, G.S., Alle, R.L., Goulder, P.J.R., Kelleher, A., Sewell, A.K., O'Callaghan, C.A., Rowland-Jones, S.L., Callan, M.F.C. and McMichael, A.J., 1998, Oligoclonal expansion of CD8+ T cells in chronic HIV infection are antigen specific. *Journal of Experimental Medicine*, **188**, 785–790.

Winkler, C., Modi, W., Smith, M.W., Nelson, G.W., Wu, X., Carrington, M., Dean, M., Honjo, T., Tashiro, K., Yabe, D., Buchbinder, S., Vittinghoff, E., Goedert, J.J., O'Brien, T.R., Jacobson, L.P., Detels, R., Donfield, S., Willoughby, A., Gomperts, E., Vlahov, D., Phair, J. and O'Brien, S., 1998, Genetic restriction of AIDS pathogenesis by an SDF-1 chemokine gene variant. *Science*, **279**, 389–393.

Wood, N.W., Sawcer, S.J., Kellar-Wood, H.F., Holmans, P., Clayton, D., Robertson, N. and Compston, D.A., 1995, The T-cell receptor beta locus and susceptibility to multiple sclerosis. *Neurology*, **45**, 1859–1863.

Yamagata, K., Nakajima, H., Hanafusa, T., Noguchi, T., Miyazaki, A., Miyagawa, J., Sada, M., Amemiya, H., Tanaka, T., Kono, N. and Tarui, D., 1989, Aspartic acid at position 57 of DQ beta chain does not protect against type 1 (insulin-dependent) diabetes mellitus in Japanese subjects. *Diabetologia*, **32**, 762–764.

Yednock, T.A., Cannon, C., Fritz, L., Sanchez-Madrid, F., Steinman, L. and Karin, N., 1992, Prevention of experimental autoimmune encephalomyelitis by antibodies against 4b1 integrin. *Nature*, **356**, 63–66.

Zack-Howard, O.M., Ben-Baruch, A. and Oppenheim, J.J., 1996, Chemokines: progress toward identifying molecular targets for therapeutic agents. *Trends in Biotechnology*, **14**, 46–51.

Zhao, T.M., Whitaker, S.E. and Robinson, M.A., 1994, A genetically determined insertion/deletion related polymorphism in human TCRB includes functional variable gene segments. *Journal of Experimental Medicine*, **180**, 1405–1414.

Zheng, B., Xue, W. and Kelso, G., 1994, Locus specific somatic hypermutation in germinal center T cells. *Nature*, **372**, 556.

Zhong, F., McCombs, C.C., Olson, J.M., Elston, R.C., Stevens, F.M., McCarthy, C.F. and Michalski, J.P., 1996, An autosomal screen for genes that predispose to celiac disease in the western counties of Ireland. *Nature Genetics*, **14**, 329–333.

Zhou, M., Sayad, A., Simmons, W.A., Jones, R.C., Maika, S.D., Satumtira, N., Dorris, M.L., Gaskell, S.J., Bordoli, R.S., Sartor R.B., Slaughter, C.A., Richardson, J.A., Hammer, R.E. and Taurog, J.D., 1998, The specificity of peptides bound to human histocompatibility leukocyte antigen (HLA)-B27 influences the prevalence of arthritis in HLA-B27 transgenic rats. *Journal of Experimental Medicine*, **188**, 877–886.

Zou, Y.-R., Kottmann, A.H., Kuroda, M., Taniuchi, I. and Littman, D.R., 1998, Function of the chemokine receptor CXCR4 in haematopoiesis and in cerebellar development. *Nature*, **393**, 595–599.

8

POLYMORPHISMS IN HEART DISEASE

R. Mark Payne

8.1 Background

The rapid explosion of biochemical techniques has lent itself well to the molecular analysis of diseases in large populations. Genes associated with cardiovascular disease were among the first to be explored and with good reason: atherosclerotic heart disease remains the prime cause of morbidity and mortality in all advanced cultures today. Combined with this has been the more recent identification and understanding of familial transmission of heart disease of all types, and how polymorphisms causing or associated with many of these diseases can remain in a population. Many gene defects associated with arrhythmias, cardiac structure, cardiac contractile function, and cardiac vascular supply have been identified and become vitally important for two reasons: (1) identification and counseling with regard to risk factors; and (2) new therapies and interventions are based on understanding the molecular mechanism of the defects. This chapter will thus focus on recent advances in understanding cardiovascular disease and on the known single-gene defects responsible for these diseases. The author acknowledges that many of the more recent gene defects not explored here will no doubt become important with time.

Multiple polymorphic loci have been linked directly to heart disease, or to a susceptibility to heart disease. Limited phenotypic response of the cardiovascular system accounts for similar pathology from different gene defects, and many of these polymorphisms have variable phenotypic expression in other systems apart from the heart. These single-gene defects fall into one of four broad classes based on their phenotypic expression, including polymorphisms associated with: (1) coronary artery disease such as lipid disorders and hypertension; (2) defects in cardiovascular structure; (3) cardiomyopathy; and (4) defects in cardiac rhythm. Although environmental factors influence the expression of many of these diseases, it is clear that single-gene defects have been identified which cause heart disease.

8.2 Polymorphisms associated with coronary artery disease

Many polymorphisms associated with diseases of the cardiovascular system may have implications for other organ systems as well. Such diverse associations include the renin–angiotensin system, the coagulation/fibrinolysis genes, lipid disorders, and hypertension. Major polymorphisms associated with each condition will be discussed and related to other genes.

8.2.1 Polymorphisms of the angiotensin I converting enzyme

The renin–angiotensin system in humans is a cascade of enzymatic reactions that participate in cardiovascular homeostasis. Renin is an aspartyl protease produced in the kidney that converts angiotensinogen, an α_2-globulin synthesized in the liver, to angiotensin I. Angiotensin-converting enzyme (ACE) is a transmembrane, zinc-based metallopeptidase anchored in plasma membranes of endothelial cells and is responsible for cleaving two amino acid residues from the carboxy-terminus of angiotensin I to produce the highly vasoactive octapeptide, angiotensin II (Erdos and Skidgel, 1987). ACE also sequentially cleaves two amino acids from bradykinin, which suppresses the vasodepressor action of this peptide. Angiotensin II is a potent cardiovascular agent that increases heart rate, blood pressure, and contractility. It also stimulates the adrenal cortex to produce aldosterone, which increases extra-cellular volume and sodium retention. Angiotensin II has growth-promoting effects in multiple tissues by increasing expression of proto-oncogenes (*c-myc*, *c-fos*, and *c-jun*) and other growth factors (Takahashi *et al.*, 1996; Berk and Corson, 1997). In heart and vascular smooth muscle cells, angiotensin II binds to the angiotensin II type 1 receptor initiating a cascade of secondary messenger systems (Dzau, 1988). In particular, angiotensin II activates a number of vascular growth factors including those associated with proliferation pathways, such as basic fibroblast growth factor and platelet-derived growth factor, and those associated with antiproliferative path-ways, such as transforming growth factor β-1 (Itoh *et al.*, 1990, 1993). This key position of the renin–angiotensin and kallikrein–kinin systems in maintaining cardiovascular homeostasis has long suggested that these systems could be involved in the pathophysiology of coronary heart disease and an extensive search has been made to link them to heart disease.

In 1990, Rigat *et al.* described an insertion/deletion polymorphism in the ACE gene that accounted for roughly half of the serum ACE variation in a normal popu-lation (Rigat *et al.*, 1990). This polymorphism is a 287 bp *Alu* repeat sequence in intron 16 of the ACE gene and is inherited as either inserted (I) or deleted (D). Thus, individuals homozygous for the insertion polymorphism are designated as II, homozygous deletions are designated DD, and heterozygosity is designated by the ID genotype. The *Alu* repeat sequence is extremely common in the human genome, is believed to be derived from the 7SL RNA gene, and the 5′ half contains an RNA polymerase III promoter that can be active. The ACE polymorphism follows Mendelian characteristics. By measuring serum ACE levels, Rigat *et al.* (1990) showed that the II, ID, and DD genotypes were associated with ACE levels of 299, 393, and 494 μg/l, respectively, thus establishing strong linkage of the I/D poly-morphism with the gene locus controlling ACE expression.

Large clinical trials have clearly shown a beneficial effect of ACE inhibitor drugs in patients with heart failure (The SOLVD Investigators, 1991) and after myocar-dial infarction (The Acute Infarction Ramipril Efficacy (AIRE) Study Investigators, 1993; Pfeffer *et al.*, 1992), stimulating intense interest in the key role of the renin–angiotensin system in the therapy of heart disease. Furthermore, the ACE genotype has been shown to correlate with the effect of ACE inhibitor drugs. Ueda *et al.* (1998) demonstrated that the effect of enalaprilat, an ACE inhibitor, was greater and lasted longer in normotensive men homozygous for the ACE II genotype. More difficult, however, has been determining linkage of the ACE polymorphism with cardiovascular disease. The ability of angiotensin II to stimulate cardiac musle growth, as well as vascular smooth muscle growth and migration, has been demonstrated (Dominiczak *et al.*, 1996) but large population studies have shown conflicting results regarding the association of the ACE polymorphism with heart failure, infarction, or hypertension (Cambien *et al.*, 1992; Wang *et al.*, 1996; Butler *et al.*, 1997). Recently, two large, independent studies have reported linkage of the ACE locus and blood pressure (Fornage *et al.*, 1998; O'Donnell *et al.*, 1998). Both studies used the same markers, a highly polymorphic and complex tandem repeat marker located on the nearby growth hormone gene (*GH*), for which no recombination has been seen with the ACE gene. The study by O'Donnell *et al.* used sibpairs to test for linkage by relating the qualitative or quantitative trait differences to genotype resemblance. Fornage *et al.* used a different approach to quantify the contribution of genetic variation in the region of four genes, including the ACE gene, to variations in blood pressure. They found that the ACE gene itself influences blood pressure in a sex-specific manner. O'Donnell's group also found that the ACE locus may be a sex-specific candidate gene for hypertension. Although the statistical significance of both studies is not robust, studies drawn from the rat model of hereditary hypertension, linkage studies of normotension and hypertension in humans, and human monogenetic hypertension (type II pseudohypoaldosteronism) (Mansfield *et al.*, 1997) strongly point to the ACE locus on human chromosome 17 as a hypertension gene.

Thus, multiple studies have shown an association between the ACE (I/D) genotype and angiotensin II levels in humans. Linkage of the ACE locus with disease is less strong and may reflect a complex interaction of environment and individual genotypes. Many of the studies are compromised by population bias and unmeasured confounding variables. This is particularly true, for example, of studies that examine the ACE polymorphism and stroke, a clinical scenario in which the genotype should logically be expected to be associated with a disease. Most of these studies are retrospective and small. However, a recent large, prospective study failed to show an association between the ACE D/I polymorphism and stroke (Zee *et al.*, 1999). Because the ACE locus may also interact with other genes to affect cardiovascular outcome, such as with diabetes (Baron, 1994; Ruiz *et al.*, 1994), the possibility of multiple genes interacting must be considered in the association of the ACE DD genotype with a disease. However, taken together, these studies strongly support the notion that the ACE DD genotype influences cardiovascular disease.

8.2.2 *Paraoxonase gene polymorphisms*

Human serum paraoxonase (PON1) is a Ca^{2+}-dependent, 45-kDa glycoprotein bound to high-density lipoprotein (HDL) that is capable of hydrolyzing lipid peroxides. It is believed to be partially responsible for the protective effect of HDL against low-density lipoprotein (LDL) oxidation. It also hydrolyzes organophosphate insecticides and nerve gases and is responsible for determining the selective toxicity of these compounds in mammals (Mackness *et al.*, 1998). Of more widespread interest, however, has been the association of polymorphisms in the PON1 gene with heart disease.

Two genetic polymorphisms resulting in amino acid substitutions in the PON1 gene are responsible for determining the activity of the paraoxonase product. The first is a glutamine to arginine substitution at position 192 of PON1. The genotype is termed Q (low activity towards paraoxon) when there is a glutamine at 192, and R (high activity towards paraoxon) when there is an arginine at position 192. These substitutions are also referred to as the A allele, and B allele, respectively, and are the major determinant of the PON1 activity polymorphism. A second polymorphism occurs at amino acid 55 and substitutes a methionine (M genotype) for leucine (L genotype). This amino acid substitution at position 55 modulates paraoxon activity independently of the 192 substitution, with serum from individuals homozygous for QQ/MM having lowest paraoxonase activity, and RR/LL homozygotes having the highest activity (Mackness *et al.*, 1997).

Mackness reported in 1991 that paraoxonase inhibited the copper-catalyzed oxidation of LDL, thus suggesting a molecular mechanism whereby HDL might affect LDL oxidation (Mackness *et al.*, 1991a). Serum levels of paraoxonase were also found to be lower in patients with familial hypercholesterolemia and diabetes: these patients are at greatly increased risk for premature atherosclerotic vascular disease (Mackness *et al.*, 1991b). To examine the hypothesis that the enzyme may alter the risk for vascular disease by cleansing oxidized lipids from LDL *in vivo*, multiple studies have looked at the 192 substitution as it relates to coronary artery disease (Ruiz *et al.*, 1995; Odawara *et al.*, 1997; Sanghera *et al.*, 1997). To summarize these six published studies, three studies have reported no association with congestive heart disease (two European and one Japanese), and three have reported a positive association with heart disease (two European and one Japanese). Using one-way analysis of variance, the paraoxonase genotype was found to be associated with cholesterol levels in men (Rice *et al.*, 1997), and racial differences in the frequency of specific genotypes have been observed, such as the Arg192 substitution (B allele) being the dominant isoform among Japanese individuals (Sanghera *et al.*, 1997; Yamasaki *et al.*, 1997). Compounding the difficulty of linking the 192 polymorphism with coronary artery disease is the presence of the second polymorphism, Met→Leu at codon 55 of the PON1 gene. The 55 polymorphism was found to be highly significant for plasma concentrations and activities of paraoxonase between genotypes, LL being highest and MM lowest, and linkage disequilibrium ($P < 0.0001$) was found between the two mutations at 192 and 55 (Garin *et al.*, 1997; Mackness *et al.*, 1997). Leviev *et al.* (1997) found that the two alleles of PON1 produced differing amounts of mRNA, depending on their genotype at the 55 codon. L/L carriers had significantly higher amounts of paraoxonase than M/M allele carriers. In a large study of Asian Indians and Chinese (Sanghera *et al.*, 1998a), the less common codon 55/L was significantly higher in frequency in Indians than Chinese, and yet there was no association of the 55 polymorphism with heart disease risk or with plasma lipoprotein–lipid variation between groups. The authors concluded that the B allele of the codon 192 polymorphism was an independent risk factor for heart disease in Indians.

Finally, two other PON-like genes have been found based on homology to PON1, designated PON2 and PON3, and are linked on chromosome 7 with PON1 (Primo-Parmo *et al.*, 1996). Their function and protein products are unknown. Sanghera and colleagues (1998b) have reported a common polymorphism on the PON2 gene, Cys→Ser at codon 311, associated with an increased risk of heart disease in Asian Indians. However, they also found that the 192 polymorphism (B allele of PON1) and the 311 polymorphism (311Ser) were synergistic in contributing to the heart disease risk in this population and that this risk was independent of the lipid profile. A second polymorphism has also been found on the PON2 gene at codon 148 (Ala→Gly) and is associated with elevated fasting plasma glucose in subjects with non-insulin dependent diabetes (NIDDM) (Hegele *et al.*, 1997). Hegele *et al.* observed that in subjects with NIDDM, the PON2 codon 148 G/G homozygotes had a significantly higher fasting plasma glucose level than in subjects with 148 A/G or A/A. This suggests that PON2 may have a different physiological role than PON1, and may also contribute to heart disease based on the association of diabetes and vasculopathy.

In summary, the discovery of the PON gene family and their polymorphisms has added a new piece to the complex puzzle of heart disease. PON1 polymorphisms appear to be associated with an increased risk of developing coronary artery disease, especially in Caucasian men, but this association is linked to race as well as the other PON genes. Although a number of other genetic influences are likely to affect risk and have yet to be examined, such as the hepatic lipase gene (HDL cholesterol) and the methylenetetrahydrofolate reductase gene (homocysteine levels), it is clear that the PON gene family represents a logical and important direction of investigations for both assessing, and altering, the risk of heart disease.

8.2.3 Polymorphisms of coagulation and fibrinolysis genes

The coagulation system is important in the maintenance of normal hemostasis, and is also a key element in the generation of coronary artery disease and cerebrovascular events (Table 8.1). It represents a necessarily complex balance between those genes controlling an evolutionarily conserved mechanism for hemostasis from an injury, and the more recent development of long life-span. For example, it is now widely accepted that blood-clot formation in coronary arteries is a key event in the development of myocardial infarction. Thus, minor differences in the genes of the clotting cascade might be expected to exert a significant influence when viewed over decades of life-span and multiple generations. Most work here has examined the genes coding for plasminogen activator inhibitor (PAI-1), factor VII of the coagulation cascade, and fibrinogen. Common polymorphisms have been described in all three genes that relate to phenotypic expression of the coded protein and indicate a heritable transmission.

8.2.3.1 Plasminogen activator inhibitor-1 (PAI-1)

Plasminogen activator inhibitor-1 (PAI-1) is a serine protease with an estimated molecular weight of 50 kDa. It plays an important role in the regulation of the fibrinolytic enzyme system and thus is a key component in regulation of

Table 8.1 Components of the coagulation system[a]

	Site of synthesis	Mass (kDa)/ #Peptides	Chromosome	Gene size/ Intron	Polymorphism/ Disease association
Coagulant protein					
Fibrinogen (factor I)	Hepatocyte	333/6	4q31	50 kb 3 genes: α, β, γ	β-gene has multiple polymorphisms associated with CAD
Prothrombin (factor II)	Hepatocyte	69/1	11p11–q12	21 kb/13	G/A at 20210 nt (3′-UTR), ↑ CAD
Factor V	Hepatocyte	330/1	1q21–25	>80 kb/24	FV Q506 mutation, ↑ thrombophilia
Factor VII	Hepatocyte	55/1	13q34	2.8 kb/7	Multiple polymorphisms in promoter and coding region, ↑ CAD
Factor VIII	Hepatocyte	330/2	Xq28	186 kb/25	Microsatellite markers in introns 13 and 22 linked to hemophilia
Factor IX	Hepatocyte	56/2	Xq27	34 kb/7	No significant polymorphisms
Factor X	Hepatocyte	55/2	13q34	25 kb/7	No significant polymorphisms
Factor XI	Hepatocyte	160/2	4q35	25 kb/14	No significant polymorphisms
Factor XII	Hepatocyte	80/1	5q33–qter	12 kb/13	No significant polymorphisms
Factor XIII	Hepatocyte/ megakaryocyte	330/4 (2 pairs not identical)	A subunit: 6p24 B subunit: 1q31	160 kb/14 28 kb/11	G→A (Val34Leu) in exon 2, ↑ CAD
Tissue factor	Many cell types	45/1	1p21–22	12 kb/5	No significant polymorphisms
von Willebrand's factor	Endothelial megakaryocytes	Variable/ Variable	12p2.1	180 kb/51	RFLP in exon 12 assoc. with NIDD
Coagulant inhibitors					
Antithrombin	Hepatocyte	65/1	1q23–25	19 kb/5	No significant polymorphisms
Protein C	Hepatocyte	62/1 or 2	2q13–14	11 kb/8	No significant polymorphisms
Protein S	Hepatocyte/ endothelium	69/1	3p11.1–11.2	50 kb/14	No significant polymorphisms
Thrombomodulin	Endothelium	74/1	20p11.2	4 kb/none	Dinucleotide mutation in promoter ↑ CAD. Neutral polymorphism at Ala455Val
Tissue factor pathway inhibitor	Hepatocyte/ endothelium	40/1	2q31–q32	85 kb/8	No significant polymorphisms

[a]Chromosome location and markers as per Koeleman *et al.* (1997). CAD, coronary artery disease.

thrombosis. It is a rapid inhibitor of tissue-type plasminogen activator (tPA) and, in conditions where serum levels of PAI-1 are increased, the half-life of functional tPA is substantially lowered, causing a decrease in the fibrinolytic potential. PAI-1 has been shown to be an independent risk factor for myocardial infarction at a young age. The PAI-1 gene is located on chromosome 7 (q21.3–q22), contains nine exons across 12.3 kb, and encodes two distinct PAI-1 mRNAs with lengths of 2.3 and 3.2 kb (Strandberg *et al.*, 1988). It is in close linkage with the cystic fibrosis gene, the paraoxonase gene, and the erythropoietin gene. Interestingly, paraoxonase is also a candidate gene for cardiovascular disease (see above), and the PAI-1 gene is strongly related to dyslipidemia. Three polymorphisms have been described in the PAI-1 gene that correlate significantly with the circulating levels of PAI-1. These include an intronic $(CA)_n$ dinucleotide repeat polymorphism (Dawson *et al.*, 1991), a *Hind*III polymorphism (Klinger *et al.*, 1987), and a sequence length polymorphism (4G/5G) located in the promoter region (Dawson *et al.*, 1993).

Polymorphisms in the PAI-1 gene seem to be important in the determination of serum levels of PAI-1. For example, genotyping the *Hind*III and $(CA)_n$ dinucleotide repeat in young post-infarction patients and comparing them with healthy controls showed that PAI-1 activity correlated with the polymorphism (Dawson *et al.*, 1991). Higher PAI-1 levels were associated with the 1/1 genotype of the *Hind*III polymorphism than the 2/2 genotype in both patients and controls. Higher plasma levels of PAI-1 were also associated with the smaller alleles of the eight allele dinucleotide repeat polymorphism. Finally, a guanosine single base-pair insertion (GACACGTGGGG(G)AGT) in the PAI-1 promoter has been identified that is associated with plasma PAI-1 activity. Homozygosity for the 4(G) allele is associated with higher PAI-1 activity than individuals with either homozygous or heterozygous 5G alleles (Dawson *et al.*, 1993). This effect may arise because the 5G allele seems to contain an additional binding site for a DNA-binding protein that may function to decrease transcription of the PAI-1 gene, resulting in lower serum levels of PAI-1. Possession of the 4G/4G genotype is associated with higher PAI-1 levels, lower fibrinolytic activity, and a past history of myocardial infarction (Grant, 1997).

8.2.3.2   *Tissue-type plasminogen activator (tPA)*

Plasminogen activators convert the inactive proenzyme, plasminogen, into the active enzyme, plasmin, involved with fibrinolysis, tissue remodeling, and cell migration. There are two distinct types of plasminogen activators: urokinase, located on chromosome 10q24; and tissue-type plasminogen activator (tPA). tPA is the main endogenous fibrinolytic enzyme and has the highest affinity for fibrin. Acute plaque rupture with subsequent exposure of blood components to procoagulant components in the vessel wall is considered a primary event in the thrombotic pathway leading to myocardial infarction. Recombinant tPA is currently used to lyse these thrombotic lesions in both stroke and myocardial infarction. Thus, there is much interest in the plasma levels of endogenous tPA and the risk of acute MI.

The tPA gene is located in the pericentromeric region of chromosome 8 (8p12–p11.2) and is composed of 14 exons across approximately 37 000 bp (Degen *et al.*, 1986; Yang-Feng *et al.*, 1986). Its location coincides with a translocation breakpoint found in myeloproliferative disorders. Acute myeloid leukemic cells that secrete only tPA are associated with chemotherapy failure, whereas other types of leukemic cells usually secrete both types of plasminogen activators and are sensitive to chemotherapy. The gene and 5′ flanking region contain 28 copies of *Alu* repetitive DNA and a single *Kpn*I repeat sequence.

Several studies have examined polymorphisms of the tPA gene relative to clinical condition, i.e. myocardial infarction (MI). A second *Alu* repeat sequence is found in the eighth intron as a dimorphic insertion (I)/deletion (D) polymorphism similar to that found in the ACE gene (Ludwig *et al.*, 1992). This *Alu* repeat sequence is a unique and highly stable polymorphic event in humans that has significant allelic frequency variation among human populations, making it particularly good for studies of genetic drift during human evolution. However, this *Alu* I/D polymorphism is also responsible for the *Eco*RI, *Pst*I, *Hinc*II, *Taq*I, and *Xmn*I restriction fragment length polymorphisms (RFLPs) found with the tPA gene, which means that these RFLPs are detecting the same polymorphism (Tishoff *et al.*, 1996). Thus, various data sets using these RFLPs can be combined. The first large population study of this Alu repeat sequence found that the genotypes I/I or I/D were associated with a 2.2-fold increase in risk of non-fatal MI when compared with the D/D genotype, and that this insertion allele was independent of plasma concentrations of tPA or other known risk factors (van der Bom *et al.*, 1997). As a cross-sectional study, however, these results were limited to cases of non-fatal MI. A longitudinal study would be needed to determine whether the I/D polymorphism is a marker for fatal MI. In a similar study, Steeds *et al.* (1998) used a substantially larger cohort of MI patients, recruited at the time of diagnosis, to examine the I/D polymorphism. In contrast to the van der Bom study, they were unable to show an association between the insertion allele and MI. However, this study differs from the first study in the diagnosis and method of recruitment, age, and ethnicity of the patients, making it difficult to compare the results. As with the ACE gene polymorphisms, these results may also underscore the powerful influence of other risk factors that have been poorly characterized or understood, and again emphasize the need for large prospective data/sample collection strategies that allow analysis of multiple variables.

8.2.3.3 *Platelet glycoprotein IIIa*

Platelets have a central role in the pathogenesis of thromboembolic disease. Platelet aggregation under normal circumstances requires an intact fibrinogen receptor. This receptor consists of two glycoproteins, GPIIb and GPIIIa, forming a heterodimer that functions as a receptor for fibrinogen, von Willebrand factor, and fibronectin on activated platelets. It is dysfunctional in the bleeding disorder Glanzmann's thrombasthenia (Jin *et al.*, 1996). Point mutations in the genes that encode GPIIb and GPIIIa result in other disorders of platelet binding, and recent evidence suggests that some of these mutations may represent independent risk factors for heart disease (Nurden, 1997). The genes for both GPIIb and GPIIIa are located on chromosome 17. GPIIIa contains 14 exons across 46 kb of gene sequence and encodes an 84.5-kDa glycoprotein. The gene for GPIIb is more complex and consists of 30 exons across 17.2 kb. This gene encodes a continuous reading frame for 1039 amino acids which forms two unequally sized disulfide-linked subunits to create the GPIIb glycoprotein.

Formation of a platelet-rich thrombus at the site of a fissured atherosclerotic plaque is a primary cause of myocardial ischemia and infarction. An amino acid substitution of proline for leucine-33 in GPIIIa, termed PlA2, was initially reported to be associated with a twofold increase in the risk of coronary ischemia (Weiss *et al.*, 1996). Following up on this, Walter *et al.* (1997) found that among men having coronary stents placed for treatment of coronary artery disease, those patients having the PlA2 allele had a significantly higher risk of occlusion of the stent following the procedure. Other investigators, however, have not found an association between the PlA2 polymorphism of the GPIIIa gene and an increased risk of heart disease (Ridker *et al.*, 1997). A valid criticism of these studies is that the populations, controls, and interventions, e.g., aspirin therapy, are not the same, making comparisons between studies difficult. Because this polymorphism is a relatively new finding, and because of its importance as a risk factor that may be treatable in coronary artery disease, large multicenter studies have been called for to investigate polymorphisms of both genes, as well as establish factors that may influence expression of these genes, such as receptor density on the platelet surface.

8.2.4 Lipid disorders

Risk factors for heart disease have been extensively investigated for many years and have been reviewed recently (Gensini *et al.*, 1998). Except for age, dyslipidemia is the most important predictive factor for coronary artery disease. Although the specific contributions of each are often arguable, it is clear that both genetic and environmental factors play a part in determining the risk profile for coronary artery disease, and the reader is referred to more extensive reviews for an in-depth discussion of this interaction (Pasternak *et al.*, 1996; Murata *et al.*, 1998). However, it is also clear that the lipid profile is a powerful predictor of coronary artery disease with a strong, independent, and positive association between cholesterol levels and risk of coronary events. For example, a 1% increase in the LDL cholesterol level may lead to a 2–3% increase in coronary artery disease risk. Many genes control lipid levels, and polymorphisms within these genes and their products are extremely common. Polymorphisms associated with the LDL receptors, apolipoproteins E and A, and the lipoprotein lipase gene will be discussed.

8.2.4.1 Low-density lipoprotein (LDL) receptor

The function of the LDL receptor is to bind and remove low-density lipoproteins, the major carriers of cholesterol, from the circulation. Familial hypercholesterolemia is one of the most common single-gene disorders, affecting approximately 1 in 500 people, and is characterized by reduced or absent levels of LDL receptors in the liver (Nicholls *et al.*, 1998). This causes a marked rise in circulating cholesterol in the heterozygous state, and individuals homozygous for the LDL receptor gene defect, approximately 1 in 1 000 000 individuals in North America, have extraordinarily high levels of cholesterol and exhibit early death from coronary artery disease. The LDL gene is located on chromosome 19p13.2, and consists of 18 exons. Well over 300 mutations have been described in this area, including point mutations and large deletions, which cause marked elevations in the cholesterol levels of affected individuals (Day *et al.*, 1997; Soutar *et al.*, 1998). In addition, a similar condition may be caused by a mutation in the ligand for the LDL receptor, apolipoprotein B. This usually consists of a single amino acid substitution ARG^{3500} to GLN and requires genetic screening to identify familial defective apolipoprotein B from familial hypercholesterolemia (Tybjaerg-Hansen and Humphries, 1992).

8.2.4.2 *Apolipoprotein E (apoE)*

Apolipoprotein E (apoE) plays a central role in the metabolism of cholesterol and triglycerides. ApoE is a polypeptide comprising 229 amino acids found in chylomicrons, very low-density lipoproteins (VLDLs), and high-density lipoproteins (HDLs). In these particles, apoE serves as a ligand for their uptake by lipoprotein receptors. The gene is relatively small and consists of four exons across 3600 bp on chromosome 19q13.2, and is closely linked to the apolipoprotein CI/CII genes (de Knijff and Havekes, 1996). The structural gene for apoE is polymorphic with three major isoforms, apoE$_2$, apoE$_3$, and apoE$_4$, differing in isoelectric point by one charge unit based on one amino acid substitution: apoE$_4$ is the most basic and apoE$_2$ is the most acidic. Synthesis of these three common isoforms is under the control of three independent codominant alleles, APOE*2, APOE*3, and APOE*4, and accounts for six common phenotypes (Contois *et al.*, 1996). These isoforms interact differently with specific lipoprotein receptors thereby altering circulating lipid levels. For example, the APOE*2 allele is associated with lower LDL cholesterol and higher triglyceride concentrations, whereas the APOE*4 allele is associated with elevated total cholesterol and LDL cholesterol levels. It is estimated that as much as 60% of the variation in cholesterol level is genetically determined, and that apoE polymorphism accounts for 15% of this variation. Two prospective studies have provided convincing evidence that risk of coronary artery disease is allele-specific, with the APOE*4 allele conferring higher risk of myocardial infarction in men (Wilson *et al.*, 1994; Stengard *et al.*, 1995). Specific apoE isoforms have also been linked to type III hyperlipoproteinemia, stroke, and Alzheimer's disease.

8.2.4.3 *Apolipoprotein A (apoA)*

Lipoprotein A is a dominantly inherited lipoprotein-like particle consisting of a low-density lipoprotein molecule in which the apoprotein apolipoprotein B 100 is linked to another large protein, apolipoprotein A (apoA). It is not bound by the LDL receptor and analysis of the protein and cDNA for apoA reveals a high degree of homology to plasminogen. The apoA gene is located on chromosome 6q2.6–q2.7 and more than 30 alleles at this highly polymorphic locus determine a size polymorphism for the protein apoA. High levels of lipoprotein A have been established as a risk factor for atherosclerosis, and this risk factor is strongly controlled by the apolipoprotein A gene (Cressman *et al.*, 1992; Sigurdsson *et al.*, 1992). Approximately 45–70% of the wide variability of the lipoprotein A plasma concentrations in Caucasian populations can be explained by the apoA size polymorphisms (Kronenberg *et al.*, 1996). Several prospective studies have indicated high lipoprotein A levels as an independent risk factor for heart disease (Sigurdsson *et al.*, 1992). Furthermore, there is an association between low-molecular-weight phenotypes of apoA protein in the lipoprotein A and coronary artery disease, as well as cerebrovascular disease and end-stage renal disease. Future studies will need to include a clear assessment of the apoA phenotype along with measurements of the lipoprotein A phenotypes, since both are factors in atherogenesis.

8.2.4.4 *Lipoprotein lipase (LPL)*

This enzyme has a central role in lipid metabolism and is an obvious candidate for contributing to the risk of atherosclerotic disease. LPL activity is highly regulated by both insulin levels and serum lipid concentrations, and functions to hydrolyze triacylglycerol in chylomicrons and VLDL with apolipoprotein CII as an essential cofactor (Fisher *et al.*, 1997). The gene for LPL is located on chromosome 8p22, and spans about 30 kb with 10 exons encoding a 475 amino acid glycoprotein of approximately 50 kDa in size. Several polymorphic variants have been reported within the non-coding regions but they appear to have little functional role or association with disease (Jemaa *et al.*, 1995). Within the coding region of the LPL gene, three common amino acid polymorphisms have been described: N291S (A→G at bp 1127), S447X (C→G at bp 1595), and D9N (G→A at bp 280). The N291S and D9N polymorphisms both occur at a carrier frequency of about 5% and have been shown to be associated with increased plasma triacylglycerol and decreased HDL cholesterol levels. These effects seem to be more pronounced in obesity and are associated with increased risk of heart disease (Mailly *et al.*, 1996). However, since both variants are found in normolipidemic individuals from the general population, it is unlikely that these polymorphisms cause dyslipidemia. The S447X polymorphism is very common, with a carrier frequency of about 20%, and the C→T conversion results in a premature stop codon at Ser447 which shortens the LPL protein by two amino acids. It is associated with a more favorable lipid profile and carriers of the S447X phenotype in the general population have higher apoAI levels, lower VLDL cholesterol, and lower triglyceride levels (Zhang *et al.*, 1995). A final polymorphism has also been described in the promoter for the LPL gene, a −93T→G transition, which confers increased promoter activity *in vitro*, and is associated with lower plasma triglyceride levels (Hall *et al.*, 1997). This polymorphism is present at a carrier frequency of about 3.4% in Caucasian men, but is much higher at 61% in Afro-Caribbean men and may confer a beneficial advantage to the lipid profile.

In summary, genes regulating the lipid profile and atherogenic risk factors in humans have been identified and expressed. They interact with each other and with the environment in complex and often unpredictable fashion, and tend to be highly polymorphic. Larger prospective population studies will need to include genotyping and DNA banking in order to clearly identify risk mutations that can be used in directing diagnosis and therapy. The potential for impact on human outcome from this research is enormous.

8.3 Polymorphisms associated with structural heart disease

Defects in cardiac structure are extremely common and are among the most frequent birth defects (Hoffman, 1995). An estimated 1.5 million births worldwide, and

600 000 children in the United States, are affected with cardiovascular diseases of which structural defects are the most common (Moller *et al.*, 1993). Ventricular septal defects and patent ductus arteriosus constitute the bulk of these defects. Because of the enormous advances in diagnosis and treatment of these defects in the past three decades, the need for better prenatal and familial genetic diagnostic abilities has never been more acute.

8.3.1 *Marfan's syndrome*

Marfan's syndrome is a pleotrophic connective tissues disease involving skeletal, cardiovascular, and ocular systems, and has an incidence of at least one in 5000. It is transmitted in an autosomal dominant fashion, and premature death occurs because of the cardiovascular anomalies, primarily dilation and rupture of the aortic wall (Pyeritz and McKusick, 1979). A more severe form appears early during neonatal life and has a high mortality because of severe heart disease.

The defective gene that causes Marfan's syndrome is the fibrillin gene (FBN-1), located on chromosome 15q21, and contains 65 exons across its 110 kb span (Sakai *et al.*, 1986). This gene encodes a 350-kDa glycoprotein product that serves as a scaffold for the deposition of tropoelastin, and is a major component of the extracellular 10–14-nm microfibrils which give elastic strength to the vascular walls and heart valves. The fibrillin monomer contains large numbers of epidermal growth factor-like motifs that can bind calcium, and a small number of motifs resembling the binding protein for transforming growth factor β. In 1991, Dietz and colleagues described a single base-pair substitution in FIB-1 resulting in a non-conservative amino acid substitution (Arg→Pro) that caused Marfan's syndrome (Dietz *et al.*, 1991). Because fibrillin is part of a complex of proteins, the mechanism by which the abnormal gene product causes the phenotype of Marfan's syndrome is most likely to be due to a dominant negative effect, that is, the presence of a defective copy of the gene product causes abnormal function in the other normal copies of the same gene product. This also suggests that polymorphisms in seemingly unrelated genes may have an impact on microfibril construction.

Since then, well over 50 polymorphisms have been described in the FBN1 gene and can be classified into one of four categories: (1) truncation of the gene product from deletion of a portion of the gene; (2) exon skipping from a mutation within an exon; (3) mutations within introns that result in abnormal splicing and exon skipping; (4) missense mutations that result in non-conservative amino acid substitutions in the epidermal growth factor-like domains and are predicted to alter calcium binding or disulfide-bond formation (Dietz and Pyeritz, 1995). Single amino acid substitutions constitute the largest group of polymorphisms in the FBN1 gene. All families identified with Marfan's syndrome have had mutations in the FBN1 gene. However, each of these mutations within the FBN1 gene are unique to the affected family, suggesting that *de novo* mutations in this gene are common. Furthermore, the phenotype shows a great deal of variability in expression, and thus the clinical indications for genotyping are used inconsistently. For example, a four-generation family kindred was described with mutations in the FBN1 gene (D1229E) that caused ectopia lentis and skeletal abnormalities, but had no cardiac involvement (Kainulainen *et al.*, 1994). This high degree of polymorphism within the FBN1 gene greatly complicates the task of widespread molecular screening for Marfan's syndrome. However, four highly informative microsatellite polymorphisms can be used to distinguish between the many copies of the FBN1 gene that segregate in any given family. Thus, by determining which allele cosegregates with the affected individual, molecular diagnosis can be achieved for asymptomatic, or equivocally affected, individuals (Pereira *et al.*, 1994).

Other fibrillin genes have been identified which also have polymorphic expression. For example, fibrillin-2 is encoded by FBN2 on chromosome 5q23, and mutations have been associated with congenital contractural arachnodactyly, a Marfan's-like syndrome with autosomal dominant inheritance but without cardiac involvement (Tsipouras *et al.*, 1992). Because at least five other proteins are found in the microfibrils, it is highly probable that mutations in these other genes may be a source of phenotypic changes in the extracellular matrix and connective tissues, and will make the microfibrils highly polymorphic in their expression.

8.3.2 *Supravalvar aortic stenosis*

Obstruction to blood flow from the left side of the heart, particularly the area just above the aortic valve, is rare but has been noted to occur in three settings: (1) in approximately 40% of the cases, a familial pattern with high penetrance and autosomal dominant inheritance is found; (2) sporadic occurrence without any familial associations in 25% of the cases; and (3) as part of a syndrome involving stenoses of major arteries and mental retardation with characteristic facies known as Williams' syndrome (approximately 35%). Williams' syndrome is also usually sporadic. Following two large families with dominantly transmitted supravalvar aortic stenosis, Keating and colleagues linked this disease to the elastin gene located on chromosome 7q11.23 (Curran *et al.*, 1993). In particular, a balanced translocation from chromosome 6 to 7, t(6:7)(p21.1;q11.23), interrupted the elastin gene in exon 28, at the carboxy-terminal coding region of the 3' end of the gene. This truncated elastin gene product lacked sequences necessary for desmosine cross-links, an important disulfide bond, and the microfibril-associated glycoprotein-binding site. Similar to the FBN1 gene mutations, the presence of one altered elastin gene product produced a dominant negative effect on elastin fibril formation.

Knowing that supravalvar aortic stenosis is an important pathology in Williams' syndrome, Ewart *et al.* (1993) explored the hypothesis that monosomy at the elastin gene locus was involved in Williams' syndrome. Both by Southern blotting and fluorescent *in situ* hybridization, they showed that hemizygosity at the elastin locus caused Williams' syndrome and was associated with many families with dominantly inherited supravalvar aortic stenosis. Furthermore, they found that a large deletion of the elastin gene region is usually present in these patients and involves many other unknown genes. This raised the possibility that Williams' syndrome is part of a contiguous-gene syndrome, involving many genes at the elastin locus that may contribute to behavioral, developmental, and cardiac abnormalities of Williams' syndrome. Analysis of more than 100 patients with Williams' syndrome shows deletions at the elastin locus in more than 95% of the patients (Payne *et al.*, 1995), and the cosmid markers for FISH analysis at the 7q11.23 locus are now widely used in cytogenetic screening to allow definitive diagnosis in young infants who may not have developed the obvious clinical features of Williams' syndrome.

8.3.3 *Alagille syndrome*

Alagille syndrome is a dominantly inherited disease with high penetrance and major abnormalities in heart, liver, eye, and vertebrae. Although rare, with an incidence of one in 70 000 live births, it is the second most common cause of intrahepatic cholestasis in infancy. The cardiac defects affect the pulmonary arterial bed with peripheral stenosis. The responsible gene has been identified recently as the human *Jagged 1* (*JAG1*) gene, located on chromosome 20p12 (Li *et al.*, 1997; Oda *et al.*, 1997). *JAG1* encodes a ligand for the transmembrane receptor NOTCH 1, and its 5.5 kb mRNA is encoded by 26 exons. The *JAG1* protein contains 16 epidermal-like growth factor repeat sequences (EGF), a cysteine-rich region, and a highly conserved DSL domain so named because it is conserved in Delta, Serrate, and Lag-2, the ligand for the *C. elegans* NOTCH homologue. Point mutations in the *JAG1* gene cause Alagille syndrome in families, and include frameshifts and non-sense mutations, exon deletions, and truncation (Yuan *et al.*, 1998). Over 15 different mutations have been described among as many probands, making it difficult to identify new cases that are mildly affected. However, the highly polymorphic marker *D20S1154* identifies a CA dinucleotide repeat from intron 19, and should be useful for the detection of submicroscopic deletions (Oda *et al.*, 1997). Activation of the NOTCH family of proteins by ligands, such as *Jagged 1* affects the expression of lineage-specific genes, which leads to inhibition of differentiation along particular pathways. Thus, mutations in the *JAG1* gene affect multiple seemingly unrelated tissues by altering expression of an early gene program, a process most readily recognized as a "field defect".

8.4 Polymorphisms associated with cardiomyopathies

Disorders of cardiac function are an important cause of morbidity and mortality in both children and adults throughout the world. The pathogenesis of many of the primary cardiomyopathies is unknown, whereas most secondary cardiomy-opathies arise from well-understood etiologies, such as hypertension and ischemic heart disease (discussed earlier). Polymorphisms have been associated with both hypertrophic cardiomyopathies, as well as dilated cardiomyopathies, and frequently relate to either disorders of the contractile element of the heart, the sarcomere, or disorders of cardiac energy metabolism, respectively. The true incidence of poly-morphism in cardiomyopathies is difficult to establish because phenotypic expression may be subtle and the inheritance pattern is frequently autosomal recessive. However, estimates of dilated cardiomyopathies range from two to eight per 100 000 in the USA and Europe (Manolio *et al.*, 1992). Although polymorphisms of the mitochondrial genome are an increasingly recognized source of cardiomy-opathies, these defects have been discussed in an earlier chapter and will not be pursued here.

8.4.1 Myocardial energy production

Myocardial energy metabolism is derived primarily through the stepwise mitochondrial β-oxidation of fatty acids to acetyl-CoA (Neely *et al.*, 1972). This process involves a large number of nuclear-encoded genes that are targeted and imported into mitochondria. In the past 15 years defects in more than 15 proteins responsible for transport of fatty acids into mitochondria, or within the fatty acid β-oxidation spiral, have been identified (Largilliere *et al.*, 1995; Sims *et al.*, 1995; Treem *et al.*, 1996). Defects in fatty-acid oxidation are among the most common metabolic diseases and result in inherited cardiac and skeletal myopathy, hypoketotic hypoglycemia, and sudden death in infants (Kelly and Strauss, 1994). The estimated incidence of medium chain acyl-dehydrogenase deficiency (MCAD) may be as high as one in 10 000 live births (Coates, 1992), and fatty-acid oxidation defects may account for as many as 5% of all unexplained infant deaths (Boles *et al.*, 1998). All known errors in fatty-acid oxidation are autosomal recessive defects, and mutations in the genes encoding two members of the acyl-CoA dehydrogenase family, short-chain and medium-chain acyl-CoA dehydrogenase, were among the first to be identified (Naito *et al.*, 1989; Kelly *et al.*, 1990). Investigations from both the Tanaka and Strauss laboratories determined that the molecular basis of MCAD was surprisingly homogeneous: a single substitution of guanine for adenine at position 985 of the mRNA coding region results in substitution of glutamic acid for lysine in the mutant protein. Matsubara and colleagues determined that the carrier state for the $985^{G \to A}$ substitution ranged from 12 per 479 live births in Britain to 5 per 536 newborns in the USA (Matsubara *et al.*, 1991) and that the MCAD locus on chromosome 1 was highly polymorphic, making it informative for linkage mapping (Kidd *et al.*, 1990). This knowledge of such a homogeneous mutation makes molecular diagnostic screening feasible. For example, the presentation of sudden death in an infant would suggest a defect in MCAD and a postmortem molecular screening can rapidly detect the $985^{G \to A}$ substitution. Postmortem identification of the defect would allow screening of asymptomatic siblings and relatives.

Mutations in the gene coding for short-chain acyl-CoA dehydrogenase (SCAD), a mitochondrial flavoenzyme that catalyzes the initial reaction in short-chain fatty-acid β-oxidation, are more heterogeneous, making it more difficult to detect polymorphisms using a rapid molecular screening tool (Naito *et al.*, 1990). The SCAD gene contains 10 exons across 13 kb and is located on the distal part of chromosome 12 (Corydon *et al.*, 1997). Four polymorphisms have been determined in the SCAD mRNA of which three are silent, whereas the $625^{G \to A}$ variant is associated with ethylmalonic aciduria. Cardiomyopathies are less common with defects in either of these genes, and the most common presentation is sudden death in an infant following a fasting situation. Thus, the likelihood is very high that most individuals with these polymorphisms will not experience prolonged starvation during childhood, and these autosomal defects will remain common in the gene pool.

Other polymorphisms in genes responsible for fatty-acid oxidation have been associated with cardiomyopathy, sudden death, acute hepatic encephalopathy, and skeletal myopathy. For example, the very long-chain acyl-CoA dehydrogenase (VLCAD) gene catalyzes the first step in β-oxidation. Multiple mutations have been described of which one, a homozygous mutation in the consensus dinucleotide of the donor splice site ($g^{+1} \rightarrow a$), was associated with skipping of the prior exon (exon 11). A second patient was a compound heterozygote with a missense mutation at $1837^{C \rightarrow T}$ changing an arginine to a tryptophan at residue 613 on one allele, and a single base-pair deletion at the intron–exon 6 boundary on the other allele (Strauss *et al.*, 1995). The gene for VLCAD has been localized to 17p11.2–p11.1305, and typically, both alleles will contain mutations in affected patients (Andresen *et al.*, 1996). Hardly any of the patients have contained identical mutations suggesting that the mutational heterogeneity in VLCAD deficiency is very large. Finally, a complex gene of the β-oxidation pathway for fatty acids has been described containing polymorphisms associated with severe disease in both pregnant women and their affected infants. The long-chain 3-hydroxyacyl-CoA dehydrogenase gene (LCHAD) catalyzes the third step in β-oxidation, and the enzyme activity is present on the C-terminal portion of the α-subunit of the mitochondrial trifunctional protein. Strauss and colleagues used single-stranded conformational variance of the LCHAD exons to determine the molecular basis of three families with children presenting with sudden death or hepatic encephalopathy, and in which the mothers suffered acute fatty liver of pregnancy (Sims *et al.*, 1995). They identified a homozygous mutation at $1528^{G \rightarrow C}$ of the α-subunit of the trifunctional protein on both alleles. A third child was a compound heterozygote with the same mutation on one allele, and a different mutation ($1132^{C \rightarrow T}$) creating a premature stop codon on the second allele. These findings were confirmed in other studies that also localized the gene encoding the α-subunit of the trifunctional protein on chromosome 2p24.1–23.3 (IJlst *et al.*, 1996). Here, Wanders and colleagues found an 87% allele frequency of the $1528^{G \rightarrow C}$ mutation in 34 patients, making this a valuable test for prenatal diagnosis.

In summary, multiple polymorphic loci in the genes for β-oxidation of fatty acids have been associated with disease, primarily cardiovascular, but also having hepatic and neurological phenotypes as well. The heterogeneic pattern of these mutations currently makes it difficult to reliably use these polymorphisms in genetic testing. However, with further definition of these single-gene defects, rapid analysis of both carriers and affected individuals should be possible.

8.4.2 Polymorphisms of contractile proteins

The recent discovery of mutations in the contractile proteins causing cardiomyopathy has led to an explosion in our knowledge of heart disease. Because of this, the genes encoding these proteins are, perhaps, among the best-defined genes of cardiac function. Multiple polymorphisms have been found in the genes of the basic contractile unit, the sarcomere, and are associated with either dilated or hypertrophic cardiomyopathy. Many of these mutations have been expressed in transgenic animals with reproduction of the human disease phenotype (Spindler *et al.*, 1998; Tardiff *et al.*, 1998). Although genetic studies have provided great insight into the heterogeneity of familial cardiomyopathies, it is clear that examination of large population samples must be performed before unambiguous conclusions can be made.

Hypertrophic cardiomyopathy (HCM) has been defined as a primary disease of the cardiac muscle in the absence of any other mitigating factors, such as hypertension, that would predispose to a cardiomyopathy. HCM clearly demonstrates autosomal dominant inheritance, with about 60% of the cases having a familial pattern of inheritance, and 40% being sporadic. Penetrance is somewhat variable but affected individuals have a 50% chance of transmission to their offspring.

Mutations in over seven different genes encoding proteins of the myofibrillar apparatus have been associated with HCM: β-myosin heavy chain gene (*MYH7*), ventricular myosin essential light chain (*MYL3*), cardiac troponin T (*TNNT2*), ventricular myosin regulatory light chain (*MYL2*), cardiac troponin I (*TNNI3*), cardiac myosin-binding protein C (*MYBPC3*), and α-tropomyosin (*TPM1*) (Table 8.2). In addition to the locus heterogeneity of these polymorphisms, allelic heterogeneity is evident because multiple mutations have been found in all of these genes. In a recent comprehensive review of HCM, Bonne *et al.* (1998) identified over 50 mutations in the β-myosin heavy chain alone that have been associated with hypertrophic cardiomyopathy. This gene is located on chromosome 14, is composed of 40 exons that span >32 kb, and encodes a 1935 amino acid protein. It is the major contractile (myosin) isoform of the human ventricle and is, perhaps, the best defined of the contractile genes in the heart. Three functional domains have been defined; a globular amino-terminal head which corresponds to the motor domain containing the ATP-binding site and the actin-binding site, the neck region which consists of a long 8.5-nm α-helix and associates with the light chains, and the carboxy-terminal rod region which is responsible for the assembly of myosin into thick filaments. Mutations of the β-myosin heavy chain tend to cluster in four regions of the

Table 8.2 Known polymorphisms in the proteins of the sarcomere

Protein	*Gene*	*Gene locus*	*Phenotype*
β-Myosin heavy chain (β-MyHC)	MYH7	14q11.2–q12	Highly polymorphic, often very malignant
Cardiac troponin T (cTnT)	TNNT2	1q3	Milder expression
α-Tropomyosin (α-TM)	TPM1	15q22	Milder expression, no sudden death reported
Myosin binding protein C (cMyBP-C)	MYBPC3	11p11.2	Milder expression
Ventricular myosin essential light chain (MLC-1s/v)	MYL3	3p21.2–p21.3	Very mild thickening of left ventricle
Ventricular myosin regulatory light chain (MLC-2s/v)	MYL2	12q23–q24.3	Mild, as well as malignant, expression
α-Myosin cardiac heavy chain	MYH6	14q11–q12	Trinucleotide repeat in flanking region
Cardiac troponin I (cTnI)	TNNI3	19p13.2–q13.2	Variable expression, can be malignant
Unidentified gene on locus			
Unknown gene (MacRae *et al.*, 1995)	Unknown	7q3	Wolff–Parkinson–White and HCM

polypeptide: the actin binding interface, adjacent to the region that connects two reactive cysteine residues, around the ATP binding site, or in proximity to the interface of the heavy chain with the essential light chain. Of these mutations, three "hot spot" mutations were identified at codons 403 (Geisterfer-Lowrance *et al.*, 1990), 719 (Consevage *et al.*, 1994), and 741 (Arai *et al.*, 1995) which are predominantly missense mutations located in the head, or head–rod junction, of the myosin heavy chain. Relatively few mutations are found in the myosin rod region. The *MHY7* gene also contains two polymorphic dinucleotide repeats, one in intron 24 and one in the promoter region, termed MYOII and MYOI, respectively, that allow easy linkage analysis in family studies of HCM (Fougerousse *et al.*, 1992; Schwartz *et al.*, 1992). Defects in the *MYH7* gene have been associated with HCM in 40% of the cases.

The α-myosin heavy chain (α-MHC) is the other cardiac isoform of myosin and is expressed in human atria. It is encoded by the *MYH6* gene, and is in a tandem cluster 4.5 kb upstream from the *MYH7* gene on chromosome 14q11.2–q13. A compound trinucleotide repeat sequence is present between positions 14722 and 14820, $(GAA)_{23}(GAG)_{10}$ (van den Berg *et al.*, 1995). Analysis of 55 unrelated individuals indicated that >93% were heterozygous at this locus, with 17 alleles being found to contain 21 to 50 repeats. Thus, this may become a highly informative polymorphic marker in linkage analysis for familial HCM. Polymorphic loci associated with HCM have also been identified in the α-tropomyosin gene (15q22) (Thierfelder *et al.*, 1993) and cardiac troponin T genes (1q3) (Watkins *et al.*, 1993).

There is considerable variability of disease expression (penetrance) and malignancy in HCM, even within families with the same mutation. For example, the $908^{Leu \rightarrow Val}$ missense mutation, which is considered a "conservative" mutation, has a penetrance of approximately 77% but a low incidence of severe outcome, around 5–10%, as measured by sudden death and syncope (Epstein *et al.*, 1992). However, the $403^{Arg \rightarrow Gln}$ mutation, which results in a charge change, has a high incidence of penetrance (100%) and is associated with the classic presentation for HCM of sudden death and syncope. It is considered a malignant mutation within families. This heterogeneity of clinical presentation underscores the limited phenotypic responses of cardiac tissue to a wide variety of molecular polymorphisms, but also reflects the complex interaction of other genes and the environment.

8.4.3 X-linked cardiomyopathies

Two sex-linked diseases, Becker's and Duchenne's muscular dystrophies, are among the most common causes of skeletal and cardiac myopathies, with an incidence of one in 3500 males (Walton, 1982). The inheritance pattern is X-linked recessive, although spontaneous mutations occur quite frequently (30% of new cases) and the pathological hallmark is progressive degeneration and atrophy of skeletal muscle. Duchenne's muscular dystrophy tends to be more severe with early progressive muscle weakness, mental retardation, and death in the second decade of life. Becker's muscular dystrophy has a milder course with significant skeletal muscle strength maintained throughout life. Cardiac involvement is common in both forms and is the most common cause of death. The severity of the cardiomyopathy is independent of the skeletal muscle myopathy, and includes hypertrophic ventricular dilation with progressive congestive heart failure, and conduction abnormalities and arrhythmias.

The defective gene is the same for both disorders. Dystrophin is a large membrane-associated protein, approximately 427 kDa, that forms part of a glycoprotein oligomeric complex with at least 10 other proteins. These proteins are thought to form three complexes, the sarcoglycan, dystroglycan, and syntrophin complexes, which, among other functions, serve as a transmembrane glycoprotein complex to link dystrophin to laminin in the extracellular basal lamina. This links the intracellular contractile elements to the extracellular basal lamina and protects the delicate cell membrane from mechanical damage. The complex is associated with the sarcolemmal membranes and T tubules in both cardiac and skeletal muscle, as well as the membrane surface of the cardiac Purkinje fiber (Ervasti and Campbell, 1991). The huge dystrophin gene is located on the short arm of the X chromosome (Xp21), is approximately 2.5 Mb and encodes a 14-kb mRNA. Use of multiple promoters and alternative splicing at the 3' end of the mRNA gives rise to multiple isoforms of the normal dystrophin gene product, some of which are expressed in a tissue-specific fashion.

Approximately two-thirds of the genomic mutations that cause these muscular dystrophies involve gross gene rearrangements including deletions and, less frequently, duplications. The remaining gene defects are point mutations (Baranzini *et al.*, 1997). Disease severity seems to correlate best with the degree of gene disruption and the resultant decrease in gene product: mutations that cause a shift in the reading frame result in severe disease (Duchenne's), whereas milder disease (Becker's) is associated with deletions that maintain the reading frame producing smaller, but semi-functional, dystrophin proteins (Chelly *et al.*, 1990). Finally, mutations in the cardiac-specific promoter of the dystrophin gene may result in abnormal cardiac dystrophin and a familial cardiomyopathy. These same patients can have normal levels of dystrophin in brain and skeletal muscle because the dystrophin gene has multiple tissue-specific promoters that regulates its expression.

8.4.4 *Myotonic dystrophy*

This neuromuscular disease is the most common form of adult muscular dystrophy with an incidence of about one in 8000. The disease progressively worsens during the life of the affected individual (potentiation), and shows increasing clinical severity across successive generations in muscular dystrophy pedigrees (genetic anticipation). Cardiac abnormalities are common with this dystrophy and include dilated cardiomyopathy and conduction disorders. Sudden cardiac death is the leading cause of mortality in these patients. The inheritance is autosomal dominant and there is a marked variability in expression, both within and between families.

The 13-kb gene for myotonic dystrophy has been mapped to chromosome 19q13.3 (Shaw *et al.*, 1993). The gene is transcribed into a 2.7-kb mRNA that encodes a protein kinase of approximately 55 kDa in size whose function is unknown. The genetic defect in myotonic dystrophy is an unstable trinucleotide repeat polymorphism, $(CTG)_n$, in the 3'-untranslated region that undergoes expansion, causing disease (Brook *et al.*, 1992; see also Chapter 10). A low copy number of these repeats, between 5 and 40 repeats, is associated with normal individuals.

However, in myotonic dystrophy, affected individuals may have over 1000 repeat sequences, resulting in unstable, or decreased, mRNA transcripts and low expression of the protein kinase. The mechanism by which the trinucleotide repeat causes disease is unknown but one model of myotonic dystrophy suggests that expansion of the affected allele creates a gain-of-function mutation resulting from the inappropriate recognition of the CTG repeats by binding proteins (Philips *et al.*, 1998). Myotonic dystrophy can also occur with fewer repeats but only with alleles of maternal origin. Genetic diagnosis of myotonic dystrophy is by measurement of the trinucleotide expansion, and prenatal diagnosis is now possible for this polymorphism (Geifman-Holtzman and Fay, 1998).

8.5 Polymorphisms associated with cardiac arrhythmias

Cardiac arrhythmias cause more than 300 000 sudden deaths each year in the USA. Disturbance of rhythm in cardiac function may be either a secondary event, such as ventricular fibrillation from myocardial ischemia or infarction, or a primary disease in which an arrhythmia is associated with no other phenotype. Much progress has been made recently in defining single-gene mutations associated with familial arrhythmias (Priori *et al.*, 1999a). Because of this, our understanding of the genetic basis associated with isolated arrhythmias has increased tremendously and provided a logical foundation for therapeutic intervention (Wang *et al.*, 1998). Familial occurrence of three arrhythmias in particular, long QT syndrome, arrhythmogenic right ventricular dysplasia, and congenital heart block, have led to defining either polymorphisms or a gene locus linked with the disease.

8.5.1 *Long QT syndrome*

Long QT syndrome was the first arrhythmia in which a gene defect was identified. It is a familial disease that includes the findings of syncope, seizures, and sudden death with autosomal dominant inheritance. Penetrance and expression are both variable and may depend on the gene and polymorphism (Priori *et al.*, 1999b). Symptomatic patients have characteristic findings on their EKGs which include a prolonged corrected QT interval, slow heart rate, and repolarization abnormalities in the cardiac cycle. The arrhythmia may be triggered by emotions or startling noises and is typically a ventricular tachycardia that can degenerate into a malignant cardiac arrhythmia, such as ventricular fibrillation and death. Mortality is high in symptomatic patients.

Four genes on separate chromosomes and over 40 mutations have been linked to the long QT syndrome: a fifth locus has been identified on chromosome 4 but the identity of the gene is unknown. These are usually missense mutations and are frequently distinctive within a family. Because they are often not associated with an overtly morbid phenotype, or may not even be expressed, these polymorphisms can remain in the genotype of large families across many generations.

These four genes have been well characterized and encode cardiac ion channels (Splawski *et al.*, 1998). The KVLQT1 gene on chromosome 11p15.5 is a 400-kb gene composed of 16 exons and was the first gene linked to long QT syndrome. It is also linked to the Harvey *ras*-1 locus, which initially suggested the *ras* locus as a candidate for long QT syndrome. KVLQT1 is completely linked to the polymorphic markers *TH* and *D11S1318* located near the KVLQT1 locus (Keating *et al.*, 1991). KVLQT1 encodes the α-subunit of the I_{Ks} potassium channel protein and combines with another ancillary protein, minimal potassium-channel subunit (minK), encoded by the KCNE1 gene on 21q22.1, to form a multimeric potassium channel that functions in myocellular repolarization. The KCNE1 gene is approximately 20 kb in size, encoding a 1.7-kb mRNA across three exons. Defects in either the KVLQT1 or KCNE1 gene cause a dominant-negative mutation whereby one copy of the polymorphic gene product disrupts the entire channel complex. Humans heterozygous for the KVLQT1 or KCNE1 mutation have long QT syndrome, whereas the homozygous state of either gene, or a compound heterozygote, is associated with long QT syndrome and congenital sensory deafness (Splawski *et al.*, 1997). Using the single-stranded conformational polymorphism technique, a 200-bp region in the KVLQT1 gene was identified as polymorphic and contained a single nucleotide (G) insertion at nucleotide 282 that caused a frameshift and premature stop codon. Other investigators have found multiple polymorphic sites within the KVLQT1 gene, including deletions and splice-site mutations (Ackerman *et al.*, 1998; Kanters *et al.*, 1998).

The SCN5A gene on chromosome 3p21–24 encodes the inward cardiac sodium channel termed I_{Na} (Wang *et al.*, 1995). Mutations in this gene cause a "gain of function" in which there is a persistent late inward sodium current due to defective inactivation. In two unrelated families, an intragenic region responsible for the sodium channel inactivation is deleted (ΔKPQ), causing abnormal cardiac repolarization. This mutation suggests that pharmacological therapies targeted towards inhibiting the gain of function may be successful. The fourth known gene causing long QT syndrome is the HERG gene on 7q35–q36. This 19 kb gene has 15 exons and encodes the inward rectifying I_{Kr} potassium channel protein. As with the other two potassium channel genes, mutations of the HERG gene cause a loss of function, with a dominant-negative effect when the mutant protein is co-expressed with the native protein to form the tetrameric channel structure.

8.5.2 *Familial heart block*

Two forms of progressive, familial heart block have been described based on their EKG findings, both of which are inherited in an autosomal dominant fashion. The typical symptoms are dramatic and include syncope, slow heart rate, and sudden death. The genes for both forms are not known; however, the gene locus for progressive familial heart block, type I (PFHB-I) has been linked to chromosome 19q13.2–q13.3 through analysis of several large South African families (Brink *et al.*, 1995). The candidate gene has been localized to within 10 cM of the kallikrein locus and has been confirmed in other families. Genes residing in this region include myotonic dystrophy, creatine kinase-M, apolipoprotein C2, troponin T, and the histidine-rich Ca^{2+}-binding protein (a luminal sarcoplasmic reticulum-associated protein). The first three of these genes have been excluded as the causative genes. However, this is a gene-rich area of chromosome 19, making it highly likely that a single-gene polymorphism will be identified that causes this form of heart block.

8.5.3 *Arrhythmogenic right ventricular dysplasia*

This is a familial cardiomyopathy that is characterized by fibrofatty replacement of the right ventricular myocardium and life-threatening ventricular arrhythmias that originate from the right ventricle. The disease is progressive, predominantly inherited as an autosomal dominant condition, and is especially common in north-eastern Italy (prevalence, 1:1000). To date, five loci for this cardiomyopathy have been identified, two of which are in close proximity to each other on chromosome 14 (14q23–q24 and 14q12–q22). A third locus is present at 1q42–q43, and the fourth at 2q32.1–q32.2. A fifth locus has been recently mapped to a 9.3-cM region on 3p23, between markers D3S3610 and D3S3659 (Ahmad *et al.*, 1998). Possible candidate genes in this region include a DND-binding protein, a protein-tyrosine phosphatase, and a *raf* serine-threonine protein kinase. The complexity in gene identification for this disease is compounded by the difficulty in making a clinical diagnosis: current techniques are insensitive and the phenotype is frequently non-specific with the first symptom being sudden death and the diagnosis made at autopsy. Thus, it is vitally important to be able to rule out these genes in other family members where sudden death was the presenting symptom in an otherwise normal subject.

8.6 Summary

A characteristic of many of the gene defects and polymorphisms discussed in this chapter has been identification based on transmission of a dominant phenotype, such as the long QT syndrome, and sudden death or syncope. This is understandable because of the greater ease in ascertainment of affected family members with dominantly inherited heart disease as compared to recessive disease transmission where the phenotype is rare, seems to be sporadic in its appearance, and requires a large family to investigate. This will change as more genes are found that contribute to the formation and function of the heart. An example of this might be polymorphisms in transcription factors that control heart formation, or in promoter and enhancer regions of genes that would modify cardiac function. A second factor that has played a large part in finding genes associated with heart disease has been the development of better tools for diagnosis. More subtle phenotypes are now identified, markedly increasing the power of linkage studies. Finally, as more genes are found that contribute to cardiac function and disease, greater collaboration among investigators will allow the larger studies necessary for analysis of risk factors and banking of DNA. These are clearly exciting times for exploration of the heart.

Acknowledgment

This work was supported by an Established Investigator Award from the American Heart Association.

References

Ackerman, M.J., Schroeder, J.J., Berry, R., Schaid, D.J., Porter, C.J., Michels, V.V. and Thibodeau, S.N., 1998, A novel mutation in KVLQT1 is the molecular basis of inherited long QT syndrome in a near-drowning patient's family. *Pediatr. Res.*, **44**, 148–153.

Ahmad, F., Li, D., Karibe, A., Gonzalez, O., Tapscott, T., Hill, R., Weilbaecher, D., Blackie, P., Furey, M., Gardner, M., Bachinski, L.L. and Roberts, R., 1998, Localization of a gene responsible for arrhythmogenic right ventricular dysplasia to chromosome 3p23. *Circulation*, **98**, 2791–2795.

Andresen, B.S., Bross, P., Vianey-Saban, C., Divry, P., Zabot, M.T., Roe, C.R., Nada, M.A., Byskov, A., Kruse, T.A., Neve, S., Kristiansen, K., Knudsen, I., Corydon, M.J. and Gregersen, N., 1996, Cloning and characterization of human very-long-chain acyl-CoA dehydrogenase cDNA, chromosomal assignment of the gene and identification in four patients of nine different mutations within the VLCAD gene. *Human Molec. Genet.*, **5**, 461–472.

Arai, S., Matsuoka, R., Hirayama, K., Sakurai, H., Tamura, M., Ozawa, T., Kimura, M., Imamura, S., Furutani, Y., Joh-o, K., Kawana, M., Takao, A., Hosoda, S. and Momma, K., 1995, Missense mutation of the beta-cardiac myosin heavy-chain gene in hypertrophic cardiomyopathy. *Am. J. Med. Genet.*, **58**, 267–276.

Baranzini, S.E., Lenk, U., Szijan, I. and Speer, A., 1997, Four new polymorphisms in the human dystrophin gene from an Argentinian population. *Muscle and Nerve*, **20**, 1451–1453.

Baron, A.D., 1994, Hemodynamic actions of insulin. *Am. J. Physiol.*, **267**, E187–E202.

Berk, B.C. and Corson, M.A., 1997, Angiotensin II signal transduction in vascular smooth muscle: role of tyrosine kinases. *Circ. Res.*, **80**, 607–616.

Boles, R.G., Buck, E.A., Blitzer, M.G., Platt, M.S., Cowan, T.M., Martin, S.K., Yoon, H., Madsen, J.A., Reyes-Mugica, M. and Rinaldo, P., 1998, Retrospective biochemical screening of fatty acid oxidation disorders in postmortem livers of 418 cases of sudden death in the first year of life. *J. Pediat.*, **132**, 924–933.

Bonne, G., Carrier, L., Richard, P., Hainque, B. and Schwartz, K., 1998, Familial hypertrophic cardiomyopathy: from mutations to functional defects. *Circ. Res.*, **83**, 580–593.

Brink, P.A., Ferreira, A., Moolman, J.C., Weymar, H.W., van der Merwe, P.L. and Corfield, V.A., 1995, Gene for progressive familial heart block type I maps to chromosome 19q13. *Circulation*, **91**, 1633–1640.

Brook, J.D., McCurrach, M.E., Harley, H.G., Buckler, A.J., Church, D., Aburatani, H., Hunter, K., Stanton, V.P., Thirion, J.P. and Hudson, T., 1992, Molecular basis of myotonic dystrophy: expansion of a trinucleotide (CTG) repeat at the 3′ end of a transcript encoding a protein kinase family member. *Cell*, **68**, 799–808.

Butler, R., Morris, A.D. and Struthers, A.D., 1997, Angiotensin-converting enzyme gene polymorphism and cardiovascular disease. *Clin. Sci.*, **93**, 391–400.

Cambien, F., Poirier, O., Lecerf, L., Evans, A., Cambou, J.P., Arveiler, D., Luc, G., Bard, J.M., Bara, L. and Ricard, S., 1992, Deletion polymorphism in the gene for angiotensin-converting enzyme is a potent risk factor for myocardial infarction. *Nature*, **359**, 641–644.

Chelly, J., Gilgenkrantz, H., Lambert, M., Hamard, G., Chafey, P., Recan, D., Katz, P., de la Chapelle, A., Koenig, M. and Ginjaar, I.B., 1990, Effect of dystrophin gene deletions on mRNA levels and processing in Duchenne and Becker muscular dystrophies. *Cell*, **63**, 1239–1248.

Coates, P.M., 1992, Historical perspective of medium-chain acyl-CoA dehydrogenase deficiency: a decade of discovery, in Coates, P.M. and Tanaka, K. (Eds) *New Developments in Fatty Acid Oxidation*, pp. 409–423, New York: Wiley-Liss.

Consevage, M.W., Salada, G.C., Baylen, B.G., Ladda, R.L. and Rogan, P.K., 1994, A new missense mutation, Arg719Gln, in the beta-cardiac heavy chain myosin gene of patients with familial hypertrophic cardiomyopathy. *Human Molec. Genet.*, **3**, 1025–1026.

Contois, J.H., Anamani, D.E. and Tsongalis, G.J., 1996, The underlying molecular mechanism of apolipoprotein E polymorphism: relationships to lipid disorders, cardiovascular disease, and Alzheimer's disease. *Clinics Lab. Med.*, **16**, 105–123.

Corydon, M.J., Andresen, B.S., Bross, P., Kjeldsen, M., Andreasen, P.H., Eiberg, H., Kolvraa, S. and Gregersen, N., 1997, Structural organization of the human short-chain acyl-CoA dehydrogenase gene. *Mammalian Genome*, 8, 922–926.

Cressman, M.D., Heyka, R.J., Paganini, E.P., O'Neil, J., Skibinski, C.I. and Hoff, H.F., 1992, Lipoprotein(a) is an independent risk factor for cardiovascular disease in hemodialysis patients. *Circulation*, 86, 475–482.

Curran, M.E., Atkinson, D.L., Ewart, A.K., Morris, C.A., Leppert, M.F. and Keating, M.T., 1993, The elastin gene is disrupted by a translocation associated with supravalvular aortic stenosis. *Cell*, 73, 159–168.

Dawson, S., Hamsten, A., Wiman, B., Henney, A. and Humphries, S., 1991, Genetic variation at the plasminogen activator inhibitor-1 locus is associated with altered levels of plasma plasminogen activator inhibitor-1 activity. *Arteriosclerosis and Thrombosis*, 11, 183–190.

Dawson, S.J., Wiman, B., Hamsten, A., Green, F., Humphries, S. and Henney, A.M., 1993, The two allele sequences of a common polymorphism in the promoter of the plasminogen activator inhibitor-1 (PAI-1) gene respond differently to interleukin-1 in HepG2 cells. *J. Biol. Chem.*, 268, 10739–10745.

Day, I.N., Whittall, R.A., O'Dell, S.D., Haddad, L., Bolla, M.K., Gudnason, V. and Humphries, S.E., 1997, Spectrum of LDL receptor gene mutations in heterozygous familial hypercholesterolemia. *Human Mut.*, 10, 116–127.

de Knijff, P. and Havekes, L.M., 1996, Apolipoprotein E as a risk factor for coronary heart disease: a genetic and molecular biology approach. *Curr. Opin. Lipidol.*, 7, 59–63.

Degen, S.J., Rajput, B. and Reich, E., 1986, The human tissue plasminogen activator gene. *J. Biol. Chem.*, 261, 6972–6985.

Dietz, H.C. and Pyeritz, R.E., 1995, Mutations in the human gene for fibrillin-1 (FBN1) in the Marfan syndrome and related disorders. *Human Molec. Genet.*, 4 Spec No., 1799–1809.

Dietz, H.C., Cutting, G.R., Pyeritz, R.E., Maslen, C.L., Sakai, L.Y., Corson, G.M., Puffenberger, E.G., Hamosh, A., Nanthakumar, E.J. and Curristin, S.M., 1991, Marfan syndrome caused by a recurrent de novo missense mutation in the fibrillin gene. *Nature*, 352, 337–339.

Dominiczak, A.F., Devlin, A.M., Lee, W.K., Anderson, N.H., Bohr, D.F. and Reid, J.L., 1996, Vascular smooth muscle polyploidy and cardiac hypertrophy in genetic hypertension. *Hypertension*, 27, 752–759.

Dzau, V.J., 1988, Circulating versus local renin–angiotensin system in cardiovascular homeostasis. *Circulation*, 77, I4–13.

Epstein, N.D., Cohn, G.M., Cyran, F. and Fananapazir, L., 1992, Differences in clinical expression of hypertrophic cardiomyopathy associated with two distinct mutations in the beta-myosin heavy chain gene. A 908Leu→Val mutation and a 403Arg→Gln mutation. *Circulation*, 86, 345–352.

Erdos, E.G. and Skidgel, R.A., 1987, The angiotensin I-converting enzyme. *Lab. Investig.*, 56, 345–348.

Ervasti, J.M. and Campbell, K.P., 1991, Membrane organization of the dystrophin-glycoprotein complex. *Cell*, 66, 1121–1131.

Ewart, A.K., Morris, C.A., Atkinson, D., Jin, W., Sternes, K., Spallone, P., Stock, A.D., Leppert, M. and Keating, M.T., 1993, Hemizygosity at the elastin locus in a developmental disorder, Williams' syndrome. *Nature Genet.*, 5, 11–16.

Fisher, R.M., Humphries, S.E. and Talmud, P.J., 1997, Common variation in the lipoprotein lipase gene: effects on plasma lipids and risk of atherosclerosis. *Atherosclerosis*, 135, 145–159.

Fornage, M., Amos, C.I., Kardia, S., Sing, C.F., Turner, S.T. and Boerwinkle, E., 1998, Variation in the region of the angiotensin-converting enzyme gene influences interindividual differences in blood pressure levels in young white males. *Circulation*, 97, 1773–1779.

Fougerousse, F., Dufour, C., Roudaut, C. and Beckmann, J.S., 1992, Dinucleotide repeat polymorphism at the human gene for cardiac beta-myosin heavy chain (MYH6). *Human Molec. Genet.*, 1, 64.

Garin, M.C., James, R.W., Dussoix, P., Blanche, H., Passa, P., Froguel, P. and Ruiz, J., 1997, Paraoxonase polymorphism Met-Leu54 is associated with modified serum concentrations of the enzyme. A possible link between the paraoxonase gene and increased risk of cardiovascular disease in diabetes. *J. Clin. Invest.*, 99, 62–66.

Geifman-Holtzman, O. and Fay, K., 1998, Prenatal diagnosis of congenital myotonic dystrophy and counseling of the pregnant mother: case report and literature review. *Am. J. Med. Genet.*, 78, 250–253.

Geisterfer-Lowrance, A.A., Kass, S., Tanigawa, G., Vosberg, H.P., McKenna, W., Seidman, C.E. and Seidman, J.G., 1990, A molecular basis for familial hypertrophic cardiomyopathy: a beta cardiac myosin heavy chain gene missense mutation. *Cell*, 62, 999–1006.

Gensini, G.F., Comeglio, M. and Colella, A., 1998, Classical risk factors and emerging elements in the risk profile for coronary artery disease. *Europ. Heart J.*, 19(Suppl A), A53–A61.

Grant, P.J., 1997, Polymorphisms of coagulation/fibrinolysis genes: gene environment interactions and vascular risk. *Prostaglandins, Leukotrienes and Essential Fatty Acids*, 57, 473–477.

Hall, S., Chu, G., Miller, G., Cruickshank, K., Cooper, J.A., Humphries, S.E. and Talmud, P.J., 1997, A common mutation in the lipoprotein lipase gene promoter, –93T/G, is associated with lower plasma triglyceride levels and increased promoter activity *in vitro*. *Arteriosclerosis, Thromb. Vasc. Biol.*, 17, 1969–1976.

Hegele, R.A., Connelly, P.W., Scherer, S.W., Hanley, A.J., Harris, S.B., Tsui, L.C. and Zinman, B., 1997, Paraoxonase-2 gene (PON2) G148 variant associated with elevated fasting plasma glucose in noninsulin-dependent diabetes mellitus. *J. Clin. Endocrinol. Metab.*, 82, 3373–3377.

Hoffman, J.I., 1995, Incidence of congenital heart disease: I. Postnatal incidence. *Pediat. Cardiol.*, 16, 103–113.

IJlst, L., Ruiter, J.P., Hoovers, J.M., Jakobs, M.E. and Wanders, R.J., 1996, Common missense mutation G1528C in long-chain 3-hydroxyacyl-CoA dehydrogenase deficiency. Characterization and expression of the mutant protein, mutation analysis on genomic DNA and chromosomal localization of the mitochondrial trifunctional protein alpha subunit gene. *J. Clin. Invest.*, 98, 1028–1033.

Itoh, H., Pratt, R.E. and Dzau, V.J., 1990, Atrial natriuretic polypeptide inhibits hypertrophy of vascular smooth muscle cells. *J. Clin. Invest.*, 86, 1690–1697.

Itoh, H., Mukoyama, M., Pratt, R.E., Gibbons, G.H. and Dzau, V.J., 1993, Multiple autocrine growth factors modulate vascular smooth muscle cell growth response to angiotensin II. *J. Clin. Invest.*, 91, 2268–2274.

Jemaa, R., Fumeron, F., Poirier, O., Lecerf, L., Evans, A., Arveiler, D., Luc, G., Cambou, J.P., Bard, J.M. and Fruchart, J.C., 1995, Lipoprotein lipase gene polymorphisms: associations with myocardial infarction and lipoprotein levels, the ECTIM study. Etude Cas Temoin sur l'Infarctus du Myocarde. *J. Lipid Res.*, 36, 2141–2146.

Jin, Y., Dietz, H.C., Montgomery, R.A., Bell, W.R., McIntosh, I., Coller, B. and Bray, P.F., 1996, Glanzmann thrombasthenia. Cooperation between sequence variants in cis during splice site selection. *J. Clin. Invest.*, 98, 1745–1754.

Kainulainen, K., Karttunen, L., Puhakka, L., Sakai, L. and Peltonen, L., 1994, Mutations in the fibrillin gene responsible for dominant ectopia lentis and neonatal Marfan syndrome. *Nature Genet.*, 6, 64–69.

Kanters, J.K., Larsen, L.A., Orholm, M., Agner, E., Andersen, P.S., Vuust, J. and Christiansen, M. 1998, Novel donor splice site mutation in the KVLQT1 gene is associated with long QT syndrome. *J. Cardiovasc. Electrophysiol.*, **9**, 620–624.

Keating, M., Atkinson, D., Dunn, C., Timothy, K., Vincent, G.M. and Leppert, M., 1991, Linkage of a cardiac arrhythmia, the long QT syndrome, and the Harvey ras-1 gene. *Science*, **252**, 704–706.

Kelly, D.P. and Strauss, A.W., 1994, Inherited cardiomyopathies. *New Engl. J. Med.*, **330**, 913–919.

Kelly, D.P., Whelan, A.J., Ogden, M.L., Alpers, R., Zhang, Z.F., Bellus, G., Gregersen, N., Dorland, L. and Strauss, A.W., 1990, Molecular characterization of inherited medium-chain acyl-CoA dehydrogenase deficiency. *Proc. Natl Acad. Sci., USA*, **87**, 9236–9240.

Kidd, J.R., Matsubara, Y., Castiglione, C.M., Tanaka, K. and Kidd, K.K., 1990, The locus for the medium-chain acyl-CoA dehydrogenase gene on chromosome 1 is highly polymorphic. *Genomics*, **6**, 89–93.

Klinger, K.W., Winqvist, R., Riccio, A., Andreasen, P.A., Sartorio, R., Nielsen, L.S., Stuart, N., Stanislovitis, P., Watkins, P. and Douglas, R., 1987, Plasminogen activator inhibitor type 1 gene is located at region q21.3–q22 of chromosome 7 and genetically linked with cystic fibrosis. *Proc. Natl Acad. Sci., USA*, **84**, 8548–8552.

Koeleman, B.P., Reitsma, P.H., Bakker, E. and Bertina, R.M., 1997, Location on the human genetic linkage map of 26 genes involved in blood coagulation. *Thrombosis and Haemostasis*, **77**, 873–878.

Kronenberg, F., Utermann, G. and Dieplinger, H., 1996, Lipoprotein(a) in renal disease. *Am. J. Kidney Dis.*, **27**, 1–25.

Largilliere, C., Vianey-Saban, C., Fontaine, M., Bertrand, C., Kacet, N. and Farriaux, J.P., 1995, Mitochondrial very long chain acyl-CoA dehydrogenase deficiency – a new disorder of fatty acid oxidation. *Arch. Dis. Childhood Fetal and Neonatal Edition*, **73**, F103–F105.

Leviev, I., Negro, F. and James, R.W., 1997, Two alleles of the human paraoxonase gene produce different amounts of mRNA. An explanation for differences in serum concentrations of paraoxonase associated with the (Leu-Met54) polymorphism. *Arteriosclerosis, Thrombosis and Vascular Biol.*, **17**, 2935–2939.

Li, L., Krantz, I.D., Deng, Y., Genin, A., Banta, A.B., Collins, C.C., Qi, M., Trask, B.J., Kuo, W.L., Cochran, J., Costa, T., Pierpont, M.E., Rand, E.B., Piccoli, D.A., Hood, L. and Spinner, N.B., 1997, Alagille syndrome is caused by mutations in human Jagged1, which encodes a ligand for Notch1. *Nature Genet.*, **16**, 243–251.

Ludwig, M., Wohn, K.D., Schleuning, W.D. and Olek, K., 1992, Allelic dimorphism in the human tissue-type plasminogen activator (TPA) gene as a result of an Alu insertion/deletion event. *Human Genet.*, **88**, 388–392.

Mackness, M.I., Arrol, S. and Durrington, P.N., 1991a, Paraoxonase prevents accumulation of lipoperoxides in low-density lipoprotein. *FEBS Letters*, **286**, 152–154.

Mackness, M.I., Harty, D., Bhatnagar, D., Winocour, P.H., Arrol, S., Ishola, M. and Durrington, P.N., 1991b, Serum paraoxonase activity in familial hypercholesterolaemia and insulin-dependent diabetes mellitus. *Atherosclerosis*, **86**, 193–199.

Mackness, B., Mackness, M.I., Arrol, S., Turkie, W. and Durrington, P.N., 1997, Effect of the molecular polymorphisms of human paraoxonase (PON1) on the rate of hydrolysis of paraoxon. *Brit. J. Pharmacol.*, **122**, 265–268.

Mackness, B., Durrington, P.N. and Mackness, M.I., 1998, Human serum paraoxonase. *Gen. Pharmacol.*, **31**, 329–336.

MacRae, C.A., Ghaisas, N., Kass, S., Donnelly, S., Basson, C.T., Watkins, H.C., Anan, R., Thierfelder, L.H., McGarry, K. and Rowland, E., 1995, Familial hypertrophic cardiomyopathy with Wolff–Parkinson–White syndrome maps to a locus on chromosome 7q3. *J. Clin. Invest.*, **96**, 1216–1220.

Mailly, F., Fisher, R.M., Nicaud, V., Luong, L.A., Evans, A.E., Marques-Vidal, P., Luc, G., Arveiler, D., Bard, J.M., Poirier, O., Talmud, P.J. and Humphries, S.E., 1996, Association between the LPL-D9N mutation in the lipoprotein lipase gene and plasma lipid traits in myocardial infarction survivors from the ECTIM Study. *Atherosclerosis*, **122**, 21–28.

Manolio, T.A., Baughman, K.L., Rodeheffer, R., Pearson, T.A., Bristow, J.D., Michels, V.V., Abelmann, W.H. and Harlan, W.R., 1992, Prevalence and etiology of idiopathic dilated cardiomyopathy (summary of a National Heart, Lung, and Blood Institute workshop). *Am. J. Cardiol.*, **69**, 1458–1466.

Mansfield, T.A., Simon, D.B., Farfel, Z., Bia, M., Tucci, J.R., Lebel, M., Gutkin, M., Vialettes, B., Christofilis, M.A., Kauppinen-Makelin, R., Mayan, H., Risch, N. and Lifton, R.P., 1997, Multilocus linkage of familial hyperkalaemia and hypertension, pseudohypoaldosteronism type II, to chromosomes 1q31–42 and 17p11–q21. *Nature Genet.*, **16**, 202–205.

Matsubara, Y., Narisawa, K., Tada, K., Ikeda, H., Yao, Y.Q., Danks, D.M., Green, A. and McCabe, E.R., 1991, Prevalence of K329E mutation in medium-chain acyl-CoA dehydrogenase gene determined from Guthrie cards. *Lancet*, **338**, 552–553.

Moller, J.H., Allen, H.D., Clark, E.B., Dajani, A.S., Golden, A., Hayman, L.L., Lauer, R.M., Marmer, E.L., McAnulty, J.H. and Oparil, S., 1993, Report of the task force on children and youth. American Heart Association. *Circulation*, **88**, 2479–2486.

Murata, M., Kawano, K., Matsubara, Y., Ishikawa, K., Watanabe, K. and Ikeda, Y., 1998, Genetic polymorphisms and risk of coronary artery disease. *Sem. Thromb. Hemostasis*, **24**, 245–250.

Naito, E., Indo, Y. and Tanaka, K., 1990, Identification of two variant short chain acyl-coenzyme A dehydrogenase alleles, each containing a different point mutation in a patient with short chain acyl-coenzyme A dehydrogenase deficiency. *J. Clin. Invest.*, **85**, 1575–1582.

Naito, E., Ozasa, H., Ikeda, Y. and Tanaka, K., 1989, Molecular cloning and nucleotide sequence of complementary DNAs encoding human short chain acyl-coenzyme A dehydrogenase and the study of the molecular basis of human short chain acyl-coenzyme A dehydrogenase deficiency. *J. Clin. Invest.*, **83**, 1605–1613.

Neely, J.R., Rovetto, M.J. and Oram, J.F., 1972, Myocardial utilization of carbohydrate and lipids. *Prog. Cardiovasc. Dis.*, **15**, 289–329.

Nicholls, P., Young, I.S. and Graham, C.A., 1998, Genotype/phenotype correlations in familial hypercholesterolaemia. *Curr. Opin. Lipidol.*, **9**, 313–317.

Nurden, A.T., 1997, Platelet glycoprotein IIIa polymorphism and coronary thrombosis. *Lancet*, **350**, 1189–1191.

Oda, T., Elkahloun, A.G., Pike, B.L., Okajima, K., Krantz, I.D., Genin, A., Piccoli, D.A., Meltzer, P.S., Spinner, N.B., Collins, F.S. and Chandrasekharappa, S.C., 1997, Mutations in the human Jagged1 gene are responsible for Alagille syndrome. *Nature Genet.*, **16**, 235–242.

Odawara, M., Tachi, Y. and Yamashita, K., 1997, Paraoxonase polymorphism (Gln192–Arg) is associated with coronary heart disease in Japanese noninsulin-dependent diabetes mellitus. *J. Clin. Endocrinol. Metab.*, **82**, 2257–2260.

O'Donnell, C.J., Lindpaintner, K., Larson, M.G., Rao, V.S., Ordovas, J.M., Schaefer, E.J., Myers, R.H. and Levy, D., 1998, Evidence for association and genetic linkage of the angiotensin-converting enzyme locus with hypertension and blood pressure in men but not women in the Framingham Heart Study. *Circulation*, **97**, 1766–1772.

Pasternak, R.C., Grundy, S.M., Levy, D. and Thompson, P.D., 1996, 27th Bethesda Conference: matching the intensity of risk factor management with the hazard for coronary disease events. Task Force 3. Spectrum of risk factors for coronary heart disease. *J. Am. Coll. Cardiol.*, **27**, 978–990.

Payne, R.M., Johnson, M.C., Grant, J.W. and Strauss, A.W., 1995, Toward a molecular understanding of congenital heart disease. *Circulation*, 91, 494–504.

Pereira, L., Levran, O., Ramirez, F., Lynch, J.R., Sykes, B., Pyeritz, R.E. and Dietz, H.C., 1994, A molecular approach to the stratification of cardiovascular risk in families with Marfan's syndrome. *New Engl. J. Med.*, 331, 148–153.

Pfeffer, M.A., Braunwald, E., Moye, L.A., Basta, L., Brown, E.J.J., Cuddy, T.E., Davis, B.R., Geltman, E.M., Goldman, S. and Flaker, G.C., 1992, Effect of captopril on mortality and morbidity in patients with left ventricular dysfunction after myocardial infarction. Results of the survival and ventricular enlargement trial. The SAVE Investigators. *New Engl. J. Med.*, 327, 669–677.

Philips, A.V., Timchenko, L.T. and Cooper, T.A., 1998, Disruption of splicing regulated by a CUG-binding protein in myotonic dystrophy. *Science*, 280, 737–741.

Primo-Parmo, S.L., Sorenson, R.C., Teiber, J. and La Du, B.N., 1996, The human serum paraoxonase/arylesterase gene (PON1) is one member of a multigene family. *Genomics*, 33, 498–507.

Priori, S.G., Barhanin, J., Hauer, R.N., Haverkamp, W., Jongsma, H.J., Kleber, A.G., McKenna, W.J., Roden, D.M., Rudy, Y., Schwartz, K., Schwartz, P.J., Towbin, J.A. and Wilde, A.M., 1999a, Genetic and molecular basis of cardiac arrhythmias: impact on clinical management: Parts I and II. *Circulation*, 99, 518–528.

Priori, S.G., Napolitano, C. and Schwartz, P.J. 1999b, Low penetrance in the long-QT syndrome: clinical impact. *Circulation*, 99, 529–533.

Pyeritz, R.E. and McKusick, V.A., 1979, The Marfan syndrome: diagnosis and management. *New Engl. J. Med.*, 300, 772–777.

Rice, G.I., Ossei-Gerning, N., Stickland, M.H. and Grant, P.J., 1997, The paraoxonase Gln-Arg 192 polymorphism in subjects with ischaemic heart disease. *Coronary Artery Dis.*, 8, 677–682.

Ridker, P.M., Hennekens, C.H., Schmitz, C., Stampfer, M.J. and Lindpaintner, K., 1997, PIA1/A2 polymorphism of platelet glycoprotein IIIa and risks of myocardial infarction, stroke, and venous thrombosis. *Lancet*, 349, 385–388.

Rigat, B., Hubert, C., Alhenc-Gelas, F., Cambien, F., Corvol, P. and Soubrier, F., 1990, An insertion/deletion polymorphism in the angiotensin I-converting enzyme gene accounting for half the variance of serum enzyme levels. *J. Clin. Invest.*, 86, 1343–1346.

Ruiz, J., Blanche, H., Cohen, N., Velho, G., Cambien, F., Cohen, D., Passa, P. and Froguel, P., 1994, Insertion/deletion polymorphism of the angiotensin-converting enzyme gene is strongly associated with coronary heart disease in non-insulin-dependent diabetes mellitus. *Proc. Natl Acad. Sci.*, USA, 91, 3662–3665.

Ruiz, J., Blanche, H., James, R.W., Garin, M.C., Vaisse, C., Charpentier, G., Cohen, N., Morabia, A., Passa, P. and Froguel, P., 1995, Gln-Arg192 polymorphism of paraoxonase and coronary heart disease in type 2 diabetes. *Lancet*, 346, 869–872.

Sakai, L.Y., Keene, D.R. and Engvall, E., 1986, Fibrillin, a new 350-kD glycoprotein, is a component of extracellular microfibrils. *J. Cell Biol.*, 103, 2499–2509.

Sanghera, D.K., Saha, N., Aston, C.E. and Kamboh, M.I., 1997, Genetic polymorphism of paraoxonase and the risk of coronary heart disease. *Arteriosclerosis, Thrombosis Vasc. Biol.*, 17, 1067–1073.

Sanghera, D.K., Aston, C.E., Saha, N. and Kamboh, M.I., 1998a, DNA polymorphisms in two paraoxonase genes (PON1 and PON2) are associated with the risk of coronary heart disease. *Am. J. Human Genet.*, 62, 36–44.

Sanghera, D.K., Saha, N. and Kamboh, M.I., 1998b, The codon 55 polymorphism in the paraoxonase 1 gene is not associated with the risk of coronary heart disease in Asian Indians and Chinese. *Atherosclerosis*, 136, 217–223.

Schwartz, K., Beckmann, J., Dufour, C., Faure, L., Fougerousse, F., Carrier, L., Hengstenberg, C., Cohen, D., Vosberg, H.P. and Sacrez, A., 1992, Exclusion of cardiac myosin heavy chain and actin gene involvement in hypertrophic cardiomyopathy of several French families. *Circulation Res.*, **71**, 3–8.

Shaw, D.J., McCurrach, M., Rundle, S.A., Harley, H.G., Crow, S.R., Sohn, R., Thirion, J.P., Hamshere, M.G., Buckler, A.J. and Harper, P.S., 1993, Genomic organization and transcriptional units at the myotonic dystrophy locus. *Genomics*, **18**, 673–679.

Sigurdsson, G., Baldursdottir, A., Sigvaldason, H., Agnarsson, U., Thorgeirsson, G. and Sigfusson, N., 1992, Predictive value of apolipoproteins in a prospective survey of coronary artery disease in men. *Am. J. Cardiol.*, **69**, 1251–1254.

Sims, H.F., Brackett, J.C., Powell, C.K., Treem, W.R., Hale, D.E., Bennett, M.J., Gibson, B., Shapiro, S. and Strauss, A.W., 1995, The molecular basis of pediatric long chain 3-hydroxyacyl-CoA dehydrogenase deficiency associated with maternal acute fatty liver of pregnancy. *Proc. Natl Acad. Sci., USA*, **92**, 841–845.

Soutar, A.K., 1998, Update on low density lipoprotein receptor mutations. *Curr. Opin. Lipidol.*, **9**, 141–147.

Spindler, M., Saupe, K.W., Christe, M.E., Sweeney, H.L., Seidman, C.E., Seidman, J.G. and Ingwall, J.S., 1998, Diastolic dysfunction and altered energetics in the alphaMHC403/+ mouse model of familial hypertrophic cardiomyopathy. *J. Clin. Invest.*, **101**, 1775–1783.

Splawski, I., Timothy, K.W., Vincent, G.M., Atkinson, D.L. and Keating, M.T., 1997, Molecular basis of the long-QT syndrome associated with deafness. *New Engl. J. Med.*, **336**, 1562–1567.

Splawski, I., Shen, J., Timothy, K.W., Vincent, G.M., Lehmann, M.H. and Keating, M.T., 1998, Genomic structure of three long QT syndrome genes: KVLQT1, HERG, and KCNE1. *Genomics*, **51**, 86–97.

Steeds, R., Adams, M., Smith, P., Channer, K. and Samani, N.J., 1998, Distribution of tissue plasminogen activator insertion/deletion polymorphism in myocardial infarction and control subjects. *Thrombosis and Haemostasis*, **79**, 980–984.

Stengard, J.H., Zerba, K.E., Pekkanen, J., Ehnholm, C., Nissinen, A. and Sing, C.F., 1995, Apolipoprotein E polymorphism predicts death from coronary heart disease in a longitudinal study of elderly Finnish men. *Circulation*, **91**, 265–269.

Strandberg, L., Lawrence, D. and Ny, T., 1988, The organization of the human-plasminogen-activator-inhibitor-1 gene. Implications on the evolution of the serine-protease inhibitor family. *Europ. J. Biochem.*, **176**, 609–616.

Strauss, A.W., Powell, C.K., Hale, D.E., Anderson, M.M., Ahuja, A., Brackett, J.C. and Sims, H.F., 1995, Molecular basis of human mitochondrial very-long-chain acyl-CoA dehydrogenase deficiency causing cardiomyopathy and sudden death in childhood. *Proc. Natl Acad. Sci., USA*, **92**, 10496–10500.

Takahashi, T., Kawahara, Y., Okuda, M. and Yokoyama, M., 1996, Increasing cAMP antagonizes hypertrophic response to angiotensin II without affecting Ras and MAP kinase activation in vascular smooth muscle cells. *FEBS Letters*, **397**, 89–92.

Tardiff, J.C., Factor, S.M., Tompkins, B.D., Hewett, T.E., Palmer, B.M., Moore, R.L., Schwartz, S., Robbins, J. and Leinwand, L.A., 1998, A truncated cardiac troponin T molecule in transgenic mice suggests multiple cellular mechanisms for familial hypertrophic cardiomyopathy. *J. Clin. Invest.*, **101**, 2800–2811.

The Acute Infarction Ramipril Efficacy (AIRE) Study Investigators, 1993, Effect of ramipril on mortality and morbidity of survivors of acute myocardial infarction with clinical evidence of heart failure. *Lancet*, **342**, 821–828.

The SOLVD Investigators, 1991, Effect of enalapril on survival in patients with reduced left ventricular ejection fractions and congestive heart failure. *New Engl. J. Med.*, **325**, 293–302.

Thierfelder, L., MacRae, C., Watkins, H., Tomfohrde, J., Williams, M., McKenna, W. Bohm, K., Noeske, G., Schlepper, M., Bowcock, A., Vosberg, H.P., Seidman, J.G. and Seidman, C.E., 1993, A familial hypertrophic cardiomyopathy locus maps to chromosome 15q2. *Proc. Natl Acad. Sci., USA*, **90**, 6270–6274.

Tishkoff, S.A., Ruano, G., Kidd, J.R. and Kidd, K.K., 1996, Distribution and frequency of a polymorphic Alu insertion at the plasminogen activator locus in humans. *Human Genet.*, **97**, 759–764.

Treem, W.R., Shoup, M.E., Hale, D.E., Bennett, M.J., Rinaldo, P., Millington, D.S., Stanley, C.A., Riely, C.A. and Hyams, J.S., 1996, Acute fatty liver of pregnancy, hemolysis, elevated liver enzymes, and low platelets syndrome, and long chain 3-hydroxyacyl-coenzyme A dehydrogenase deficiency. *Am. J. Gastroenterol.*, **91**, 2293–2300.

Tsipouras, P., Del Mastro, R., Sarfarazi, M., Lee, B., Vitale, E., Child, A.H., Godfrey, M., Devereux, R.B., Hewett, D. and Steinmann, B., 1992, Genetic linkage of the Marfan syndrome, ectopia lentis, and congenital contractural arachnodactyly to the fibrillin genes on chromosomes 15 and 5. The International Marfan Syndrome Collaborative Study. *New Engl. J. Med.*, **326**, 905–909.

Tybjaerg-Hansen, A. and Humphries, S.E., 1992, Familial defective apolipoprotein B-100: a single mutation that causes hypercholesterolemia and premature coronary artery disease. *Atherosclerosis*, **96**, 91–107.

Ueda, S., Meredith, P.A., Morton, J.J., Connell, J.M.C. and Elliott, H.L., 1998, Ace (I/D) genotype as a predictor of the magnitude and duration of the response to an ACE inhibitor drug (enalaprilat) in humans. *Circulation*, **98**, 2148–2153.

van den Berg, M.H., Meijer, H. and Geraedts, J.P., 1995, A trinucleotide repeat combination polymorphism in the cardiac alpha myosin heavy chain (MYH6) gene. *Human Genet.*, **95**, 723–724.

van der Bom, J.G., de Knijff, P., Haverkate, F., Bots, M.L., Meijer, P., de Jong, P.T., Hofman, A., Kluft, C. and Grobbee, D.E., 1997, Tissue plasminogen activator and risk of myocardial infarction. The Rotterdam Study. *Circulation*, **95**, 2623–2627.

Walter, D.H., Schachinger, V., Elsner, M., Dimmeler, S. and Zeiher, A.M., 1997, Platelet glycoprotein IIIa polymorphisms and risk of coronary stent thrombosis. *Lancet*, **350**, 1217–1219.

Walton, J., 1982, *Skeletal Muscle Pathology*, Edinburgh: Churchill Livingstone.

Wang, Q., Shen, J., Splawski, I., Atkinson, D., Li, Z., Robinson, J.L., Moss, A.J., Towbin, J.A. and Keating, M.T., 1995, SCN5A mutations associated with an inherited cardiac arrhythmia, long QT syndrome. *Cell*, **80**, 805–811.

Wang, Q., Chen, Q. and Towbin, J.A., 1998, Genetics, molecular mechanisms and management of long QT syndrome. *Annals of Med.*, **30**, 58–65.

Wang, X.L., McCredie, R.M. and Wilcken, D.E., 1996, Genotype distribution of angiotensin-converting enzyme polymorphism in Australian healthy and coronary populations and relevance to myocardial infarction and coronary artery disease. *Arteriosclerosis, Thrombosis and Vasc. Biol.*, **16**, 115–119.

Watkins, H., MacRae, C., Thierfelder, L., Chou, Y.H., Frenneaux, M., McKenna, W., Seidman, J.G. and Seidman, C.E., 1993, A disease locus for familial hypertrophic cardiomyopathy maps to chromosome 1q3. *Nature Genet.*, **3**, 333–337.

Weiss, E.J., Bray, P.F., Tayback, M., Schulman, S.P., Kickler, T.S., Becker, L.C., Weiss, J.L., Gerstenblith, G. and Goldschmidt-Clermont, P.J., 1996, A polymorphism of a platelet glycoprotein receptor as an inherited risk factor for coronary thrombosis. *New Engl. J. Med.*, **334**, 1090–1094.

Wilson, P.W., Myers, R.H., Larson, M.G., Ordovas, J.M., Wolf, P.A. and Schaefer, E.J., 1994, Apolipoprotein E alleles, dyslipidemia, and coronary heart disease. The Framingham Offspring Study. *J. Am. Med. Assoc.*, **272**, 1666–1671.

Yamasaki, Y., Sakamoto, K., Watada, H., Kajimoto, Y. and Hori, M., 1997, The Arg192 isoform of paraoxonase with low sarin-hydrolyzing activity is dominant in the Japanese. *Human Genet.*, **101**, 67–68.

Yang-Feng, T.L., Opdenakker, G., Volckaert, G. and Francke, U., 1986, Human tissue-type plasminogen activator gene located near chromosomal breakpoint in myeloproliferative disorder. *Am. J. Human Genet.*, **39**, 79–87.

Yuan, Z.R., Kohsaka, T., Ikegaya, T., Suzuki, T., Okano, S., Abe, J., Kobayashi, N. and Yamada, M., 1998, Mutational analysis of the Jagged 1 gene in Alagille syndrome families. *Human Molec. Genet.*, **7**, 1363–1369.

Zee, R.Y., Ridker, P.M., Stampfer, M.J., Hennekens, C.H. and Lindpaintner, K., 1999, Prospective evaluation of the angiotensin-converting enzyme insertion/deletion polymorphism and the risk of stroke. *Circulation*, **99**, 340–343.

Zhang, Q., Cavanna, J., Winkelman, B.R., Shine, B., Gross, W., Marz, W. and Galton, D.J., 1995, Common genetic variants of lipoprotein lipase that relate to lipid transport in patients with premature coronary artery disease. *Clin. Genet.*, **48**, 293–298.

9

GENETIC POLYMORPHISMS IN THE MITOCHONDRIAL GENOME

Gerard T. Berry and Paolo Fortina

9.1 Introduction

Mitochondria are cytoplasmic double-membrane-structured organelles, which provide most of the energy of eukaryotic cells. Each human cell contains several hundreds of mitochondria and thousands of closed-circular information-dense 16 569-base-pair (bp) human mitochondrial DNA (mtDNA). The human mtDNA contains a total of 37 genes: 13 polypeptide-encoding genes (mRNAs) which are essential to the mitochondrial energy-generating oxidative phosphorylation (OXPHOS) system, 22 transfer RNAs (tRNAs) and the small (12S, MTRNR1) and large (16S, MTRNR2) ribosomal RNAs (rRNAs) necessary for its protein synthesis (Anderson, 1981; Attardi, 1982; Wallace, 1995; Wallace *et al.*, 1995, 1996; Kogelnik *et al.*, 1996). The OXPHOS system is located within the inner membrane and involves five enzymes, named complex I to V, that are responsible for generating most of the ATP required by cells. The ATP is produced through a cascade of coordinated redox reactions derived from a variety of metabolic pathways in which electrons are sequentially transported from complex I to complex IV. As electrons traverse each of three reaction centers, protons are pumped out across the mitochondrial inner membrane generating a gradient that is used by the ATP synthase or complex V to synthesize ATP from ADP. Hence, mutations of the mtDNA manifest as disorders of electron transfer and oxidative phosphorylation (Shoffner and Wallace, 1995; Wallace *et al.*, 1996).

The two strands of the circular mtDNA chromosome have an asymmetric distribution of Gs and Cs generating heavy (H)- and light (L)-strands. Gene products encoded by the L-strand include one polypeptide (MTND6) and nine tRNAs whereas gene products of the H-strand include 12 polypeptides, 2 rRNAs and 13 tRNAs. Seven of the polypeptides are subunits of complex I (NADH-dehydrogenase-ubiquinone reductase) and are named MTND 1, 2, 3, 4, 4L, 5 and 6; cytochrome *b* (MTCYB) is part of complex III (ubiquinone-cytochrome *c* reductase); three polypeptides, COI (MTCOI), COII (MTCO2) and COIII (MTCO3), are the catalytic subunits of complex IV (cytochrome *c* oxidase); and finally, ATPase-6 (MTATP6) and ATPase-8 (MTATP8) are components of complex V. All the subunits of complex II (succinate dehydrogenase-ubiquinone reductase) are nuclear DNA (nDNA)-encoded (Leckschat *et al.*, 1993; Hirawake *et al.*, 1994; Morris *et al.*, 1994;

Au *et al.*, 1995). In order to produce functionally active complexes, the mtDNA-encoded subunits of each complex must interact with several subunits, which are encoded by nDNA and then imported into mitochondria from the cytoplasm (Glick and Schatz, 1991).

The human mtDNA has an extraordinarily compact gene organization. All coding sequences lack significant non-translated flanking regions and are adjacent to each other, or separated by only a few nucleotides. The only two non-coding sequences of functional significance are the one kilobase (kb) region, defined as the D-loop, containing the origin of replication of the H-strand as well as the promoters for L- and H-strand transcription and the 30-nucleotide (nt) region in the leading L-strand replication. This region, which is surrounded by a cluster of five tRNA genes, is able to form a stable hairpin structure, and serves as the origin of replication of the L-strand (OL) (Chang *et al.*, 1984, 1985, 1986). In addition, some of the polypeptide-coding sequences are incomplete, missing one or two nucleotides within the termination codon, which are then completed by post-transcriptional polyadenylation (Wallace *et al.*, 1996). Figure 9.1 illustrates the structure and gene organization of human mtDNA.

9.2 Unique genetic properties of mtDNA

9.2.1 *Polyplasmy*

Mitochondria are polyploid. Each human cell has hundreds of mitochondria, each containing multiple mtDNA molecules (Wallace, 1982). At cell division, mitochondria and their genomes are distributed randomly to daughter cells.

9.2.2 *Semiautonomous genetic system*

Although mitochondria are an integral component of the cell, they retain a semi-autonomous genetic system with associated replication, transcription and translation processes consistent with the mtDNA having an endosymbiotic origin. This autonomy was demonstrated by showing that resistance to the mitochondrial ribosome inhibitor chloramphenicol could be transferred between cells by using a cytoplasmic hybrid system of cultured eukaryotic cells (Blanc *et al.*, 1981; Kearsey and Craig, 1981; Wallace, 1981).

9.2.3 *Genetic code*

mtDNA has a slightly different genetic code from the mammalian nuclear genetic code. The 22 tRNAs can translate the entire genetic code using modified codon rules. For example, the universal AGA and AGG, arginine codons, are read as stop codons, while UGA codon is read as tryptophan rather than as a stop codon and AUA is read as methionine (Barrell *et al.*, 1979; Anderson *et al.*, 1981, 1982; Wallace, 1982).

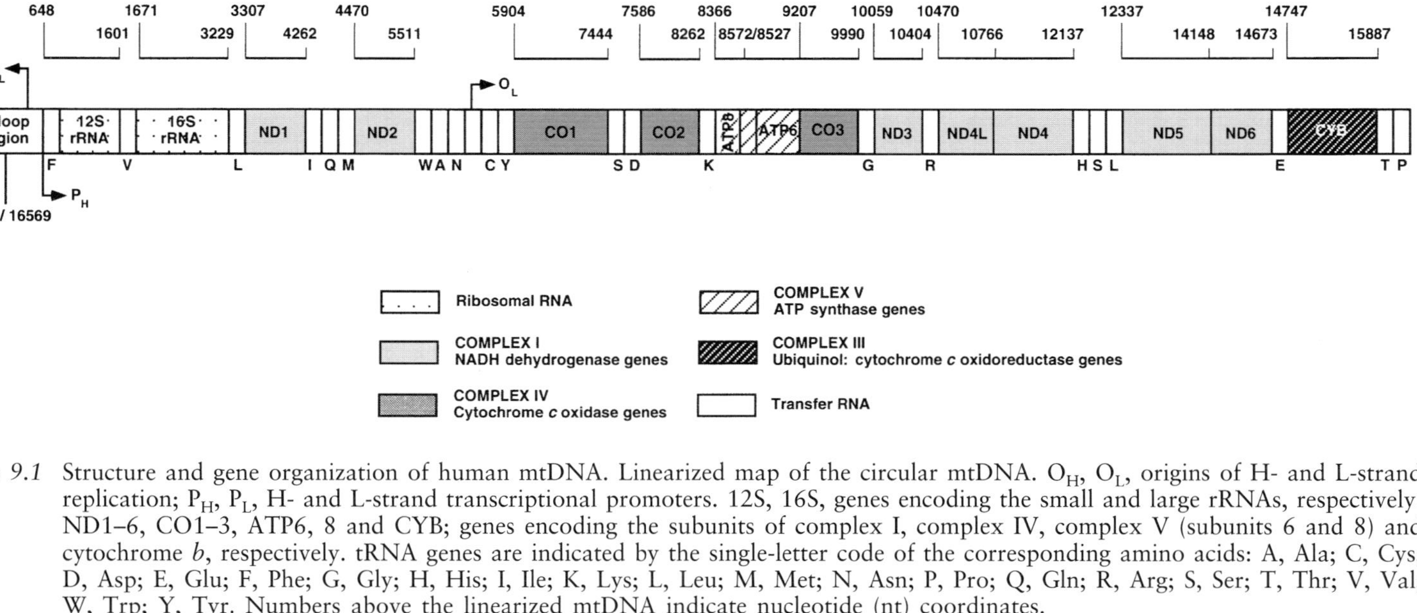

Figure 9.1 Structure and gene organization of human mtDNA. Linearized map of the circular mtDNA. O_H, O_L, origins of H- and L-strand replication; P_H, P_L, H- and L-strand transcriptional promoters. 12S, 16S, genes encoding the small and large rRNAs, respectively. ND1–6, CO1–3, ATP6, 8 and CYB; genes encoding the subunits of complex I, complex IV, complex V (subunits 6 and 8) and cytochrome *b*, respectively. tRNA genes are indicated by the single-letter code of the corresponding amino acids: A, Ala; C, Cys; D, Asp; E, Glu; F, Phe; G, Gly; H, His; I, Ile; K, Lys; L, Leu; M, Met; N, Asn; P, Pro; Q, Gln; R, Arg; S, Ser; T, Thr; V, Val; W, Trp; Y, Tyr. Numbers above the linearized mtDNA indicate nucleotide (nt) coordinates.

9.2.4   Maternal inheritance

At fertilization, the egg and the sperm donate an equal amount of nDNA information to the progeny. Conversely, human mtDNA is strictly maternally inherited (Giles *et al.*, 1980; Case and Wallace, 1981). Although a few hundred sperm mtDNAs contribute to the zygote, sperm mtDNA is eventually eliminated, suggesting a molecular mechanism for controlling the paternal mtDNA inheritance (Chen *et al.*, 1995; Kaneda *et al.*, 1995). Therefore, the mode of transmission for a pathogenic mtDNA mutation differs from Mendelian inheritance. A mother will pass her mtDNA-carrying mutation to her male and female offspring, but only her daughters will transmit it on to their progeny. Finally, since the mtDNA is uniparentally inherited, it very rarely, if ever, undergoes recombination.

9.2.5   Replicative segregation and heteroplasmy

mtDNA undergoes mitotic and meiotic segregation, meaning that at cell division or during the proliferation of female eggs leading to oocytes, the proportion of normal and mutant mtDNAs may shift in daughter cells (Michaels *et al.*, 1982; Shoffner and Wallace, 1990). This uneven distribution of wild-type and mutant mtDNAs, defined as heteroplasmy, may further change over repeated cell divisions, leading to either a pure mutant or pure normal homoplasmy mitochondrial genotype (Holt *et al.*, 1990; Zeviani *et al.*, 1991). Therefore, the amount of mutant sequence is an important factor in disease severity and correlates with the OXPHOS activity. The clinical implications of heteroplasmy and replicative segregation are self-evident. For example, twins with identical nuclear autosomal DNA may have different mitochondrial genotypes and hence different phenotypes.

9.2.6   Threshold expression

The nature of the mutation, the proportion of wild-type versus mutant mtDNAs and the relative dependence of a given tissue on mitochondrial energy production (OXPHOS system) determine the expression of a pathogenic mtDNA mutation. Consequently, a minimum critical number of mutant mtDNA is necessary to severely impair any specific organ system (Wallace, 1986). However, different organ tissues rely on mitochondrial energy to different extents depending on their vulnerability to the decline of mitochondrial ATP production. Therefore, the presence of a certain mutation in a tissue at a given percentage may be clinically silent, while the same percentage of mutant DNA can be phenotypically relevant in the central nervous system, muscle or the heart. In addition, since OXPHOS enzyme activities have been shown to decline with age in human organs and a variety of somatic mtDNA mutations accumulate over time, the extent of mtDNA damage correlates with those tissues most prone to age-related dysfunction (Singh *et al.*, 1989). The age-related decline in OXPHOS may explain why several mitochondrial diseases have a delayed onset. The simplest explanation is that as somatic mutations accumulate, inherited OXPHOS defects exacerbate the "programmed" decline until the combined defect is sufficient to result in energetic failure of the tissues.

9.2.7 *Evolution rate and mtDNA variation*

The mitochondrial genome has a mutation rate one order of magnitude higher than unique nDNA sequences (Brown *et al.*, 1979, 1982; Brown 1980; Anderson *et al.*, 1982; Wallace *et al.*, 1987; Esteal, 1991). Estimates of the average number of base pair differences between two human mitochondrial genomes range between 9.5 and 66 (Cann *et al.*, 1987; Merriwether *et al.*, 1991). This is explained by the fact that mitochondria lack an efficient DNA repair system as well as protective proteins such as histones. In addition, mtDNA is associated physically with the inner mitochondrial membrane, where highly mutagenic oxygen radicals are generated as by-products of the OXPHOS system (Bandy and Davison, 1990). Therefore, a considerable number of restriction fragment length polymorphisms, coding region nucleotide variants and other mtDNA mutations have accumulated sequentially along radiating maternal lineages. As the human population has migrated and

Table 9.1 Common continent-specific mtDNA variants

Continent	Haplogroup	mtDNA variants	Continental mtDNA (%)
Africa	L	3592+Hpa I	76
Total			76
Europe	H	7025–Alu I	39
Europe	I	1715–Dde I	
		10028+Alu I	
		4529–Hae II	
		8249+Ava II	
		16389+Bam HI/Mbo I	7
Europe	J	13704–Bst NI; 16065–Hinf I	9
	K	9052–Hae II/Hha I	8
Total			63
Asia	A	663+Hae III	12
Asia	B	8271–8281 9 bp del	
		16517+Hae III	4
Asia	F	12406–Hpa I/Hinc II	
		16517+Hae III	4
Asia	M	10394+Dde I	
		10397+Alu I	57
Asia	M-C	13259–/13262+Hinc II/Alu I	21
Asia	M-D	5176–Alu I	14
Total			77
America	A (Amerinds)	663+Hae III	44
America	A (Na-Dene)	663+Hae III; 16329–Rsa I	29
America	B (Amerinds)	8271–8281 9 bp del	
		16517+Hae III	22
America	M-C (Amerinds)	13259–/13262+Hinc II/Alu I	18
America	M-D (Amerinds)	5176–Alu I	16
Total			100
Total	Africa + Europe + Asia + America		72

Table 9.2 mtDNA mutations

HGM mutation designation	nt position	Gene/AA	Homoplasmy	Heteroplasmy
Confirmed (at least 2 independent pedigrees have been described)				
MTRNR1*DEAF1555G	A to G: 1555	12S rRNA	+	−
MTTL1*MELAS3243G	A to G: 3243	tRNA$^{Leu(UUR)}$	−	+
MTTL1*MM3251G	A to G: 3251	tRNA$^{Leu(UUR)}$	−	+
MTTL1*MELAS3256T	C to T: 3256	tRNA$^{Leu(UUR)}$	−	+
MTTL1*MMC3260G	A to G: 3260	tRNA$^{Leu(UUR)}$	−	+
MTTL1*MM3302G	A to G: 3302	tRNA$^{Leu(UUR)}$	−	+
MTND1*LHON3460A	G to A: 3460	ND1: Ala to Thr	+	−
MTND1*LHON4216C	T to C: 4216	ND1: Tyr to His	+	−
MTCOI*DEAF7445G	A to G: 7445	COI or tRNA$^{Leu(UCN)}$	+	+
MTTK*MERRF8344G	A to G: 8344	tRNA^{Lys}	−	+
MTTK*MERRF8356C	T to C: 8356	tRNA^{Lys}	−	+
MTATP6*NARP8993C	T to C: 8993	ATP6: Leu to Pro	−	+
MTATP6*NARP8993G	T to G: 8993	ATP6: Leu to Arg	−	+
MTND4*LHON11778A	G to A: 11 778	ND4: Arg to His	+	+
MTND5*LHON13708A	G to A: 13 708	ND5: Ala to Thr	+	−
MTND6*LDYT14459A	G to A: 14 459	ND6: Ala to Val	+	+
MTND6*LHON14484C	T to C: 14 484	ND6: Met to Val	+	+
MTCYB*LHON15257A	G to A: 15 257	Cy b: Asp to Asn	+	−
MTCYB*LHON15812A	G to A: 15 812	Cy b: Val to Met	+	
MTTT*LIMM15923G	A to G: 15 923	tRNA^{thr}	ND	+
Provisional (a single pedigree has been described)				
MTRNR1*ADPD956–965	5-bp ins 956–965	12S rRNA	+	−
MELAS:1642A	G to A: 1642	tRNA^{val}	−	+
MTRNR2*ADPD3196A	G to A: 3196	16S rRNA	+	+
MTTL1*MM3250C	T to C: 3250	tRNA$^{Leu(UUR)}$	−	+
MTTL1*MELAS3252G	A to G: 3252	tRNA$^{Leu(UUR)}$	−	+
MTTL1*MELAS3271C	T to C: 3271	tRNA$^{Leu(UUR)}$	−	+
MTTL1*PEM3271Δ	1-bp del: 3271	tRNA$^{Leu(UUR)}$	−	+
MTTL1*MELAS3291C	T to C: 3291	tRNA$^{Leu(UUR)}$	−	+
MTTL1*MMC3303T	C to T: 3303	tRNA$^{Leu(UUR)}$	+	+
MTND1*NIDDM3316A	G to A: 3316	ND1: Ala to Thr	+	−
MTND1*LHON3394C	T to C: 3394	ND1: Tyr to His	+	−
MTND1*ADPD3397G	A to G: 3397	ND1: Met to Val	+	−
MTND1*LHON4136G	A to G: 4136	ND1: Tyr to Cys	+	−
MTND1*LHON4160C	T to C: 4160	ND1: Leu to Pro	+	−
MTTI*FICP4269G	A to G: 4269	tRNA^{Ile}	−	+
CPEO:4285C	T to C: 4285	tRNA^{Ile}	−	+
MTTI*MICM4300G	A to G: 4300	tRNA^{Ile}	−	+
MTTI*FICP4317G	A to G: 4317	tRNA^{Ile}	ND	ND
CME:4320	C to T: 4320	tRNA^{Ile}	−	+
MTTQ:ADPD4336C	T to C: 4336	tRNA^{Gln}	+	−
MTND2*LHON4917G	A to G: 4917	ND2: Asp to Asn	+	−
MTND2*LHON5244A	G to A: 5244	ND2: Gly to Ser	−	+
MTTW*DEMCH5549A	G to A: 5549	tRNA^{Trp}	−	+
MTTN*CPEO5692G	A to G: 5692	tRNA^{ASN}	−	+
MTTN*CPEO5703G	G to A: 5703	tRNA^{ASN}	−	+
MTCOI*LHON7444A	G to A: 7444	COI ter to Lys	+	−

Table 9.2 continued

HGM *mutation* designation	*nt position*	*Gene/AA*	*Homo-plasmy*	*Hetero-plasmy*
MTTS1*AMDF7472+C	1 bp ins (C): 7472	tRNA[Ser(UCN)]	+	+
MTTS1*MERME7512C	T to C: 7512	tRNA[Ser(UCN)]	+	+
EM: 8272–8289 insertion	9 bp (C)$_5$TCTA ins (8272–8289)	Non-coding region	ND	+
SN: 8851C	T to C: 8851	ATP6: Trp to Arg	–	+
MTATP6*LHON9101C	T to C: 9101	ATP6: Ile to Thr	+	–
MTATP6*FBSN9176C	T to C: 9176	ATP6: Leu to Pro	+	+
MTCO3*LHON9438A	G to A: 9438	CO3: Gly to Ser	+	–
CrMy: 9487–9501 del	15-bp del: 9487–9501	CO3	–	+
MTCO3*LHON9738T	G to T: 9738	CO3: Ala to Thr	+	–
MTCO3*LHON9804A	G to A: 9804	CO3: Ala to Thr	+	–
MTCO3*PEM9957C	T to C: 9957	CO3: Phe to Leu	–	+
MTTG*MHCM9997C	T to C: 9997	tRNA[Gly]	ND	+
MTTG*CIPO10006G	A to G: 10006	tRNA[Gly]	ND	ND
SD: 10044G	A to G: 10044	tRNA[Gly]	–	ND
LHON+SD: 11696G	A to G: 11696	ND4: Val to Ile	+	+
MTTS1*CIPO12246G	C to G: 12246	tRNA[Ser(AGY)]	ND	ND
MTTL2*CPEO12311C	T to C: 12311	tRNA[Leu(CUN)]	+	+
CPEO: 12315A	G to A: 12315	tRNA[Leu(CUN)]	–	+
MTND5*LHON13730A	G to A: 13730	ND5: Gly to Glu	–	+
MTTE*MDM14709G	A to G: 14709	tRNA[Glu]	–	+
MTTP*MM15990T	C to T: 15990	tRNA[Pro]	–	+

Refer to Wallace *et al.* (1996) and MITOMAP for primary references for each of the listed mutations. According to the Report of the Committee on the Human Mitochondrial Genome, five descriptors designate mutations: (1) chromosome (MT); (2) mtDNA gene name; (3) clinical phenotype; (4) nucleotide position; and (5) mutant base. An asterisk separates designators 2 and 3. HGM, Human Gene Map; AD, Alzheimer's disease; ADPD, Alzheimer's disease plus Parkinson's disease; CIPO, chronic intestinal pseudo-obstruction with myopathy and ophthalmoplegia; CME, cardiomyopathy and encephalopathy; CPEO, chronic progressive external ophthalmoplegia; CrMy, cramps and myalgia; DEAF, maternally inherited deafness or aminoganglyoside-induced deafness; DMDF, diabetes mellitus and deafness; EM, encephalomyopathy; FICP, fatal infantile cardiomyopathy with a MELAS-associated cardiomyopathy; LDYT, Leber's hereditary optic neuropathy plus dystonia; LHON, Leber's hereditary optic neuropathy; LHON+SD, Leber's hereditary optic neuropathy and spastic dystonia; LIMM, lethal infantile mitochondrial myopathy; MELAS, mitochondrial encephalomyopathy, lactic acidosis and stroke-like episodes; MERFF, myoclonic epilepsy and ragged-red fiber disease; MM, mitochondrial myopathy; MMC, mitochondrial myopathy and cardiomyopathy; NARP, neurogenic muscle weakness, ataxia and retinitis pigmentosa; PEM, progressive encephalomyopathy; SD, sudden death. ND, not determined.

colonized different geographical region of the world, these lineages have consequently diverged. Continent-specific polymorphic restriction sites and their estimated frequencies are shown in Table 9.1. The continental origin of roughly 70% of the African, European, Asian and Native American mtDNAs can be assessed using these variants. Analysis of mtDNA variation in human populations has provided a large amount of information about human origins, evolution and patterns of migrations as well as variability of human genetic disease patterns (Shoffner, 1996; Jacobs, 1997; Zeviani *et al.*, 1997; DiMauro and Hirano, 1998; Morris *et al.*, 1998; Shah *et al.*, 1998).

Table 9.3 mtDNA deletions

Gene deleted	Deletion size (bp)	Coordinates (nt–nt)
D loop-	–6	105–112
MTTFH–MTND2	–4681	470–5152
MTTFH–MTND2	–4939	502–5443
MTTM–	–3895	547–4443
MTRNR2–MTND2	–3610	1836–5447
MTRNR2–MTND6	–10987	3173–14161
MTND1–MTND1	–264	3323–3588
MTTQ–MTCYB	–10422	4398–14822
MTTC–MTND5	–8136	5786–13923
MTTC–MTND5	–6973	5793–12767
MTND5–MTTY	–6825	5835–12661
MTCO1–MTND6	–8400	6023–14424
MTCO1–MTATP6	–3104	6074–9179
MTCO1–MTND5	–7723	6075–13799
MTCO1–MTND5	–7279	6226–13456
MTCO1–MTND5	–7864	6238–14103
MTCO1–MTND5	–7663	6325–13989
MTCO1–MTND5	–7664	6329–13994
MTCO1–MTND5	–7663	6330–13994
MTCO1–MTND5	–7715	6380–14096
MTCO1–MTND5	–7669	6465–14135
MTCO1–MTND6	–7402	7193–14596
MTCO1–MTND5	–6037	7438–13476
MTTS1–MTTT	–8476	7449–15926
MTTS1–MTND4	–3512	7491–11004
MTTS1–MTND5	–5268	7493–12762
MTND6–MTTS1	–6926	7501–14428
MTCO2–MTCYB	–7804	7635–15440
MTCO2–MTCYB	–7767	7669–15437
MTCO2–MTCYB	–4666	7697–12364
MTCO2–MTND5	–6016	7777–13794
MTCO2–MTCYB	–6990	7808–14799
MTCO2–MTCYB	–7565	7815–15381
MTCO2–MTND5	–6305	7829–14135
MTCO2–MTND5	–6063	7841–13905
MTCO2–MTND5	–6063	7841–13905
MTCO2–MTCO3	–1902	7845–9748
MTCO2–MTCYB	–7521	7974–15496
MTCO2–MTATT	–8042	8032–16075
MTCO2–MTCYB	–7128	8210–15339
MTCO2–MTND5	–5777	8213–13991
Intergen-region	–9	8271–8281
MTTK–MTND5	–5491	8278–13770
MTTK–MTCYB	–6750	8304–15055
MTND5–MTATP8	–4467	8426–12894
MTND5–MTATP8	–4977	8468–13446
MTND5–MTATP8	–4977	8469–13447
MTATP8–MTCYB	–6903	8517–15421
MTATP6–MTND5	–5196	8563–13758
MTATP6–MTND6	–6032	8563–14596
MTATP8–MTCYB	–4665	8570–13236

Table 9.3 continued

Gene deleted	Deletion size (bp)	Coordinates (nt–nt)
MTATP8–MTCYB	–7153	8573–15727
MTATP6–MTCYB	–7150	8580–15731
MTATP6–MTTP	–7374	8582–15957
MTATP6–MTCYB	–7038	8623–15662
MTATP6–MTND5	–5261	8624–13886
MTATP6–MTND5	–4881	8631–13513
MTATP6–MTTP	–7446	8637–16084
MTATP6–MTTP	–7436	8648–16085
MTATP6–MTND5	–5015	8707–13723
MTATP6–MTCYB	–7031	8823–15855
MTATP6–MTCYB	–6067	8828–14896
MTATP6–MTTP	–7079	8992–16072
MTATP6–MTND5	–4671	9144–13816
MTATP6–MTND6	–5100	9180–14281
MTATP6–MTND5	–3717	9191–12909
MTCO3–MTCYB	–6377	9238–15576
MTCO3–MTND5	–4507	9357–13865
MTCO3–MTND5	–3539	9515–13055
MTCO3–MTND5	–3397	9574–12972
MTTG–MTTT	–5901	9995–15897
MTTG–MTCYB	–5025	10050–15076
MTND6–MTND3	–4534	10058–14593
MTND3–MTTT	–5790	10154–15945
MTND3–MTND6	–4265	10169–14435
MTND3–MTND5	–3562	10190–13753
MTND3–MTND5	–2461	10367–12829
MTND3–MTCYB	–5199	10370–15570
MTND4L–MTTT	–5325	10587–15913
MTND4L–MTND5	–2607	10598–13206
MTND4L–MTCYB	–4190	10665–14856
MTND4L–MTCYB	–4191	10676–14868
MTND4L–MTND5	–3379	10744–14124
MTND4–MTCYB	–4420	10941–15362
MTND4–MTCYB	–4884	10961–15846
MTND4–MTND5	–2747	11232–13980
MTND4–MTCYB	–4417	11368–15786
MTND4–MTND6	–2309	12102–14412
MTND4–MTND6	–2310	12103–14414
MTND4–MTND6	–2308	12113–14422
MTTH–MTCYB	–3151	12203–15355
MTND4–MTCYB	–4884	10952–15837

9.3 Human diseases caused by mtDNA mutations

Mutations in the mtDNA can be classified into two major groups: (1) point mutations that have been identified in the protein-coding genes, tRNA or rRNA
genes, as shown in Table 9.2; and (2) rearrangements, some of which are listed in
Table 9.3 (Giles *et al.*, 1980; Case and Wallace, 1981; Holt *et al.*, 1988; Zeviani
et al., 1988; Poulton and Holt, 1994). Identification of both classes may be a

complex task due to heteroplasmy and depends on the sensitivity of the procedure employed. In addition, mtDNA lesions associated with Mendelian traits have been described (Zeviani *et al.*, 1990; Cormier *et al.*, 1991; Poulton *et al.*, 1991; Ballinger *et al.*, 1992; Rotig *et al.*, 1992; Bernes *et al.*, 1993; Suomalainen *et al.*, 1995, 1997). The broad spectrum of resulting degenerative diseases may involve all major tissues including the central nervous and endocrine systems, heart, muscle, kidney, liver, pancreas and hematopoietic tissue.

9.3.1 *mtDNA point mutations*

Point mutations are usually maternally inherited, consistent with the maternal inheritance of mtDNA. According to the Report of the Committee on the Human Mitochondrial Genome, mutations are designated by the gene name, followed by an asterisk, a clinical phenotype designator, the nucleotide position, and the mutant base (http://gen.emory.edu/mitomap.html). Among the missense mutation diseases, the best studied are Leber's hereditary optic neuropathy (LHON) and neuropathy, ataxia, and retinitis pigmentosa (NARP).

LHON was the first human disease linked to an inherited mutation in mtDNA and is characterized by maternal inheritance and late-onset blindness caused by optic nerve degeneration occurring most predominantly in young men. Although the strong male bias suggests interaction between an X-linked gene and an mtDNA defect, recent studies indicate this is unlikely (Chen *et al.*, 1989; Sweeney *et al.*, 1992; Newman, 1993). LHON begins typically with acute or subacute painless, central and monolateral visual loss. The involvement of the second eye usually occurs within a few weeks. In most patients, deterioration of visual function is severe and permanent, with recovery generally restricted to a few central degrees. A triad of fundal abnormalities is reported to be pathognomonic of LHON: (1) circumpapillary telangiectatic microangiopathy; (2) pseudoedema of the disc; and (3) absence of staining on fluorescein angiography. Fundal signs are frequently not present even in the acute phase of LHON, so their absence does not exclude a diagnosis of LHON. In the majority of cases, visual dysfunction appears to be the only evident manifestation of the disease. However, in some pedigrees, cardiac, skeletal and neurological abnormalities have been described.

A number of single-base changes in seven mitochondrial genes encoding subunits of the respiratory chain complexes I, III and IV have been reported associated with typical forms of LHON. Some of these changes are considered "primary" or class I mutations (3460A, 14459A, 11778A and 14484C) since they are restricted to LHON families. They are not found in controls and are thought to be sufficient to cause disease (Wallace *et al.*, 1988a; Howell *et al.*, 1991; Huoponen *et al.*, 1991, 1993; Johns *et al.*, 1992; Mackey and Howell, 1992; Mashima *et al.*, 1993; Brown and Wallace, 1994; Jun *et al.*, 1994; Shoffner *et al.*, 1995). Others have been categorized as "secondary" or class II mutations (3394, 4025, 4136, 4216, 4917, 5244, 7444, 12 811, 13 637, 13 708, 13 967, 14 459, 15 257, 15 812) since they are also present in normal controls. Class II mutations are often found in association with primary or other secondary mutations and are thought to exacerbate LHON expression. The only exception is the nucleotide change at position 4136, which is thought to be an intragenic suppressor mutation ameliorating the neurological abnormalities and complex I deficiency. Primary mutations and most secondary mutations result in amino acid substitutions in respiratory chain complex I and cytochrome *b* of complex III. Although these changes are generally associated with reduced biochemical activities, the correlation with development of vision loss is not clear, and further studies are required to provide insight into the pathogenesis of the disease. Non-typical cases of LHON characterized by unexplained optic neuropathy, including those with late-onset, negative family history, normal funduscopic appearance or presenting in females have also been described (Weiner *et al.*, 1993).

Leigh syndrome is a progressive neurodegenerative disorder, usually of infancy, characterized by developmental delay, loss of developmental milestones, ataxia, seizures, peripheral neuropathy, myopathy, optic atrophy or retinal pigmentary lesions and persistent or intermittent elevation of blood and/or cerebrospinal fluid lactate levels (Leigh, 1951; Pincus, 1972). Neuroradiological findings include bilateral, symmetrical lesions of the basal ganglia, thalamus and brain stem (Schwartz *et al.*, 1981). Neuropathological examination reveals widespread necrotic and cavitary lesions involving the entire neuraxis with relative neuronal sparing (Pincus, 1972). The findings are similar to those seen in Wernicke encephalopathy. Some investigators believe that the diagnosis can only be made after neuropathological examination reveals these characteristic lesions including scattered areas of capillary proliferation. There is no effective treatment and patients usually die in infancy or early childhood. Both nuclear and mitochondrial DNA gene mutations have been demonstrated to result in Leigh disease. Deficiency of pyruvate dehydrogenase enzyme activity (DeVivo *et al.*, 1976) as well as primary lesions involving the respiratory chain (Willems *et al.*, 1977) can result in this disease process. Although the NARP mutations were originally associated with a neurological disorder involving peripheral neuropathy, ataxia and retinitis pigmentosa in older children, they have also been associated with Leigh disease (Tatuch *et al.*, 1992). In these instances, the disease is maternally inherited. Mutations 8993G, 8993C and 9176C, which always display a degree of heteroplasmy correlating with the severity of the disease, alter the mtATP6 gene (Holt *et al.*, 1990; Tatuch *et al.*, 1992; Shoffner *et al.*, 1992; Santorelli *et al.*, 1993, 1994; DeVries *et al.*, 1993; Trounce *et al.*, 1994; Thyagarajan *et al.*, 1995). The 8993G mutation has been linked to the inhibition of proton translocation of ATP synthase and the 8993C transition has also been described in other NARP patients.

Base substitutions altering rRNA and tRNA genes have been identified in patients with a wide variety of clinical presentations. The mildest mutations tend to be homoplasmic, occur in specific mtDNA lineages, and are associated with late-onset diseases such as Alzheimer's disease (AD), Parkinson's disease (PD), and neurosensory hearing loss. Approximately 5% of Caucasian AD patients carry the 4336C mutation (Hutchin and Cortopassi, 1995), which appears to define an mtDNA lineage predisposed to both AD and PD. Patients within this lineage can harbor additional contributory mutations such as the 3397G missense mutation and a five-nucleotide insertion between nucleotides 956 and 965 (Shoffner *et al.*, 1993).

A homoplasmic mutation in the 12S rRNA gene, 1555G, has been found to correlate with neurosensory hearing loss. In addition, exposure to aminoglycoside antibiotics or other environmental or nuclear genetic factors may precipitate deafness in individuals carrying the 1555G mutation (Hutchin *et al.*, 1993; Prezant *et al.*, 1993; Estivill *et al.*, 1998).

Certain mutations cause quite consistent phenotypes with the severity being determined by the age of the individual and the proportion of mutant to normal mtDNA in the affected tissues (DiMauro and Moraes, 1993). Age may play a deleterious role when a respiratory defect has been inherited. One example of a condition resulting from a mutation in a mitochondrial tRNA gene is myoclonus epilepsy with ragged-red fibers (MERRF). This is a devastating, maternally inherited, neuromuscular disorder characterized by myoclonic epilepsy, muscle weakness and wasting, cerebellar ataxia, deafness and dementia. The most commonly observed mutation is an A to G transition at nucleotide position 8344 in the tRNA Lys gene (Wallace *et al.*, 1988b; Shoffner *et al.*, 1990; Noer *et al.*, 1991; Enriquez *et al.*, 1995; Masucci *et al.*, 1995). Another mutation reported in the same gene at position 8356 and a T to C mutation at position 7512 in the tRNA Ser (UCN) gene have been observed in a family with a MERRF/MELAS overlap syndrome (Nakamura *et al.*, 1995).

Other tRNA mutations give several distinctive clinical presentations, the most notable being mitochondrial encephalomyopathy, lactic acidosis and stroke-like episodes (MELAS). MELAS is defined by stroke-like episodes, corroborated by focal brain lesions, and lactic acidosis and/or presence of RRF in muscle biopsy. Elevated lactate levels have also been reported. Dementia, recurrent headache and vomiting, focal or generalized seizures as well as deafness are often associated signs. MELAS was first associated with an A to G transition at position 3243, a heteroplasmic point mutation in the tRNA Leu (UUR) sequence (Goto *et al.*, 1990; Goto, 1995). As in MERRF syndrome, genetic heterogeneity has been confirmed in MELAS by the identification of other MELAS-associated point mutations. These include nucleotide changes at positions 3252, 3256, 3271 and 3291 in the tRNA Leu (UUR) gene and position 9957 in the CO III gene (Goto *et al.*, 1991; Morten *et al.*, 1993; Manfredi *et al.*, 1995). In addition, the A3243G mutation has been detected in patients with Leigh syndrome as well as in chronic progressive external ophthalmoplegia (CPEO, discussed in Section 9.3.2), with myopathy alone, and in pedigrees with maternally inherited diabetes mellitus and deafness (Reardon *et al.*, 1992; Moraes *et al.*, 1993; Mariotti *et al.*, 1994).

9.3.2 *mtDNA rearrangements*

Naturally occurring rearrangements in mtDNA have been associated with ocular myopathies including CPEO, Kearns–Sayre syndrome (KSS), Pearson's marrow/pancreas syndrome and maternally inherited, adult-onset diabetes and deafness (Holt *et al.*, 1988; Rötig *et al.*, 1988; Zeviani *et al.*, 1988; Moraes *et al.*, 1989; Ballinger *et al.*, 1992; Poulton *et al.*, 1995; Smith *et al.*, 1995). KSS and CPEO are multisystem disorders regularly presenting with ophthalmoplegia and ptosis. KSS is the more severe, presenting with an onset prior to age 20, and manifesting at least one of the following: cardiac conduction defects, cerebellar ataxia, or elevated cerebrospinal fluid protein. CPEO is characterized by bilateral ptosis of the eyelid, and ophthalmoplegia usually associated with variable degrees of proximal muscle weakness and wasting and exercise intolerance. The CPEO-plus syndrome includes other system involvement and overlap with KSS. The pediatric-onset Pearson syndrome manifests as bone marrow failure and pancytopenia, frequently associated with exocrine pancreatic insufficiency, hepatic and renal failure, and other neuromuscular problems. Pearson syndrome patients generally die young. However, infants surviving into childhood may develop the clinical features of KSS (Rötig *et al.*, 1990; McShane *et al.*, 1991). Therefore, the three phenotypes described may represent different expressions of the same molecular defect. More than half of all KSS and CPEO cases are due to spontaneous mtDNA rearrangements such as duplications and deletions of mtDNA. A common breakpoint is usually shared. Finally, large-scale rearrangements have been described in patients with different phenotypes such as the Wolfram syndrome characterized by diabetes insipidus, diabetes mellitus, optic atrophy and deafness, diffuse leukodystrophy and infantile chronic diarrhea (Rötig *et al.*, 1993; Cormier-Daire *et al.*, 1994; Barrientos *et al.*, 1996). These observations suggest that KSS, CPEO, Pearson syndrome and diabetes mellitus with deafness may represent a continuum of clinical severity that may reflect the pathogenicity of these mtDNA rearrangements. Most of the mtDNA deletions range from 1.3 to 7.6 kb in length, are localized between the end of the D-loop region and the origin of the L-strand replication, and span more than one gene, including both mRNA and tRNA genes. DNA sequence analysis indicates that deletions are usually flanked by direct repeats of variable length. Therefore, it is suggested that mtDNA rearrangements can occur through a mechanism of slippage-mispairing of the single mtDNA strands during replication (Schon *et al.*, 1989).

9.4 Origin and maintenance of mtDNA mutations

Several factors are thought to play important roles in the genesis of mtDNA mutations. These include: (1) the large number of mtDNA molecules per cell; (2) the high rate of mtDNA replication; (3) the physical proximity of mtDNA molecules to the electron transport chain, an important generator of cytosolic oxygen free-radical species; and (4) the possibility of a DNA repair system that is less effective than the nuclear repair system (Torroni and Wallace, 1994). However, the contribution of mtDNA repair mechanisms that are under control of nuclear genes to the prevalence of mtDNA mutations is unknown. When a mutation develops, the mtDNA molecule may or may not replicate. If it does undergo successful replication, the degree to which the mutant molecules contribute to the mtDNA pool will depend on random drift during replicative segregation. The randomness of this process dictates that both the ratio of mutant to wild-type mtDNA within a mitochondrion and the number of mitochondria with high or low mutant mtDNA molecule copy numbers are variable. However, the degree to which the ratio of mutant to total mtDNA species can rise within a cell or tissue will depend on cell survival. If a high percentage of mutant mtDNA does not support a critical rate of oxidative phosphorylation in a cell that normally depends on aerobic metabolism, the cell will die. Cells may be able to survive with the mutant molecules if their presence does not seriously compromise cell function, including ATP synthesis.

Mutant mtDNA species may be passed from one generation to the next if they are present in an oocyte that undergoes fertilization. In the developing embryo and fetus, random segregation will, of course, determine the degree of heteroplasmy in cells within different tissues. Heteroplasmy refers to the coexistence of normal and mutant mtDNAs within the same individual, organ, tissue or cell.

Based on our current knowledge, most human diseases caused by mtDNA mutations are uncommon. The mtDNA mutations that cause MELAS, MERFF, NARP and LHON, for example, are rarely detected among most populations around the world. At the present time, because of the rarity of these mutations, it is not possible to assess whether the incidence of mtDNA mutations varies in different geographical areas of the world. A corollary is that, currently, it is not possible to accurately assess how often mutations occur against the normal background of mtDNA polymorphisms in different populations and ethnic groups. By this, we mean sequence differences in mtDNA among different world populations that do not produce disease but can be readily analyzed as an mtDNA haplotype.

The only exception to this statement is the observation that the 11 778 and 14 484 mutations, which cause LHON, are highly associated with the European-specific J haplotype (Torroni *et al.*, 1997). This haplotype defines the polymorphic background for certain groups such as an Italian population. The J-haplotype mtDNA background may play a role in the penetrance of these two primary LHON mutations. However, this is a population genetics phenomenon. There is no direct evidence that the mtDNA background plays a role in heteroplasmy ratios. Even in related individuals with the same mtDNA genetic background, the ratio of mutant to normal mtDNA in different tissues may vary greatly. In addition, in some mtDNA disorders the mutation is homoplasmic and apparently independent of background mtDNA polymorphisms. Different ratios of mutant to normal mtDNA are due to random drift during replicative segregation, but for certain diseases the drift is so skewed that the mutation appears homoplasmic in certain tissues. A "bottlenecking" phenomenon during oogenesis (Hauswirth and Laipis, 1982; Howell *et al.*, 1992) has been proposed to explain the relatively rapid fixation of a mutation in an individual as well as the mechanism that generates homoplasmic states. In this model, only a small fraction of the total mtDNA molecules in a female germline pool, and in this case the mutant genomes, are transmitted to an offspring. In an extreme case, the heteroplasmic mother would give birth to a child homoplasmic for an mtDNA mutation. In addition, it has been suggested that mutations such as duplications in the mtDNA control region may confer preferential replication, especially in comparison with mtDNA molecules with disease-producing mutations in tRNA or polypeptide respiratory chain genes. Whether replicative preference would also be shown over wild-type molecules is a matter of debate.

9.5   mtDNA variation in human populations

The study of polymorphic sequence differences in mtDNA has provided a large body of information on human evolution, the origins of different populations and patterns of human migration. A large volume of data has accumulated that defines the individual mtDNA polymorphisms characterizing mtDNA haplotypes. Relatively speaking, the mtDNA molecule is easy to study since it is small, contains only 16 569 nucleotides and has a high copy number compared to nuclear genes. It has been estimated that there are 1000 copies of mtDNA for each nuclear gene. Thus, mitochondrial haplotype analysis of random non-disease-causing mutations accumulated as markers over thousands of years provides a simple method of fingerprinting human populations. mtDNA genomic material is maternally inherited and is not subject to meiotic recombination events, commonplace in nuclear genes. These characteristics add to the utility of mtDNA haplotype analyses for recording the evolution of the molecule in man.

Simply put, the molecular evolution of mtDNA depends on the accumulation of random mutations. As mutations accumulate over time, newer mutations become associated with previously acquired mutations. mtDNA copy number and rate of replication contribute to a relatively fast mutational rate for mtDNA and aid in establishing diversity among human haplotypes. Thus, mtDNA polymorphisms are easily applied to tracing the ancient steps of man.

The method used to determine haplotype in an individual requires purified mtDNA from cells such as leukocytes, oligonucleotide primers and the polymerase chain reaction (PCR) for amplifying mtDNA fragments, a panel of restriction endonucleases and a set of hybridization probes for fragment identification. These enzymes permit recognition of variations in mtDNA structure and, thus, construction of a haplotype map. This method has been used to generate data on mtDNA polymorphisms in African, European, Asian and American Indian populations (Torroni *et al.*, 1993a,b, 1994a,b,c,d, 1996; Stone and Stoneking, 1993; Chen *et al.*, 1995; Francalacci *et al.*, 1996).

Most mtDNA variations are continent-specific or limited to the principal ethnic groups of a specific continent. These variations can be described by a limited number of well-defined haplogroups. The study of mtDNA samples from Africans reveals a principal mtDNA lineage (L) defined by a specific coding region mutation. This lineage represents the majority of individuals in subSaharan African populations and all of the African Pygmies. This massive mtDNA lineage is not found in non-Africans, except in some individuals whose families experienced recent racial admixture. Investigators estimate that this lineage originated in Africa between 100 000 and 130 000 years ago.

Data from Asian subjects are more complicated. One lineage (M) accounts for approximately 55% of all Asians and is Asia-specific. Within the M lineages, there are sublineages C, D, Q and F. The other Asian lineages are A, B and F. The M lineage is calculated to have originated between 65 000 and 90 000 years ago. Analysis of mtDNA in American Indians confirms their initial migration from Asia. Their mtDNA is generally one of four Asian lineages or sublineages, A, B, C or D. Nearly all of the Na-Dene populations of Alaska and Western Canada, as well as the Apache and Navajo groups, share only lineage A (Torroni *et al.*, 1992). In contrast, the Amerinds, comprising most of the Native American populations, have all four haplogroups A, B, C and D. It is estimated that these Asian groups migrated to the Americas 12 000 to 44 000 years ago. However, these estimates are controversial and not accepted by all scholars in the field (Greenberg, 1986).

There are at least nine distinct haplogroups evident in European populations. Lineage U is shared between Africans and Europeans. Lineage H is the most common among all Europeans and may have originated 41 000 to 51 000 years ago. Lineage K is most common in Ashkenazi Jews but is also very common in Sicilians and Swedes. In Caucasians from North America, haplogroups H, I, J and K are most common. North American haplogroups are not present in African, Asian or Native American populations.

The analysis of mtDNA in people from around the world has provided a most valuable tool for studying the origins of man, human migration patterns, drift and, sometimes, unambiguous recognition of the derivation of a particular race or ethnic group. In addition, this type of study permits us to characterize benign polymorphisms as opposed to disease-producing mutations. The latter are detected only on rare occasions.

Acknowledgments

Data used in preparing this manuscript including text, figures and tables were derived with permission from the MITOMAP: Mitochondrial Human Genome Database at Emory University in Atlanta (http://www.gen.emory.edu/mitomap.html) in July 1998.

The authors wish to thank Dr Antonio Torroni, Dipartimento di Genetica e Biologia Molecolare, Università "La Sapienza", Roma, and Dr Massimo Zeviani, Dipartimento di Biochimica e Genetica, Istituto Neurologico Carlo Besta, Milano, Italy, for helpful suggestions and critical discussions.

References

Anderson, S., Bankier, A.T., Barrell, B.G., de Bruijn, M.H.L., Coulson, A.R., Drouin, J., Eperon, I.C., Nierlich, D.P., Roe, B.A., Sanger, F., Schreier, P.H., Smith, A.J.H., Staden, R. and Young, I.G., 1981, Sequence and organization of the human mitochondrial genome. *Nature*, **290**, 457–465.

Anderson, S., deBruijn, M.H.L., Coulson, A.R., Eperon, I.C., Sanger, F. and Young, I.G., 1982, Complete sequence of bovine mitochondrial DNA. Conserved features of the mammalian mitochondrial genome. *J. Molec. Biol.*, **156**, 683–717.

Attardi, G., Chomyn, A., Montoya, J. and Ojala, D., 1982, Identification and mapping of human mitochondrial genes. *Cytogenet. Cell Genet.*, **32**, 85–98.

Au, H.C., Ream-Robinson, D., Bellew, L.A., Broomfield, P.L., Saghbini, M. and Scheffler, I.E., 1995, Structural organization of the gene encoding the human iron-sulfur subunit of succinate dehydrogenase. *Gene*, **159**, 249–253.

Ballinger, S.W., Shoffner, J.M., Hedaya, E.V., Trounce, I., Polak, M.A., Koontz, D.A. and Wallace, D.C., 1992, Maternally transmitted diabetes and deafness associated with a 10.4 kb mitochondrial DNA deletion. *Nature Genet.*, **1**, 11–15.

Bandy, B. and Davison, A.J., 1990, Mitochondrial mutations may increase oxidative stress: implications for carcinogenesis and aging? *Free Radicals Biol. Med.*, **8**, 523–539.

Barrell, B.G., Bankier, A.T. and Drouin, J., 1979, A different genetic code in human mito-chondria. *Nature*, **282**, 189–194.

Barrientos, A., Casademont, J., Saiz, A., Cardellach, F., Volpini, V., Solans, A., Tolosa, E., Urbano-Marquez, A., Estivill, X. and Nunes, V., 1996, Autosomal recessive Wolfram syndrome associated with an 8.5-kb mtDNA single deletion. *Am. J. Hum. Genet.*, **58**, 963–970.

Bernes, S.M., Bacino, C., Prezant, T.R., Pearson, M.A., Wood, T.S., Fournier, P. and Fischel-Ghodsian, N., 1993, Identical mitochondrial DNA deletion in mother with progressive external ophthalmoplegia and son with Pearson marrow-pancreas syndrome. *J. Pediat.*, **123**, 598–602.

Blanc, H., Wright, C.T. and Bibb, M.J., 1981, Mitochondrial DNA of chloramphenicol-resistant mouse cells contains a single nucleotide change in the region encoding the 3′ end of the large ribosomal RNA. *Proc. Natl Acad. Sci., USA*, **78**, 3789–3793.

Brown, W.M., 1980, Polymorphism in mitochondrial DNA of humans as revealed by restriction endonuclease analysis. *Proc. Natl Acad. Sci., USA*, **77**, 3605–3609.

Brown, W.M., George, M. and Wilson, A.C., 1979, Rapid evolution of animal mitochondrial DNA. *Proc. Natl Acad. Sci., USA*, **76**, 1967–1971.

Brown, M.D. and Wallace, D.C., 1994, Molecular basis of mitochondrial DNA disease. *J. Bioenergy Biomemb.*, **26**, 273–289.

Brown, W.M., Prager, E.M., Wan, A. and Wilson, A.C., 1982, Mitochondrial DNA sequences in primates: tempo and mode of evolution. *J. Molec. Evol.*, **18**, 225–239.

Cann, R.L., Stoneking, M. and Wilson, A.C., 1987, Mitochondrial DNA and human evolution. *Nature*, **325**, 31–36.

Case, J.T. and Wallace, D.C., 1981, Maternal inheritance of mitochondrial DNA polymorphisms in cultured human fibroblasts. *Somatic Cell Genet.*, **7**, 103–108.

Chang, D.D. and Clayton, D.A., 1984, Precise identification of individual promoter for transcription of each strand of human mitochondrial DNA. *Cell*, **36**, 635–643.

Chang, D.D. and Clayton, D.A., 1985, Priming of human mitochondrial DNA replication occurs at the light-strand promoter. *Proc. Natl Acad. Sci., USA*, **82**, 351–355.

Chang, D.D., Hixson, J.E. and Clayton, D.A., 1986, Minor transcription initiation events indicate that both human mitochondrial promoters function bidirectionally. *Molec. Cell. Biol.*, **6**, 294–301.

Chen J.D., Cox I. and Denton, M.J., 1989, Preliminary exclusion of an X-linked gene in Leber's optic atrophy by linkage analysis. *Human Genet.*, **82**, 203–207.

Chen, Y.S., Torroni, A., Excoffier, L., Santachiara-Benerecetti, A.S. and Wallace, D.C., 1995, Analysis of mtDNA variation in African populations reveals the most ancient of all human continent-specific haplogroups. *Am. J. Human Genet.*, **57**, 133–149.

Cormier, V., Rotig, A., Tardieu, M., Colonna, M., Saudubray, J.M. and Munnich, A., 1991, Autosomal dominant deletions of the mitochondrial genome in a case of progressive encephalomyopathy. *Am. J. Human Genet.*, **48**, 643–648.

Cormier-Daire, V., Bonnefont, J.P., Rustin P., Maurage, C., Ogler, H., Schmitz, J., Ricour, C., Saudubray, J.M., Munnich, A. and Rotig, A., 1994, Mitochondrial DNA rearrangements with onset as chronic diarrhea with villous atrophy. *J. Pediat.*, **124**, 63–70.

DeVivo, D.C., Haymond, M.W., Obert, K.A., Nelson, J.S. and Pagliare, A.S., 1976, Defective activation of the pyruvate dehydrogenase complex in subacute necrotizing encephalomyelopathy (Leigh's disease). *Annals Neurol.*, **6**, 483–494.

DeVries, D.D., VanEngelen, B.G., Gabreels, F.J., Ruitenbeek, W. and VanOost, B.A., 1993, A second missense mutation in the mitochondrial ATPase 6 gene in Leigh's syndrome. *Annals Neurol.*, **34**, 410–412.

DiMauro, S. and Hirano, M., 1998, Mitochondria and heart disease. *Current Opinion in Cardiology*, **13**, 190–197.

DiMauro, S. and Moraes, C.T., 1993, Mitochondrial encephalomyelopathies. *Arch. Neurol.*, **50**, 1197–1208.

Enriquez, J.A., Chomyn, A. and Attardi, G., 1995, MtDNA mutation in MERRF syndrome causes defective aminoacylation of tRNALys and premature translation termination. *Nature Genet.*, **10**, 47–55.

Esteal, S., 1991, The relative rate of DNA evolution in primates. *Molec. Biol. Evol.*, **8**, 115–127.

Estivill, X., Govea, N., Barcelo, E., Badenas, C., Romero, E., Moral, L., Scozzri, R., D'Urbano, L., Zeviani, M. and Torroni, A., 1998, Familial progressive sensorineural deafness is mainly due to the mtDNA A1555G mutation and is enhanced by treatment of aminoglycosides. *Am. J. Human Genet.*, **62**, 27–35.

Francalacci, P., Bertranpetit, J., Calafell, F. and Underhill, P.A., 1996, Sequence diversity of the control region of mitochondrial DNA in Tuscany and its implications for the peopling of Europe. *Am. J. Physiol. Anthropol.*, **100**, 443–460.

Giles, R.E., Blanc, H., Cann, H.M. and Wallace, D.C., 1980, Maternal inheritance of human mitochondrial DNA. *Proc. Natl Acad. Sci., USA*, **77**, 6715–6719.

Glick, B. and Schatz, G., 1991, Import of proteins into mitochondria. *Annual Rev. Genet.*, **25**, 21–44.

Goto, Y., 1995, Clinical features of MELAS and mitochondrial DNA mutations. *Muscle Nerve*, **3**, S107–112.

Goto, Y., Nonaka, I. and Horai, S., 1990, A mutation in the tRNALeu (UUR) gene associated with the MELAS subgroup of mitochondrial encephalomyelopathies. *Nature*, **348**, 651–653.

Goto, Y., Nonaka, I. and Horai, S., 1991, A new mtDNA mutation associated with mitochondrial myopathy, encephalopathy, lactic acidosis and stroke-like episodes (MELAS). *Biochim. Biophys. Acta*, **1097**, 238–240.

Greenberg, J.H., Turner, C.G., II and Zegura, S.L., 1986, The settlement of the Americas: a comparison of the linguistic, dental, and genetic evidence. *Curr. Anthropol.*, **27**, 477–497.

Hauswirth, W.W. and Laipis, P.J., 1982, Mitochondrial DNA polymorphism in a maternal lineage of Holstein cows. *Proc. Natl Acad. Sci., USA*, **79**, 4686–4690.

Hirawake, H., Wang, H., Kuramochi, T., Kojima, S. and Kita, K., 1994, Human complex II (succinate–ubiquinone oxidoreductase): cDNA cloning of the flavoprotein (Fp) subunit of liver mitochondria. *J. Biochem.*, **116**, 221–227.

Holt, I.J., Harding, A.E. and Morgan-Hughes, J.A., 1988, Deletions of muscle mitochondrial DNA in patients with mitochondrial myopathies. *Nature*, **331**, 717–719.

Holt, I.J., Harding, A.E., Petty, R.K. and Morgan-Hughes, J.A., 1990, A new mitochondrial disease associated with mitochondrial DNA heteroplasmy. *Am. J. Human Genet.*, **46**, 428–433.

Howell, N., Kubacka, I., Xu, M. and McCullough, D.A., 1991, Leber's hereditary optic neuropathy, involvement of the mitochondrial ND1 gene and evidence for an intragenic supressor mutation. *Am. J. Human Genet.*, **48**, 935–942.

Howell, N., Halvorson, S., Kubacka, I., McCullough, D.A., Bindoff, L.A. and Turnbull, D.M., 1992, Mitochondrial gene segregation in mammals: is the bottleneck always narrow? *Human Genet.*, **90**, 117–120.

Huoponen, K., Vilkki, J., Aula, P., Nikoskelainen, E.K. and Savontaus, M.L., 1991, A new mtDNA mutation associated with Leber hereditary optic neuroretinopathy. *Am. J. Human Genet.*, **48**, 1147–1153.

Huoponen, K., Lamminen, T., Juvonen, V., Aula, P., Nikoskelainen, E. and Savontaus, M.L., 1993, The spectrum of mitochondrial DNA mutations in families with Leber's hereditary optic neuroretinopathy. *Human Genet.*, **92**, 379–384.

Hutchin, T. and Cortopassi, C., 1995, A mitochondrial DNA clone is associated with increased risk for Alzheimer disease. *Proc. Natl Acad. Sci., USA*, **92**, 6892–6895.

Hutchin, T., Haworth, I., Higashi, K., Fischel-Ghodsian, N., Stoneking, M., Saha, N., Arnos, C. and Cortopassi, G., 1993, A molecular basis for human hypersensitivity to aminoglycoside antibiotics. *Nucleic Acids Res.*, **21**, 4174–4179.

Jacobs, H.T., 1997, Mitochondrial deafness. *Annals Med.*, **29**, 483–491.

Johns, D.R., Neufeld, M.J. and Park, R.D., 1992, An ND-6 mitochondrial DNA mutation associated with Leber hereditary optic neuropathy. *Biochem. Biophys. Res. Commun.*, **187**, 1551–1557.

Jun, A.S., Brown, M.D. and Wallace, D.C., 1994, A mitochondrial DNA mutation at np 14459 of the ND6 gene associated with maternally inherited Leber's hereditary optic neuropathy and dystonia. *Proc. Natl Acad. Sci., USA*, **91**, 6206–6210.

Kaneda, H., Hayashi, J., Takahama, S., Taya, C., Lindahl, K.F. and Yonekawa, H., 1995, Elimination of paternal mitochondrial DNA in intraspecific crosses during early mouse embryogenesis. *Proc. Natl Acad. Sci., USA*, **92**, 4542–4546.

Kearsey, S.E. and Craig, I.W., 1981, Altered ribosomal RNA genes in mitochondria from mammalian cells with chloramphenicol resistance. *Nature*, **290**, 607–608.

Kogelnik, A.M., Lott, M.T., Brown, M.D., Navathe, S.B. and Wallace, D.C., 1996, MITOMAP: a human mitochondrial genome database. *Nucleic Acids Res.*, **24**, 177–179.

Leckschat, S., Ream-Robinson, D. and Scheffler, I.E., 1993, The gene for the iron sulfur protein of succinate dehydrogenase (SDH-IP) maps to human chromosome 1p35–36.1. *Somat. Cell. Molec. Genet.*, **19**, 505–511.

Leigh, D., 1951, Subacute necrotizing encephalomyelopathy in an infant. *J. Neurol. Neurosurg. Psychiatry*, **14**, 216–221.

Mackey, D. and Howell, N., 1992, A variant of Leber hereditary optic neuropathy characterized by recovery of vision and by an unusual mitochondrial genetic etiology. *Am. J. Human Genet.*, **51**, 1218–1228.

Manfredi, G., Schon, E.A., Moraes, C.T., Bonilla, E., Berry, G.T., Sladky, J.T. and DiMauro, S., 1995, A new mutation associated with MELAS is located in a mitochondrial DNA polypeptide-coding gene. *Neuromuscular Disorders*, **5**, 391–398.

Mariotti, C., Tiranti, V., Carrara, F., Dallapiccola, B., DiDonato, S. and Zeviani, M., 1994, Defective respiratory capacity and mitochondrial protein synthesis in transformant hybrids harboring the tRNA (Leu(UUR)) mutation associated with maternally inherited myopathy and cardiomyopathy. *J. Clin. Investig.*, **93**, 1102–1107.

Mashima, Y., Hiida, Y., Oguchi, Y., Kudoh, J. and Shimizu, N., 1993, High frequency of mutations at position 11778 in mitochondrial ND4 gene in Japanese families with Leber's hereditary optic neuropathy. *Human Genet.*, **92**, 101–102.

Masucci, J.P., Davidson, M., Koga, Y., Schon, E.A. and King, M.P., 1995, *In vitro* analysis of mutations causing myoclonus epilepsy with ragged-red fibers in the mitochondrial tRNALysgene: two genotypes produce similar phenotypes. *Molec. Cell. Biol.*, **15**, 2872–2881.

McShane, M.A., Hammans, S.R., Sweeney, M., Holt, I.J., Beattie, T.J., Brett, E.M. and Harding, A.E., 1991, Pearson syndrome and mitochondrial encephalomyopathy in a patient with a deletion of mtDNA. *Am. J. Human Genet.*, **48**, 39–42.

Merriwether, D.A., Clark, A.G., Ballinger, S.W., Schurr, T.G., Soodyall, H., Jenkins, T., Sherry, S.T. and Wallace, D.C., 1991, The structure of human mitochondrial DNA variation. *J. Molec. Evol.*, **33**, 543–555.

Michaels, G.S., Hauswirth, W.W. and Laipis, P.J., 1982, Mitochondrial DNA copy number in bovine oocytes and somatic cells. *Develop. Biol.*, **94**, 246–251.

Moraes, C.T., DiMauro, S., Zeviani, M., Lombes, A., Shanske, S., Miranda, A.F., Nakase, H., Bonilla, E., Werneck, L.C., Servidei, S., Nonaka, I., Koga, Y., Spiro, A.J., Brownell, K.W., Schmidt, B., Schotland, D.L., Zupanc, M., DeVivo, D.C., Schon, E.A. and Rowland, L.P., 1989, Mitochondrial DNA deletions in progressive external ophthalmoplegia and Kearns–Sayre syndrome. *New Engl. J. Med.*, **320**, 1293–1299.

Moraes, C.T., Ciacci, F., Bonilla, E., Jansen, C., Hirano, M., Rao, N., Lovelace, R.E., Rowland, L.P., Schon, E.A. and DiMauro, S., 1993, Two novel pathogenic mitochondrial DNA mutations affecting organelle number and protein synthesis. Is the tRNA (Leu(UUR)) gene an etiologic hot spot? *J. Clin. Investig.*, **92**, 2906–2915.

Morris, A.A., Farnsworth, L., Ackrell, B.A., Turnbull, D.M. and Birch-Machin, M.A., 1994, The cDNA sequence of the flavoprotein subunit of human heart succinate dehydrogenase. *Biochim. Biophys. Acta*, **1185**, 125–128.

Morris, A.A., Taanman, J.W., Blake, J., Cooper, J.M., Lake, B.D., Malone, M., Love, S., Clayton, P.T., Leonard, J.V. and Schapira, A.H., 1998, Liver failure associated with mitochondrial DNA depletion. *J. Hepatol.*, **28**, 556–563.

Morten, K.J., Cooper, J.M., Brown, G.K., Lake, B.D., Pike, D. and Poulton, J., 1993, A new point mutation associated with mitochondrial encephalomyopathy. *Human Molec. Genet.*, **2**, 2081–2087.

Nakamura, M., Nakano, S., Goto, Y., Ozawa, M., Nagahama, Y., Fukuyama, H., Akiguchi, I., Kaji, R. and Kimura, J., 1995, A novel point mutation in the mitochondrial tRNA

(Ser(UCN)) gene detected in a family with MERRF/MELAS overlap syndrome. *Biochim. Biophys. Res. Commun.*, **214**, 86–93.

Newman, N.J., 1993, Leber's hereditary optic neuropathy, new genetic considerations. *Archives Neurol.*, **50**, 540–548.

Noer, A.S., Sudoya, H., Lertrit, P., Thyagarajan, D., Utthanaphol, P., Kapsa, R., Byrne, E. and Marzuki, S., 1991, A tRNALys mutation in the mtDNA is the causal genetic lesion underlying myoclonic epilepsy and ragged-red fiber (MERRF) syndrome. *Am. J. Human Genet.*, **49**, 715–722.

Pincus, J.H., 1972, Subacute necrotizing encephalomyelopathy (Leigh's disease): a consideration of clinical features and etiology. *Develop. Med. Child Neurol.*, **14**, 87–101.

Poulton, J. and Holt, I., 1994, Mitochondrial DNA: does more lead to less? *Nature Genet.*, **8**, 313–315.

Poulton, J., Deadman, M.E., Ramacharan, S. and Gardiner, R.M., 1991, Germ-line deletions of mtDNA in mitochondrial myopathy. *Am. J. Human Genet.*, **48**, 649–653.

Poulton, J., Morten, K.J., Marchington, D., Weber, K., Brown, G.K., Rotig, A. and Bindoff, L., 1995, Duplications of mitochondrial DNA in Kearns–Sayre syndrome. *Muscle Nerve*, **3**, 154–158.

Prezant, T.R., Agapian, J.V., Bohlman, M.C., Bu, X., Oztas, S., Qiu, W.Q., Arnos, K.S., Cortopassi, G.A., Jaber, L., Rotter, J.I., Shohat, M. and Fischel-Ghodsian, N., 1993, Mitochondrial ribosomal RNA mutation associated with both antibiotic-induced and non-syndromic deafness. *Nature Genet.*, **4**, 289–294.

Reardon, W., Ross, R.J., Sweeney, M.G., Luxon, L.M., Pembrey, M.E., Harding, A.E. and Trembath, R.C., 1992, Diabetes mellitus associated with a pathogenic point mutation in mitochondrial DNA. *Lancet*, **340**, 1376–1379.

Rötig, A., Colonna, M., Blanche, S., Fischer, A., LeDeist, F., Frezal, J., Saudubray, J.M. and Munnich, A., 1988, Deletion of blood mitochondrial DNA in pancytopenia. *Lancet*, **2**, 567–568.

Rötig, A., Cormier, V., Blanche, S., Bonnefont, J.P., Ledeist, F., Romero, N., Schmitz, J., Rustin, P., Fischer, A., Saudubray, J.M., *et al.*, 1990, Pearson's marrow–pancreas syndrome: a multisystem mitochondrial disorder in infancy. *J. Clin. Investig.*, **86**, 1601–1608.

Rötig, A., Bessis, J.L., Romero, N., Cormier, V., Saudubray, J.M., Narcy, P., Lenoir, G., Rustin, P. and Munnich, A., 1992, Maternally inherited duplication of the mitochondrial genome in a syndrome of proximal tubulopathy, diabetes mellitus, and cerebellar ataxia. *Am. J. Human Genet.*, **50**, 364–370.

Rötig, A., Cormier, V., Chatelain, P., Francois, R., Saudubray, J.M., Rustin, P. and Munnich, A., 1993, Deletion of mitochondrial DNA in a case of early-onset diabetes mellitus, optic atrophy, and deafness (Wolfram syndrome). *J. Clin. Investig.*, **91**, 1095–1098.

Santorelli, F.M., Shanske, S., Macaya, A., DeVivo, D.C. and DiMauro, S., 1993, The mutation at nt 8993 of mitochondrial DNA is a common cause of Leigh's syndrome. *Annals Neurol.*, **34**, 827–834.

Santorelli, F.M., Shanske, S., Jain, K.D., Tick, D., Schon, E.A. and DiMauro, S., 1994, A T–C mutation at nt 8993 of mitochondrial DNA in a child with Leigh syndrome. *Neurology*, **44**, 972–974.

Schon, E.A., Rizzuto, R., Moraes, C.T., Nakase, H., Zeviani, M. and DiMauro, S., 1989, A direct repeat is a hotspot for large-scale deletion of human mitochondrial DNA. *Science*, **244**, 346–349.

Schwartz, W.J., Hutchinson, H.T. and Bert, B.O., 1981, Computerized tomography in subacute necrotizing encephalomyelopathy (Leigh disease). *Annals Neurol.*, **10**, 268–271.

Shah, Z.H., Migliosi, V., Miller, S.C., Wang, A., Friedman, T.B. and Jacobs, H.T., 1998, Chromosomal locations of three human nuclear genes (RPSM12, TUFM, and AFG3L1) specifying putative components of the mitochondrial gene expression apparatus. *Genomics*, **15**, 384–388.

Shoffner, J.M., 1996, Maternal inheritance and the evaluation of oxidative phosphorylation diseases. *Lancet*, **348**, 1283–1288.

Shoffner, J.M. and Wallace, D.C., 1990, Oxidative phosphorylation diseases. Disorders of two genomes. *Adv. Human Genet.*, **19**, 267–330.

Shoffner, J.M. and Wallace, D.C., 1995, Oxidative phosphorylation diseases, in Scriver C.R., Beaudet, A.L., Sly, W.S. and Valle, D. (Eds) *The Metabolic and Molecular Basis of Inherited Disease*, pp. 1535–1609, New York: McGraw-Hill.

Shoffner, J.M., Lott, M.T., Lezza, A.M., Seibel, P., Ballinger, S.W. and Wallace, D.C., 1990, Myoclonic epilepsy and ragged-red fiber disease (MERRF) is associated with a mitochondrial DNA tRNALys mutation. *Cell*, **61**, 931–937.

Shoffner, J.M., Fernhoff, M.D., Krawiecki, N.S., Caplan, D.B., Holt, P.J., Koontz, D.A., Takei, Y., Newman, N.J., Ortiz, R.G., Polak, M., Ballinger, S.W., Lott, M.T. and Wallace, D.C., 1992, Subacute necrotizing encephalopathy: oxidative phosphorylation defects and the ATPase 6 point mutation. *Neurology*, **42**, 2168–2174.

Shoffner, J.M., Brown, M.D., Torroni, A., Lott, M.T., Cabell, M.R., Mirra, S.S., Beal, M.F., Yang, C., Gearing, M., Salvo, R., Watts, R.L., Juncos, J.L., Hansen, L.A., Crain, B.J., Fayad, M., Reckord, C.L. and Wallace, D.C., 1993, Mitochondrial DNA variants observed in Alzheimer disease and Parkinson disease patients. *Genomics*, **17**, 171–184.

Shoffner, J.M., Brown, M.D., Stugard, C., Jun, A.S., Pollok, S., Haas, R.H., Kaufman, A., Koontz, D., Kim, Y., Graham, J., Smith, E., Dixon, J. and Wallace, D.C., 1995, Leber's hereditary optic neuropathy plus dystonia is caused by a mitochondrial DNA point mutation in a complex I subunit. *Annals Neurol.*, **38**, 163–169.

Singh, G., Lott, M.T. and Wallace, D.C., 1989, A mitochondrial DNA mutation as a cause of Leber's hereditary optic neuropathy. *New Engl. J. Med.*, **320**, 1300–1305.

Smith, O.P., Hann, I.M., Woodward, C.E. and Brockington, M., 1995, Pearson's marrow/pancreas syndrome: haematological features associated with deletion and duplication of mitochondrial DNA. *Brit. J. Haematol.*, **90**, 469–472.

Stone, A.C. and Stoneking, M., 1993, Ancient DNA from a pre-Columbian Amerindian population. *Am. J. Physiol. Anthropol.*, **92**, 463–471.

Suomalainen, A., Kaukonen, J., Amati, P., Timonen, R., Haltia, M., Weissenbach, J., Zeviani, M., Somer, H. and Peltonen, L., 1995, An autosomal locus predisposing to deletions of mitochondrial DNA. *Nature Genet.*, **9**, 146–151.

Suomalainen, A., Majander, A., Wallin, M., Setala, K., Kontula, K., Leinonen, H., Salmi, T., Paetau, A., Haltia, M., Valanne, L., Lonnqvist, J., Peltonen, L. and Somer, H., 1997, Autosomal dominant progressive external ophthalmoplegia with multiple deletions of mtDNA: clinical, biochemical, and molecular genetic features of the 10q-linked disease. *Neurology*, **48**, 1244–1253.

Sweeney, M.G., Davis, M.B., Lashwood, A., Brockington, M., Toscano, A. and Harding, A.E., 1992, Evidence against an X-linked locus close to DXS7 determining visual loss susceptibility in British and Italian families with Leber's hereditary optic neuropathy. *Am. J. Human Genet.*, **51**, 741–748.

Tatuch, Y., Christodoulou, J., Feigenbaum, A., Clarke, J.T.R., Wherret, J., Smith, C., Rudd, N., Petrova-Benedict, R. and Robinson, B.H., 1992, Heteroplasmic mtDNA mutation (T–G) at 8993 can cause Leigh disease when the percentage of abnormal mtDNA is high. *Am. J. Human Genet.*, **50**, 852–858.

Thyagarajan, D., Shanske, S., Vazquez-Memije, M., DeVivo, D. and DiMauro, S., 1995, A novel mitochondrial ATPase 6 point mutation in familial bilateral striatal necrosis. *Annals Neurol.*, **38**, 468–472.

Torroni, A. and Wallace, D.C., 1994, Mitochondrial DNA variation in human populations and implications for detection of mitochondrial DNA mutations of pathological significance. *J. Bioenergy Biomemb.*, **26**, 261–271.

Torroni, A., Schurr, T.G., Yang, C.C., Szathmary, E.J., Williams, R.C., Schanfield, M.S., Troup, G.A., Knowler, W.C., Lawrence, D.N. and Weiss, K.M., 1992, Native American mitochondrial DNA analysis indicates that the Amerind and the Nadene populations were founded by two independent migrations. *Genetics*, **130**, 153–162.

Torroni, A., Schurr, T.G., Cabell, M.F., Brown, M.D., Neel, J.V., Larsen, M., Smith, D.G., Vullo, C.M. and Wallace, D.C., 1993a, Asian affinities and continental radiation of the four founding Native American mtDNAs. *Am. J. Human Genet.*, **53**, 563–590.

Torroni, A., Sukenik, R.I., Schurr, T.G., Starikovskakya, Y.B., Cabell, M.F., Crawford, M.H., Comuzzie, A.G. and Wallace, D.C., 1993b, mtDNA variation of Aboriginal Siberians reveals distinct genetic affinities with Native Americans. *Am. J. Human Genet.*, **53**, 591–608.

Torroni, A., Chen, Y., Semino, O., Santachiara-Beneceretti, A.S., Scott, C.R., Lott, M.T., Winter, M. and Wallace, D.C., 1994a, Mitochondrial DNA and Y-chromosome polymorphisms in four native American populations from southern Mexico. *Am. J. Human Genet.*, **54**, 303–318.

Torroni, A., Lott, M.T., Cabell, M.F., Chen, Y.S., Lavergne, L. and Wallace, D.C., 1994b, MtDNA and the origin of the Caucasians. Identification of the ancient Caucasian-specific haplogroups, one of which is prone to a recurrent somatic duplication in the D-loop region. *Am. J. Human Genet.*, **55**, 760–776.

Torroni, A., Miller, J.A., Moore, L.G., Zamudio, S., Zhuang, J., Droma, R. and Wallace, D.C., 1994c, Mitochondrial DNA analysis in Tibet. Implications for the origin of the Tibetan population and its adaptation to high altitude. *Am. J. Physiol. Anthropol.*, **93**, 189–199.

Torroni, A., Neel, J.V., Barrantes, R., Schurr, T.G. and Wallace, D.C., 1994d, A mitochondrial DNA "Clock" for the Amerinds and its implications for timing their entry into North America. *Proc. Natl Acad. Sci., USA*, **91**, 1158–1162, 1994.

Torroni, A., Huoponen, K., Francalacci P., Petrozzi M., Morelli, L., Scozzari, R., Obinu, D., Savontaus, M.L. and Wallace, D.C., 1996, Classification of European mtDNAs from an analysis of three European populations. *Genetics*, **144**, 1835–1850.

Torroni, A., Petrozzi, M., D'Urbano, L., Sellitto, D., Zeviani, M., Carrara, F., Carducci, C., Leleuzzi, V., Carelli, V., Barboni, P., De Negri, A. and Scozzari, R., 1997, Haplotype and phylogenetic analyses suggest that one European-specific mtDNA background plays a role in the expression of Leber hereditary optic neuropathy by increasing the penetrance of the primary mutations 11778 and 14484. *Am. J. Human Genet.*, **60**, 1107–1121.

Trounce, I., Neill, S. and Wallace, D.C., 1994, Cytoplasmic transfer of the mtDNA nt 8993 TG (ATP6) point mutation associated with Leigh syndrome into mtDNA-less cells demonstrates cosegregation with a decrease in state III respiration and ADP/O ratio. *Proc. Natl Acad. Sci., USA*, **91**, 8334–8338.

Wallace, D.C., 1981, Assignment of the chloramphenicol resistance gene to mitochondrial deoxyribonucleic acid and analysis of its expression in cultured human cell. *Molec. Cell. Biol.*, **1**, 697–710.

Wallace, D.C., 1982, Structure and evolution of organelle genomes. *Microbiol. Rev.*, **46**, 208–240.

Wallace, D.C., 1986, Mitotic segregation of mitochondrial DNAs in human cell hybrids and expression of chloramphenicol resistance. *Somat. Cell. Molec. Genet.*, **12**, 41–49.

Wallace, D.C., 1995, Mitochondrial DNA variation in human evolution, degenerative disease, and aging. *Am. J. Human Genet.*, **57**, 201–223.

Wallace, D.C., Ye, J.H., Neckelmann, S.N., Singh, G., Webster, K.A. and Greenberg, B.D., 1987, Sequence analysis of cDNAs for the human and bovine ATP synthase β-subunit: mitochondrial DNA genes sustain seventeen times more mutations. *Curr. Genet.*, **12**, 81–90.

Wallace, D.C., Singh, G., Lott, M.T., Hodge, J.A., Schurr, T.G., Lezza, A.M., Elsas, L.J. and Nikoskelainen, E.K., 1988a, Mitochondrial DNA mutation associated with Leber's hereditary optic neuropathy. *Science*, **242**, 1427–1430.

Wallace, D.C., Zheng, X., Lott, M.T., Shoffner, J.M., Hodge, J.A., Kelley, R.I., Epstein, C.M., and Hopkins, L.C., 1988b, Familial mitochondrial encephalomyopathy (MERRF): genetic, pathophysiological, and biochemical characterization of a mitochondrial DNA disease. *Cell*, **55**, 601–610.

Wallace, D.C., Lott, M.T., Brown, M.D., Huoponen, K. and Torroni, A., 1995, Report of the committee on human mitochondrial DNA, in Cuticchia A.J. (Ed.) *Human Gene Mapping 1995: a Compendium*, pp. 910–954, Baltimore: Johns Hopkins University Press.

Wallace, D.C., Brown, M.D. and Lott, M.T., 1996, Mitochondrial genetics, in Rimoin, D.L., Connor, J.M., Pyeritz, R.E. and Emery, A.E.H. (Eds) *Emery and Rimoin's Principles and Practice of Medical Genetics*, 3rd edn, vol. 1, pp. 277–332, London: Churchill Livingstone.

Weiner, N.C., Newman, N.J., Lessell, S., Johns, D.R., Lott, M.T. and Wallace, D.C., 1993, Atypical Leber's hereditary optic neuropathy with molecular confirmation. *Arch. Neurol.*, **50**, 470–473.

Willems, J.L., Monnens, L.A.H., Trijbels, J.M.F., Veerkamp, J.H., Meyer, A.E.F.H., vanDam, K. and vanHaelst, U., 1977, Leigh's encephalomyelopathy in a patient with cytochrome c oxidase deficiency in muscle tissue. *Pediatrics*, **60**, 850–857.

Zeviani, M., Moraes, C.T., DiMauro, S., Nakase, H., Bonilla, E., Nakase, H., Bonilla, E., Schon, E.A. and Rowland, L.P., 1988, Deletions of mitochondrial DNA in Kearns–Sayre syndrome. *Neurology*, **38**, 1339–1346.

Zeviani, M., Bresolin, N., Gellera, C., Bordoni, A., Pannacci, M., Amati, P., Moggio, M., Servidei, S., Scarlato, G. and DiDonato, S., 1990, Nucleus-driven multiple large-scale deletions of the human mitochondrial genome: a new autosomal dominant disease. *Am. J. Human Genet.*, **47**, 904–914.

Zeviani, M., Gellera, C., Antozzi, C., Rimoldi, M., Morandi, L., Villani, F., Tiranti, V. and DiDonato, S., 1991, Maternally inherited myopathy and cardiomyopathy: association with mutation in mitochondrial DNA tRNA (Leu[UUR]). *Lancet*, **338**, 143–147.

Zeviani, M., Fernandez-Silva, P. and Tiranti, V., 1997, Disorders of mitochondria and related metabolism. *Curr. Opin. Neurol.*, **10**, 160–167.

10

GENETIC POLYMORPHISMS AFFECTING NERVOUS SYSTEM FUNCTION

Susan M. Mockus and Kent E. Vrana

10.1 Introduction

The strategic partnership of molecular biology and neuroscience has flourished over the past decade, providing an impressive array of advances in the genetics of nervous system diseases. The advent of recombinant DNA technology has provided a repertoire of new tools to study already existing, clinically relevant neuroscience problems. One of the future challenges will be to determine the molecular mechanisms by which identified genomic mutations translate into phenotypic expression and presentation. A cross-disciplinary approach engaging neurobiology, molecular biology, pharmacology, physiology, and clinical neuroscience will be vital to understanding neurogenetic diseases and to developing treatment of these genetically-based neurological diseases. This chapter surveys genetically determined neurological diseases with the intention of establishing a foundation on which investigators from cross disciplines might begin to refine their concept of neurogenetics.

There are a plethora of human diseases that involve dysfunction of the central and/or peripheral nervous systems (see Table 10.1). These diseases may be characterized as those primarily involving nervous system failure and those in which neurological symptoms manifest secondary to other underlying disease entities. Examples of the latter category are metabolic, mitochondrial, and cancer-related diseases that eventually impact normal neurological function. These diseases are covered in other chapters and hence this chapter will focus on inherited genetic polymorphisms predominately affecting nervous system function. In general, neurogenetics presents a challenge to the field of clinical and molecular neuroscience due to the heterogeneity of the many nervous system diseases. Heterogeneity can be manifested at both the genotypic and phenotypic levels (Wolf, 1997). Genotypic heterogeneity reflects the presentation of similar phenotypes even though varying mutations exist within the same gene or in different genes. The opposing scenario also occurs in which different mutations within the same gene may produce remarkably different phenotypes. Phenotypic and genotypic heterogeneity have contributed to the complexity and variability of nervous system disease classification. Therefore, the organization of neurogenetic diseases based solely on genotypic mutations versus phenotype is a superficial approach.

231

Table 10.1 Selected neurogenetic diseases

Disease	Class	Chromosome	Affected protein	Classifying phenotype	References
Huntington's disease (HD)	Trinucleotide repeat CAG	4p16.3	Huntingtin	Neurodegenerative	Ross et al., 1997; Wellington and Hayden, 1997
Spinobulbar muscular atrophy (SBMA); Kennedy disease	Trinucleotide repeat CAG	Xq11–Xq12	Androgen receptor	Neurodegenerative	Burright et al., 1997; Koshy and Zoghbi, 1997
Dentatorubral pallidoluysian atrophy (DRPLA)	Trinucleotide repeat CAG	12p2	Atrophin-1	Neurodegenerative	Ross et al., 1997
Spinocerebellar ataxia type 1 (SCA1)	Trinucleotide repeat CAG	6p22–p23	Ataxin-1	Neurodegenerative	Yagishita and Inoue, 1997
Spinocerebellar ataxia type 2 (SCA2)	Trinucleotide repeat CAG	12q24.1	Ataxin-2	Neurodegenerative	Yagishita and Inoue, 1997
Spinocerebellar ataxia Machado–Joseph disease (MJD; SCA3)	Trinucleotide repeat CAG	14q24.3	Ataxin-3	Neurodegenerative	Sudarsky and Coutinho, 1995; Yagishita and Inoue, 1997
Spinocerebellar ataxia type 6 (SCA6)	Trinucleotide repeat CAG	19p13	P/Q–Ca^{2+} channel	Neurodegenerative, Channelopathy	Bulman, 1997; Greenberg, 1997
Friedreich ataxia (FDRA)	Trinucleotide repeat GAA	9q13	Frataxin	Neurodegenerative	Koenig and Mandel, 1997
Fragile X type A (FRAXA)	Trinucleotide repeat CGG	Xq27.3	FMRP	Mental retardation	Tarleton and Saul, 1993; Timchenko and Caskey, 1996; Hoogeveen and Oostra, 1997
Myotonic dystrophy (MD)	Trinucleotide repeat CTG	19q13.3	DMPK	Neuromuscular	Timchenko and Caskey, 1996; Strong and Brewster, 1997
Familial hemiplegic migraine (FMA)	Missense mutation	19p13	P/Q–Ca^{2+} channel	Channelopathy	Terwindt et al., 1997; Doyle and Stubbs, 1998
Episodic ataxia type 2 (EA-2)	Frameshift mutation	19p13	P/Q–Ca^{2+} channel	Channelopathy	Bulman, 1997; Miller, 1997
Episodic ataxia type 1 (EA-1)	Missense mutations	12p	K$^+$ channel	Channelopathy	Sanguinetti and Spector, 1997

Table 10.1 continued

Disease	Class	Chromosome	Affected protein	Classifying phenotype	References
Charcot–Marie–Tooth disease type 1A (CMT1A)	Duplication	17p11.2	Peripheral myelin protein (PMP22)	Peripheral neuropathy Demyelination	Murakami *et al.*, 1996
Charcot–Marie–Tooth disease type 1B (CMT1B)	Missense mutations	1q21.2–q23	Myelin protein zero (P0)	Peripheral neuropathy Hypomyelination	Murakami *et al.*, 1996
Charcot–Marie–Tooth disease X-Linked (CMT1X; HMSN)	Missense, Frameshift Deletions	Xq13.1	Connexin-32 (Cx32)	Peripheral neuropathy	Bone *et al.*, 1997
Hereditary neuropathy with liability to pressure palasies (HNPP)	Deletion	17p11.2	PMP22	Peripheral neuropathy	Pareyson and Taroni, 1996
Dejerine–Sottas syndrome (DSS) A	Missense mutations	17p11.2	PMP22	Peripheral neuropathy Demyelinating	Scherer, 1997
Dejerine–Sottas syndrome (DSS) B	Missense mutations	1q21.2–q23	P0	Peripheral neuropathy	Scherer, 1997
Pelzaeus–Merzbacher disease	Deletions, Frameshift, Duplication, Missense mutations	Xq21–q22	Proteolipid protein (PLP) and DM20	Demyelination	Kobayashi *et al.*, 1996
Spastic paraplegia type 2 (SPG2)	Missense mutations	Xq21–q22	PLP; DM20	Demyelination	Kobayashi *et al.*, 1996
Hereditary spastic paraplegia (HSP)	Deletion	16q24.3	Paraplegin	Neuropathy, Mental retardation, Ataxia	Casari *et al.*, 1998
Isolated lissencephaly	Deletion	17p13.3	PAF acetylhydrolase	Mental retardation	Budarf and Emanuel, 1997
X-linked subcortical laminar heterotopia and lissencephaly (X-SCLH/LIS)	Frameshift, Missense mutations	Xq22.3–q23	Doublecortin	Mental retardation Epilepsy	des Portes *et al.*, 1998; Gleeson *et al.*, 1998

Table 10.1 continued

Disease	Class	Chromosome	Affected protein	Classifying phenotype	References
CRASH syndrome Spastic paraplegia 1 MASA syndrome X-hydrocephalus X-linked agenosis of corpus callosum	Missense mutations Deletions Frameshift Intronic-splice variants	Xq28	L1CAM	Mental retardation	Fransen *et al.*, 1997
Angelman syndrome (AS)	70% Maternal deletion	15q11–q13	UBE3A SNRPN	Mental retardation, Segmentally aneusomic syndrome (SAS)	Budarf and Emanuel, 1997
Prader Willi syndrome (PWS)	70% Paternal deletion	15q11–q13	SNRPN	Mental retardation, SAS	Budarf and Emanuel, 1997
Miller Dieker syndrome (MDS)	Deletion	17p14.3	LIS-1 gene	Mental retardation SAS, Lissencephaly	Muller *et al.*, 1994; Budarf and Emanuel, 1997
DiGeorge and velocardiofacial symdromes (DGS/VCFS)	Deletions "22q11 deletion disorders"	22q11.2	17 possible proteins implicated	Mental retardation, SAS	Budarf and Emanuel, 1997
Amytrophic lateral sclerosis (ALS)	Missense mutations	21q21	SOD1	Motor neuron degeneration	Siddique and Hentati, 1996; Louvel *et al.*, 1997
Creutzfeldt–Jakob disease (CJD)	Insertion, Missense mutations	20	Prion	Neurodegenerative	Meiner *et al.*, 1997
Fatal familial insomnia (FFI)	Missense mutations	20	Prion	Neurodegenerative	Prusiner and Scott, 1997
Gerstmann–Straussler– Scheinker disease (GSS)	Missense mutations	20	Prion	Neurodegenerative	Prusiner and Scott, 1997
Spinal muscular atrophy (SMA)	Deletion, Duplication, Missense mutations	5q11.2	SMN NAIP	Neurodegeneration of lower motor neurons	Crawford and Pardo, 1996; Melki, 1997

Table 10.1 continued

Disease	Class	Chromosome	Affected protein	Classifying phenotype	References
Alzheimer's disease (AD)	Missense mutations	21q21	Amyloid precursor protein (APP)	Neurodegenerative	Goate, 1998
Alzheimer's disease (AD)	Missense mutations	14q24	Presenilin-1 (PS-1)	Neurodegenerative	Kovacs and Tanzi, 1998
Alzheimer's disease (AD)	Missense mutations	1q31–42	Presenilin-2 (PS-2)	Neurodegenerative	Renbaum and Levy-Lehad, 1998
Alzheimer's disease (AD)	Natural allelic variants (disease modulation)	19q13	Apolipoprotein E	Neurodegenerative	Corder *et al.*, 1998
Parkinson's disease (PD)	Missense mutations	4q21–22	α-Synuclein	Neurodegenerative	Polymeropoulos *et al.*, 1997

For simplicity, this chapter attempts to divide neurogenetic diseases by genotypic or phenotypic similarity (Table 10.1). The trinucleotide repeat diseases are organized together because they share a common genotypic mutation. The trinucleotide expansion diseases can be further subdivided into: (a) polyglutamine disorders (CAG repeats) that all result in neurodegeneration, and (b) trinucleotide repeats (non-CAG repeats) that give rise to neuromuscular disorders (myotonic dystrophy) and mental retardation (fragile X). Other inherited nervous system diseases are organized based on phenotype similarities such as common neuroanatomical systems engaged (e.g. peripheral neuropathies) and mechanisms of molecular pathology (e.g. channelopathies). In general, these diseases display significant genotypic heterogeneity such that point mutations, deletions, insertions, and frameshift mutations within the same gene can all give rise to varying phenotypes. The last subdivision of neurogenetic diseases includes multivariate disorders in which several genes determine inheritance and disease expression. These diseases (e.g. Alzheimer's disease) display extensive heterogeneity and complexity at both the genotypic and phenotypic levels. The role of environmental factors in the expression of these multifactorial neurogenetic diseases will also be explored.

The last major focus of this chapter will be on molecular mechanisms of nervous system diseases. The translation of genetic mutations into aberrant proteins that lead to neurological manifestations is examined. Several of the trinucleotide repeat diseases, prion disease, and spinal muscular atrophy (SMA) will be used as examples to explore the role of molecular biology and transgenic technology in assessing how genetic polymorphisms affect nervous system function.

10.2 Dynamic mutations: trinucleotide repeat diseases

"Dynamic mutation" is used to describe the consequence of inheriting unstable, repeat DNA sequences in disease processes (Richards and Sutherland, 1992, 1997). By definition, these mutations naturally exist as a combination of repeated nucleotides within a continuous stretch of DNA. Repeats of DNA sequence are common throughout the human genome, yet a normal number of repeat units do not induce disease expression. However, tandem repeats have a propensity to expand in unit size because of their instability during mitotic or meiotic replication events. This expansion increases the number of inherited repeats from generation to generation which can result in disease expression if the repeat unit has expanded out of normally tolerated ranges (Figure 10.1). For example, in Huntington's disease (HD), a parent with 10 to 35 CAG repeats does not have the disease. However, if during replication of the parent's DNA, the repeat expands to between 36 and 120, then the offspring will express the HD phenotype. Trinucleotide repeats are the most common genetic source of nervous system diseases, with ten neurological diseases attributed to these dynamic mutations. The instability of triplet repeats causes expansion of the repeat size during DNA replication (Figure 10.1). The inherent tendency of triplet repeats to expand is four orders of magnitude greater than the mutation rate of genomic DNA sequence (La Spada, 1997).

There may exist a common mechanism by which repeats expand to biologically intolerable ranges. The proposed hypotheses include unequal homologous exchange,

Triuncleotide expansion

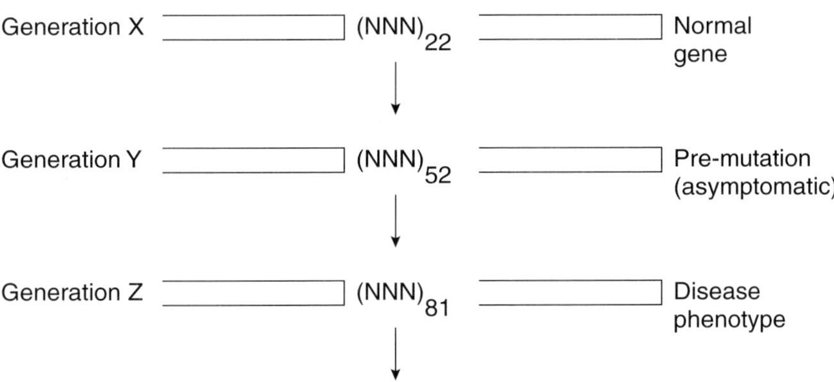

Figure 10.1 Hypothetical expansion and inheritance of a nucleotide repeat. Nucleotide repeats are common throughout the human genome. Normally, repeats within genes do not confer disease phenotypes as exemplified by generation X with 22 copies of a trinucleotide repeat (NNN). Through a variety of possible mechanisms, polymerase slippage, mismatch repair, gene conversion, unequal homologous recombination, and hairpin formation, repeats can expand in number. This expansion occurs when generation X's genetic material undergoes replication. Generation Y then inherits an increased copy number of the trinucleotide repeat (assuming that the expansion occurred during development of the germ cell). In this example, the repeat expands from 22 copies to 52 in one generation. Although generation Y has inherited an increased number of repeats, the offspring is asymptomatic. This scenario is termed a premutation and is common to trinucleotide repeat diseases such as fragile X. Generation Y's repeats can also expand when replicated such that generation Z inherits the gene which now contains 81 repeats. In this example, 81 repeats is beyond the threshold of biological tolerance and thus this individual expresses the disease phenotype. The threshold of repeat tolerance varies from disease to disease but, in general, the more repeats, the more severe the disease.

polymerase slippage, hairpin formation, gene conversion, and mismatch repair (for a review, see La Spada, 1997). Furthermore, repeat expansion provides a molecular explanation for the incidence of anticipation observed with trinucleotide repeat diseases. Anticipation describes the phenomenon of increased severity and decreased age at onset of biologically inherited diseases. In trinucleotide repeat diseases, anticipation correlates with an increased number of repeats within the affected genes. In other words, the number of repeats is often indicative of the disease severity; likewise, the age at onset is inversely proportional to the number of repeats.

Although all 10 trinucleotide repeat diseases may have common mechanisms of repeat expansion and all demonstrate anticipation, they differ in the type of trinucleotide repeat, molecular pathology, and expressed phenotype. For example, the CAG repeat diseases all result in neurodegeneration (e.g. Huntington's disease). This class of CAG repeat disease is also known as the polyglutamine diseases because the CAG repeat is found within the protein-coding region (exons) of the gene and is subsequently translated into the amino acid glutamine; the corresponding proteins have large stretches of polyglutamine tracts. Interestingly, the CAG repeats are always found in the same reading frame, thus producing polyglutamine. The non-CAG repeat diseases are Friedreich's ataxia, fragile X (types A and E), and myotonic dystrophy. These non-CAG diseases are subclassified based on phenotypes such as spinocerebellar ataxia, mental retardation, and neuromuscular disease (Figure 10.2). The trinucleotide repeat diseases are surveyed below with the primary focus on molecular pathology and transgenic animal models.

10.2.1 CAG trinucleotide repeats – overview of polyglutamine diseases

Selective neurodegeneration is a primary characteristic of all trinucleotide repeat diseases that are mediated by a CAG repeat. The major distinction between the

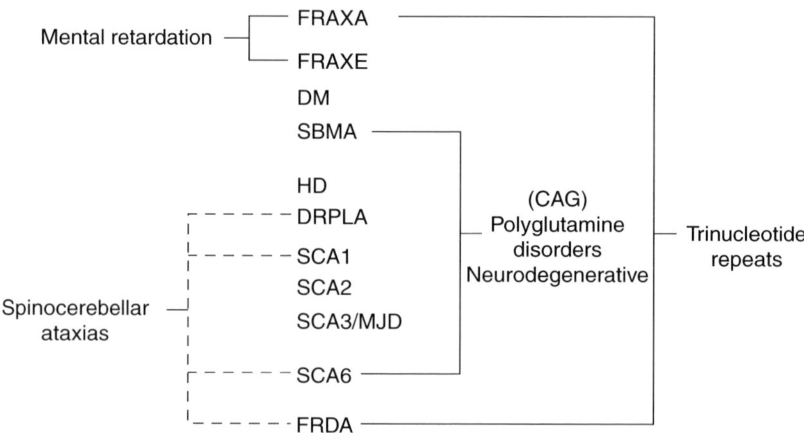

Figure 10.2 Phenotypic and genotypic similarities among the trinucleotide repeat diseases. All of the 11 diseases listed are the result of a trinucleotide repeat expansion or "dynamic mutation". Seven of these diseases (SBMA, HD, DRPLA, SCA1, SCA2, SCA3/MJD, SCA6) specifically have a CAG repeat. In all cases, this CAG repeat is translated into glutamine. Thus, the proteins corresponding to these genes have long stretches of glutamines which may or may not play a role in the neurodegenerative pathology. These seven trinucleotide repeat diseases are also called the polyglutamine disorders. Other subclasses are based on particular phenotypes such as mental retardation. In both FRAXA and FRAXE, the genotype is an expanded trinucleotide repeat and the phenotype is mental retardation. Another classifying phenotype is spinocerebellar ataxia, in which four different trinucleotide repeat diseases fall (DRPLA, SCA1, SCA6, and FRDA). Abbreviations: DRPLA, dentatorubral pallidoluysian atrophy; HD, Huntington's disease; FRAXA, fragile X type A; FRAXE, fragile X type E; FDRA, Friedreich ataxia; DM, myotonic dystrophy; SCA1, spinocerebellar ataxia type 1; SCA2, spinocerebellar ataxia type 2; SCA3/MJD, spinocerebellar ataxia type 3/Machado–Joseph disease; SCA6, spinocerebellar ataxia type 6; SBMA, spinobulbar muscular atrophy.

seven CAG repeat diseases are the particular nuclei that degenerate. Therefore, the phenotype is determined by the degenerating neuroanatomical system. It is not clear if there exists a common molecular pathogenesis that involves the translated glutamine stretch itself or if each disease pathology emerges because of the individual protein affected. There is some evidence demonstrating that the polyglutamine fragment itself can induce cellular toxicity. Furthermore, it is possible that selective neurodegeneration occurs due to the regional selectivity of proteolytic enzymes that process the mutant proteins to liberate the toxic polyglutamine fragment. This is only one hypothesis, however, and a unique neurodegenerative function for each of the seven affected proteins cannot be excluded. An alternative hypothesis is that each protein participates in a distinct and direct role in a given disease pathology.

Proposed mechanisms of common cellular toxicity are glutamate excitotoxicity, metabolic toxicity, apoptosis, and free-radical stress. An intolerable range of glutamine repeats within any protein may be sufficient to give rise to neuronal vulnerability to cellular stress. However, this may be oversimplified since it is commonly accepted that genes containing CAG repeats encode proteins that exhibit a gain of function. This gain of function could preside over the neuronal pathology. In other words, the polyglutamine stretch could specifically alter individual proteins in a way that leads to unique pathways of destruction. Furthermore, polyglutamine stretches within proteins may be required for protein–protein interactions and may also form β-pleated sheets that induce protein aggregation in a manner similar to β-amyloid in Alzheimer's disease. Overall, it remains to be determined if the particular affected protein produces a unique toxic pathway or if the polyglutamine stretch itself induces a common mechanism of neuronal toxicity.

10.2.1.1   *Huntington's disease (HD)*

Huntington's disease (HD) is a debilitating and terminal movement disorder that is inherited in an autosomal dominant fashion. The CAG repeat is located within an exon of chromosome 4p16.3. HD clinical features present when the CAG repeat expands to an abnormal range between 36 and 120 repeats. Neurodegeneration or atrophy of basal ganglia nuclei involved in motor function (caudate and putamen) predominates. More specifically, striatal medium spiny neurons, synthesizing GABA and enkephalin and projecting to the globus pallidus, are the most susceptible to degeneration. The overall hypothesized effect of striatal atrophy on the basal ganglia circuitry is decreased inhibition of the thalamus. Therefore, the thalamus becomes unregulated in the release of the excitatory amino acid, glutamate, onto the cerebral motor cortex. This overexcitation of the motor cortex may account for the observed hyperkinesia (increased movement) in HD patients. In addition to hyperkinesia, HD patients present with chorea (irregular, jerky, spastic, and involuntary movements), bradykinesia (slowness of movement), dysphagia (difficulty in swallowing), rigidity, gait abnormalities, and finally global dementia (deterioration of intellectual faculties). The age at onset is within the fourth or fifth decade and the disease progression follows anticipation in which the earlier the onset, the more severe the disease. HD eventually results in mortality 10 to 12 years following symptom onset.

The molecular pathogenesis of HD is not well understood. The HD gene encodes the *huntingtin* protein which may normally function in intracellular trafficking, cellular housekeeping, and signal transduction. Huntingtin is widely expressed in neuronal and non-neuronal cell populations; yet only neuronal cells are vulnerable to cell death in HD. In an effort to explain the regional selectivity of cell death in HD, research has focused on proteins that associate with huntingtin. A differential distribution of huntingtin-associated proteins could potentially account for the selective neuronal cell death. HAP1 (huntingtin-associated protein-1), hE2–25k (E2 ubiquitin-conjugating enzyme), HIP1 (huntingtin interactor protein 1), GAPDH (glyceraldehyde-3-phosphate dehydrogenase), and apopain (caspase, an apoptotic enzyme) are a few of the identified proteins that associate with huntingtin (for a review, see Koshy and Zoghbi, 1997; Ross *et al.*, 1997). The distribution of these associated proteins and their functional roles are still under intense investigation. Most recently, huntingtin and the brain-specific protein, HAP1, have been demonstrated to be involved in microtubule and dynein-dynactin associated intracellular transport (Li *et al.*, 1998). Similarly, HIP1 may be a cytoskeletal protein, further implicating a role for huntingtin in intracellular transport and possibly endocytosis. Although protein–protein interactions provide evidence for the normal function of huntingtin, how polyglutamine tracts may alter these protein–protein interactions remains obscure. The current hypothesis is that the polyglutamines alter huntingtin protein interactions, changing cellular processes that ultimately initiate a variety of potential cell-death mechanisms (e.g. excitotoxicity, metabolic stress, apoptosis; Ross *et al.*, 1997). These altered protein interactions could be considered the gain of function for the huntingtin protein. Further studies elucidating the role of proteins that interact with huntingtin and the possible combination of these different components will be essential to understanding the initiation of neuronal cell death by abnormal huntingtin in HD patients.

Transgenic and knockout mouse models for HD have been developed (for a review, see Burright *et al.*, 1997). Complete elimination of the mouse homolog of the human huntingtin gene, *Hdh*, is prenatally lethal. This indicates that huntingtin has a normal and essential role in embryonic development. Another murine transgenic line, R6[a], containing 116 to 156 CAG repeats, displays chorea, seizures, and resting tremors. However, neurodegeneration is not observed in this transgenic line. Therefore, this transgenic model of HD suggests that polyglutamine tracts within huntingtin can induce HD-like symptoms in mice, but that the human pathology has not been replicated.

10.2.1.2 *Spinobulbar muscular atrophy (SBMA)*

Spinobulbar muscular atrophy (SBMA; also known as Kennedy's disease) is an X-linked motor neuron disease characterized by progressive muscle weakness and atrophy of upper and lower extremities (Burright *et al.*, 1997; Koshy and Zoghbi, 1997; discussed in Chapter 6). An expanded CAG repeat was identified in the first exon of the androgen receptor gene on Xq11–Xq12 in patients with SBMA. Consistent with other polyglutamine diseases, SBMA patients undergo neurodegeneration, but it occurs selectively in the motor neurons of the anterior horn of the spinal cord, certain brainstem motor nuclei, and sensory neurons of the dorsal root ganglia. The correlation of the androgen receptor and cell death has not been established but it is hypothesized that the receptor gains a function since XY males with a deletion of the AR gene do not present with SBMA. The synthesis of transgenic mouse models of SBMA has been attempted but none have successfully produced mice with phenotypes or pathology similar to human SBMA patients.

10.2.1.3 Spinocerebellar ataxias

The autosomal dominant spinocerebellar ataxias (ADSCAs) are a clinically heterogeneous group of diseases characterized by cerebellar and brainstem dysfunction and neurodegeneration. In general, cerebellar ataxia (incoordination) predominates along with motor weakness and sometimes muscle wasting. Historically, the ataxias have been classified based on phenotypic presentation. However, this phenotypic classification is confusing since there is significant symptom overlap among the ADSCAs. Molecular genetics has aided in refining the ataxia classification system by identifying the genotypes responsible for each spinocerebellar ataxia. The genotypic classification divides the ataxias into four diseases that include spinocerebellar ataxia types 1 and 2 (SCA1 and SCA2), spinocerebellar ataxia type 3 (SCA3)/ Machado–Joseph disease (MJD), spinocerebellar ataxia type 6 (SCA6), and dentatorubral pallidoluysian atrophy (DRPLA; Table 10.1). Interestingly, all four ataxias have been attributed to trinucleotide repeats (CAG) within exons that result in polyglutamine tracts within the respective proteins (see Table 10.1). In agreement with other polyglutamine diseases, selective neurodegeneration is evident. In ASCDAs, degeneration is selective for cerebellar neurons, spinocerebellar tracts, brainstem nuclei, and other central nervous system (CNS) motor nuclei. For example, in both SCA1 and SCA2, the Purkinje cells of the cerebellar cortex selectively and severely degenerate, while in DPRLA and SCA3/MJD, the deep cerebellar nuclei (i.e. dentate) are more susceptible to cell death. The molecular pathology of the ataxias, in addition to genotype, is useful for distinguishing between the diseases (for a review, see Yagishita and Inoue, 1997). Although the proteins involved have been identified, their regional selectivity, intracellular localization, and function are just beginning to be elucidated. SCA6 is the exception in which a CAG repeat was identified in the coding region of the calcium ion channel (see Section 10.3.1).

Transgenic animal models have been developed for SCA3/MJD and SCA1 (Burright et al., 1997). In one SCA3/MJD transgenic mouse line, MJD79, a full-length copy of the MJD1 gene containing 79 CAG repeats did not induce a phenotype analogous to the human disease. However, a truncated version of the MJD1 transgene containing 79 CAG repeats in Q79C mice did produce ataxia and cerebellar atrophy. These SCA3/MJD transgenic models, in conjunction with cell-culture models, support the notion that the expanded polyglutamine tract itself is sufficient to induce cell death. However, cells may require specific proteolytic processing machinery to liberate toxic polyglutamine fragments from mutant proteins. A different scenario exists for SCA1 in which transgenic mice expressing ataxin-1 with 82 CAG repeats (but not 30 repeats) did display homologous human phenotypes and pathology. These SCA1 transgenic mice have become useful in defining the role of ataxin-1 in the maintenance of dendrites and spines.

10.2.2 Friedreich ataxia (FDRA)

Although it appears that Friedreich ataxia (FDRA) is a mitochondrial disease, it is discussed here in order to emphasize its striking contrast to the polyglutamine diseases. FDRA is one of the most common inherited neurodegenerative ataxias with symptoms (ataxia, areflexia, and dysarthria) presenting during puberty and restricting patients to wheelchairs by the late 20s and finalizing in death by the mid-30s (Bidichandani *et al.*, 1998). Unlike the polyglutamine diseases, FDRA is inherited in an autosomal recessive fashion. However, a non-Mendelian expansion of a GAA repeat does occur within the first intron of the frataxin gene. Notably, this repeat is in an intron (non-coding region) and not in an exon (mRNA-encoding region), unlike the polyglutamine diseases. Furthermore, whereas all of the polyglutamine diseases are attributed to a gain of protein function due to the dominant nature of the diseases, in FDRA it is hypothesized that there is a loss of function of the frataxin protein. The fact that only homozygous individuals express FDRA phenotypes confirms this notion while demonstrating the recessiveness of the ataxia.

The regional expression of frataxin correlates well with the cellular atrophy characteristic of FDRA. Frataxin has been localized to neurons of the spinal cord (dorsal root ganglia, Clarke's nucleus), cerebellum, medulla and pons, and cells in the pancreas and heart. Not coincidentally, symptoms of FDRA include, ataxia, neuropathy, diabetes, and cardiomyopathy. Again, this is unlike the polyglutamine diseases in which not all cell types expressing the affected protein undergo degeneration. In contrast to the polyglutamine diseases, the molecular pathology of FDRA is fairly well understood. First, there is evidence to suggest that the GAA repeat in intron 1 of the frataxin gene interferes with transcription (Bidichandani *et al.*, 1998). Reduced frataxin protein and mRNA have been observed in FDRA patients (Campuzano *et al.*, 1997). Frataxin is associated with mitochondrial membranes and is hypothesized to function in iron homeostasis. The decrease in frataxin may render cells susceptible to oxidative damage by accumulation of mitochondrial iron that undergoes Fenton reactions (iron-catalyzed generation of hydroxyl radicals). In summary, FDRA is a mitochondrial disease that results from a decreased expression of frataxin due to an intronic repeat and cell death occurs due to potential free-radical toxicity.

10.2.3 Mental retardation – fragile X type A (FRAXA)

Fragile X syndrome is a common cause of X-linked mental retardation in males and to a lesser extent females (for an excellent review of the unusual inheritance pattern of FRAXA, see Tarleton and Saul, 1993). A CGG trinucleotide repeat has been identified in the 5′ untranslated region of the fragile X mental retardation gene (FMR1). This CGG repeat influences the methylation of an adjacent CpG island in the promoter region of the FMR1 gene. It is believed that this aberrant hypermethylation inhibits transcription of the FMR1 gene. This coincides with decreased fragile X mental retardation protein (FMRP) in FRAXA patients and suggests that loss of FMRP function is responsible for the phenotype. FMRP appears to be an RNA-binding protein and may function in the transport of RNA from the nucleus to the cytoplasm. In addition, FMRP contains coiled-coils that may serve as protein interaction domains. Therefore, FMRP may also function in the transport of specific proteins. A fragile X knockout has further suggested that FMRP may play a role in dendritic spine development (Comery *et al.*, 1997). Fragile X patients could potentially benefit from gene therapy in which FMRP is reconstituted. Future studies on this mentally debilitating disease will be important in unraveling the molecular mechanisms of learning and memory and development.

10.2.4 *Neuromuscular disease – myotonic dystrophy (DM)*

Myotonic dystrophy (DM) is an autosomal dominant disease that results in muscular weakness, muscular atrophy, cardiac conduction disturbances, and sometimes mental retardation. A CTG trinucleotide repeat has been identified in the 3′ untranslated region of the DM protein kinase (DMPK) gene. It is hypothesized that the CTG repeat increases the number of binding sites of CUG-binding protein (CUG-BP) to the DMPK mRNA. The increased CUG-BP inhibits the processing and transport of DMPK mRNA. This working model agrees with decreased levels of DMPK in DM patients. The role of the DM serine/threonine protein kinase in the molecular pathology is still under intense investigation. Currently identified substrates of DMPK included specific isoforms of sodium and calcium ion channels (Timchenko and Caskey, 1996). In addition, DMPK normally phosphorylates a CUG-BP, hNab50. Decreased DMPK in DM results in hypophosphorylated CUG-BP, contributing to its accumulation in the nucleus (Roberts *et al.*, 1997). In general, CUG-BPs are involved in splicing transcripts that contain the CUG repeats (Singer, 1998). Therefore, the phosphorylation status and nuclear accumulation of CUG-BP could interfere with the posttranscriptional processing of CUG transcripts. Interestingly, one such transcript that encodes a cardiac muscle contraction protein, troponin-T, is not alternatively spliced by CUG-BP in DM-striated muscle (Philips *et al.*, 1998). It is hypothesized that other muscle-specific CUG transcripts may also have disrupted splicing in DM patients. This complex disease illuminates the novel role of RNA splicing in triplet repeat diseases (Singer, 1998).

10.3 Phenotypically related neurological diseases

This section organizes diseases that are related by the neuroanatomical system engaged and/or molecular pathology. Therefore, the classifications include channelopathies, peripheral neuropathies, mental retardation, and neurodegenerative diseases not attributed to trinucleotide repeats. Each disease category is further subdivided by genotype.

10.3.1 *Channelopathies*

Genetic defects within voltage-gated ion-channel genes (K^+, Na^+, Ca^{2+}) are causative of neuromuscular, neurological, and cardiac dysfunction. This class of ion-channel diseases has been termed the channelopathies to establish that altered ion-channel function can be directly implicated in the disease processes. Interestingly, all of the channelopathies display dominant inheritance, symptoms occur intermittently, and they are clinically diverse. Four channelopathies that specifically impair normal neurological function include familial hemiplegic migraine (FHM), spinocerebellar ataxia type 6 (SCA6), episodic ataxia type-2 (EA-2), and episodic ataxia type-1 (EA-1). The genetic defect in EA-1 is a missense mutation within the K^+-channel gene, KCNA1, which encodes the delayed rectifier K^+ channel. The remaining neurological channelopathies, EA-2, SCA6, and FMH, are actually allelic disorders of one another. This means that all three diseases are mapped to the same gene (CACNL1A4; a calcium channel) but that each is caused by a different mutation (genotypic heterogeneity). The voltage-gated ion-channel gene, CACNL1A4, encodes the α_{1A} pore-forming subunit of the P/Q-type Ca^{2+} channel. The genetic defects include missense mutations in FMH, frameshift mutations in EA-2, and dynamic mutations (CAG) in SCA6. Therefore, the resulting α_{1A} subunits contain amino acid substitutions (FMH), are truncated (EA-2), and are expanded (SCA6). It is unclear how each mutation alters the function of the ion channel but hypotheses include alterations in ion permeability and selectivity. Although the clinical presentation of FMH, EA-2, and SCA6 are different, cerebellar atrophy is common to each disease. Future studies that correlate changes in calcium ion channel function with downstream intracellular events that subsequently induce cell death will be of significant value. Transgenic animal models have not been required since naturally occurring mouse analogs, tottering (*tg*), leaner (*tg^{la}*), and lethargic (*lh*), already exist (for a review, see Doyle and Stubbs, 1998).

10.3.2 *Peripheral neuropathies/demyelination*

Peripheral neuropathy is an all-encompassing term that describes general peripheral nerve dysfunction. This section focuses on inherited peripheral neuropathies arising from genetically encoded flaws in myelin formation and/or maintenance. We distinguish these neuropathies from other diseases such as the spinocerebellar ataxias, in which neuropathies may be present but only as part of widespread neurological dysfunction. In addition, this section is restricted to those myelin diseases in which known metabolic defects cannot account for the observed disease pathology. Therefore, although porphyrias, amyloidoses (familial amyloid polyneuropathies), mitochondrial (Kearns–Sayre), phytantic acid storage diseases (Refsum's disease), and X-linked leukodystrophies (adrenoleukodystrophy) produce primary neuropathies, they will not be discussed (for a review of these neuropathies, see Harding, 1995).

10.3.2.1 *Hereditary motor and sensory neuropathies (HMSNs)*

The HMS neuropathies are poorly classified because significant genotypic heterogeneity is present. General symptoms consist of decreased nerve conduction velocities, weakness, sensory loss, and muscular atrophy. The HMS neuropathies include Charcot–Marie–Tooth (CMT) disease types 1A, 1B, and X and Dejerine–Sottas disease (DSS) types A and B. All five neuropathies result from genetic defects in the genes encoding the peripheral compact myelin proteins (peripheral myelin protein, PMP22, and myelin protein zero, P0) or the gap junction protein, connexin-32 (Table 10.1).

More specifically, CMT1A results from a duplication of the PMP22 gene and the overexpression of this myelin protein may destabilize the myelin sheath. Myelin sensitivity to PMP22 concentrations has been termed the "gene dosage effect". Thirty different mutations in another compact myelin protein, P0, have been identified in CMT1B patients. P0 also has an essential function in myelin formation and maintenance. In CMTX, 90 distinct mutations have been identified in the connexin-32 gene. Connexin-32 (Cx32) is normally involved in the formation of gap junctions that regulate the flow of ions and small molecules across the myelin sheath. Some of the Cx32 mutations inhibit gap junction formation, while the effects of other mutations are unclear.

Further complicating the CMT diseases are DSS types A and B, which are allelic variants of CMT1A and CMT1B, respectively. That is, the same genes are mutated but different clinical phenotypes present. In DSS A, missense mutations are present in PMP22 that result in phenotypes different from CMT1A. Similarly, DSS B contains missense mutations in P0 that distinguish it from CMT1B. It is hoped that future phenotype and genotype refinement of the HMSN will lead to a more simplified classification system. Naturally occurring animal models (i.e. *trembler* mice) as well as transgenic mouse models may help to clarify this complex group of demyelinating neuropathies (reviewed in Scherer, 1997).

10.3.2.2 Hereditary neuropathy with liability to pressure palsies (HNPP)

HNPP is also an allelic disorder of CMT1A. However, as opposed to the duplication of PMP22 in CMT1A, HNPP patients have a deletion of the PMP22 gene. This absence of PMP22 predisposes patients to palsies (partial muscle paralysis) after minor trauma to the peripheral nerves. In addition, tomaculi (sausage-like myelin thickenings) are numerous in HNPP. HNPP is due to a loss of function of PMP22. In summary, there are three peripheral neuropathies that consist of mutations in the compact peripheral myelin protein, PMP22. These include CMT1A (duplication), HNPP (deletion), and DSS A (missense mutations). The varying phenotypes in each disease reflect the sensitivity of peripheral myelin formation and maintenance to genetic defects in PMP22.

10.3.2.3 Pelizaeus–Merzbacher disease

Although Pelizaeus–Merzbacher disease is an X-linked dysmyelinating disorder specific to the CNS, it is briefly mentioned under peripheral neuropathies only because it also involves genetic defects in a major myelin protein, proteolipid protein (PLP). PLP is structurally and functionally analogous to PMP22 and so is considered the CNS equivalent of PMP22. The PLP gene is alternatively spliced to produce DM20 protein. PLP is responsible for myelin compaction in the CNS while DM20 is essential for oligodendrocyte survival. Pelizaeus–Merzbacher disease is allelic to spastic paraplegia type 2 (SPG2). It is hypothesized that Pelizaeus–Merzbacher patients have a loss of function of both PLP and DM20 while SPG2 patients have a mutant PLP but normal DM20 protein. This accounts for the variable clinical presentation of SPG2 versus Pelizaeus–Merzbacher disease. Conveniently, like the channelopathies and HMSNs, naturally occurring mouse models of Pelizaeus–Merzbacher disease already exist (*rumpshaker* and *jimpy*; reviewed in Kobayashi *et al.*, 1996). These animal models will have tremendous utility in further characterizing the normal function of PLP and myelination in the CNS.

10.3.2.4 *Hereditary spastic paraplegia (HSP)*

Hereditary spastic paraplegia (HSP) is clinically diverse because symptoms range from peripheral neuropathies to mental retardation to ataxia to neurodegeneration. Based on this, HSP would fit well into any of the subsections within this chapter. HSP has recently been identified as a mitochondrial disease that results in progressive weakness and spasticity of the lower limbs. Mutations in 16q24.3 which encodes paraplegin have been identified (Casari *et al.*, 1998). The current hypothesis is that paraplegin is a mitochondrial metalloprotease that may disrupt oxidative phosphorylation by unidentified mechanisms. This disease exemplifies the common theme of mitochondrial impairment in neurodegeneration.

10.3.3 **Mental retardation**

10.3.3.1 *CRASH syndrome – L1 related diseases*

CRASH is an acronym that represents a set of consensus clinical features that include **C**orpus callosum hypoplasia, mental **R**etardation, **A**dducted thumbs, **S**pastic paraplegia, and **H**ydrocephalus (Fransen *et al.*, 1997). This acronym lumps five phenotypically related diseases under one syndrome, CRASH. All five of these diseases, X-linked hydrocephalus, MASA syndrome, spastic paraplegia type 1 (SPG1), agenesis of corpus callosum (ACC), and mental retardation with clasped thumbs (MR-CT), have mutations in the L1 cell adhesion molecule (L1CAM). Therefore, CRASH syndrome can be considered a disease of the L1CAM. The molecular structure of L1CAM is similar to other cell adhesion molecules in that it is a transmembrane glycoprotein that consists of six immunoglobulin domains, five fibronectin domains, a transmembrane domain, and a cytoplasmic domain. L1CAM is expressed by neurons and Schwann cells and may have a variety of pivotal functions that include myelination, axon outgrowth, and neuronal migration. Furthermore, L1CAM has been implicated in long-term potentiation and long-term memory storage, possible linking L1CAM to mental retardation. Over 75 mutations have been identified in the L1CAM gene that causes CRASH syndrome (Fransen *et al.*, 1997). It is hypothesized that other, yet unidentified, mentally retarding diseases may also be mapped to L1CAM.

10.3.3.2 *Lissencephaly*

Lissencephaly (smooth brain or agyria) is attributed to aberrant neuronal migration during cortical development. Similar to CRASH syndrome, patients exhibit profound mental retardation. There is an autosomal lissencephaly entitled isolated lissencephaly (ILS) as well as an X-linked form called X-linked subcortical laminar heterotopia and lissencephaly (X-SCLH/LIS). The subcortical laminar heterotopia is also referred to as "double cortex" because migrating neurons come to a halt and produce an abnormal banding pattern. A deletion in the LIS-1 gene has been identified in ILS patients. This gene encodes platelet-activating factor (PAF) acetylhydrolase which may normally influence calcium fluxes. The protein involved in the pathology of X-SCLH/LIS has recently been termed doublecortin (des Portes *et al.*, 1998; Gleeson *et al.*, 1998). Doublecortin may be a member of the tyrosine kinase signal transduction pathway that controls neuronal migration. It is unknown whether doublecortin and PAF acetylhydrolase interact with one another. Both proteins are associated with an arrest in neuronal migration that consequently results in severe mental retardation. X-SCLH/LIS has an associated epilepsy to provide a differential diagnosis from ILS.

In both CRASH syndrome and lissencephaly, there is aberrant neuronal migration during development that causes mental impairment. However, entirely different subclasses of proteins have been implicated. Transgenic animal models of these diseases will be vital to further elucidating how neurons migrate to specific destinations during development. This could lead to new therapeutic interventions while providing valuable insights into other disease which may also involve aberrant neuronal migration such as schizophrenia.

10.3.3.3   Segmentally aneusomic syndrome (SAS)

Segmentally aneusomic syndrome (SAS) is a recently established term used to describe inherited diseases that are a consequence of gene imbalances such as duplications and deletions. The SASs include, but are not limited to, Angelman syndrome (AS), Prader Willi syndrome (PWS), Miller Dieker syndrome (MDS), Smith Mayenis syndrome (SMS), and DiGeorge/velocardiofacial syndromes (DGS/VCFS), which all produce some degree of mental retardation. This is not intended to be a comprehensive list of all the possible SASs. This group of mentally debilitating SASs is extremely complex because many might be multi-locus disorders and/or display genetic abnormalities such as deletions, duplications, uniparental disomies (UPDs), and genetic imprinting defects. Briefly, UPD is a situation in which two copies of one parental chromosome are present and no copies of the other parent's chromosome are inherited. For example, 1–4% of AS patients have two paternal chromosome 15 and no maternal chromosome 15. Genetic imprinting abnormalities may then occur on the paternal copies in which the genes are modified during meiosis. This basically means that the only copies of the inherited gene (paternal in this example) are genetically defective. Therefore, the deck is stacked against the offspring in two ways; (1) one parent's genetic information is not transmitted, and (2) the only parental genetic information that is received is highly susceptible to genetic mutations. For a detailed discussion of how these unusual genetic abnormalities may direct phenotypic expression of specific SASs, see Budarf and Emanuel (1997).

10.3.4   Neurodegenerative diseases

This section describes three diseases (prion disease, familial amyotrophic lateral sclerosis, and spinal muscular atrophy) that all result in neurodegeneration by unidentified mechanisms. However, unlike the CAG polyglutamine diseases, these diseases are not due to genomic trinucleotide repeats. Although the complexity of these disorders is still under investigation, it appears that specific genes have been identified as contributors to the molecular pathogenesis of these diseases. This collection of diseases exemplify recurring themes of protein aggregation in cell death and selective neuronal cell death. It is interesting that protein aggregation is a common theme in neurodegenerative disorders. The "aggregate-forming" neurodegenerative disorders include prion, Alzheimer's, Parkinson's, and Huntington's diseases as well as spinocerebellar ataxia 1 and 3 (Welch and Gambetti, 1998). Future efforts that investigate common mechanisms by which proteins form insoluble aggregates that lead to neurodegeneration will be of tantamount importance.

10.3.4.1 *Prion disease*

Prion disease or transmissible spongiform encephalopathy has three modes of acquisition; infection, sporadic occurrence, and genetic inheritance (Prusiner and Scott, 1997). The excitement and controversy over infectious prion disease stems from the fact that the infectious agent is a protein that does not contain nucleic acid. In humans, infection by prions occurs rarely through the consumption of contaminated beef products (mad cow disease). Kuru is also an infectious form of prion disease, but its occurrence has declined since the cessation of cannibalism in Papua New Guinea. Infectious prion disease does not appear to be widespread since the majority of human cases are sporadic with an elusive etiology. In addition to sporadic and infectious prion disease, 10% of human prion cases are attributed to genetic defects within the prion protein gene (PRNP) on chromosome 20. The genetic human prion diseases consist of familial Creutzfeldt–Jakob disease (*f*CJD), Gerstmann–Straussler–Scheinker disease (GSS), and fatal familial insomnia (FFI; for a review of clinical and pathological features, see Collinge and Palmer, 1997).

All three inherited human prion diseases, *f*CJD, GSS, and FFI, have mutations in the PRNP gene that encodes prion protein. The normal prion protein is designated as PRP^C. The physiological function of PRP^C is undetermined especially since *Prp* knockout mice have no gross phenotypic abnormalities (for a review of transgenic prion models, see Gabizon and Taraboulos, 1997). PRP^C protein is sensitive to proteolysis and contains several α-helices. In inherited *f*CJD, GSS, and FFI, PRP^C has a propensity to spontaneously convert to PRP^{SC}. In this case, PRP^{SC} refers to a "scrapie" form of PRP, scrapie being an animal version of spongiform encephalopathy. It is PRP^{SC} that is pathologically insulting due to its resistance to proteolysis and an adopted β-sheet conformation that promotes protein aggregation and deposition. PRP^{SC} then autocatalytically converts PRP^C to form more PRP^{SC} which establishes a cyclic and exponential "autoinfection".

10.3.4.2 *Familial amyotrophic lateral sclerosis (FALS)*

Amyotrophic lateral sclerosis (ALS) is a progressive motor neuron disease that manifests itself as muscular weakness, atrophy, and spasticity. Mortality is high in ALS, with most patients succumbing to respiratory failure. Pathogenic hallmarks of ALS include degeneration of upper and lower motor neurons. This is reminiscent of trinucleotide repeat disorders, in which selective neuronal cell death occurs. However, like prion diseases, the majority of ALS cases are sporadic. Only 10%

of all ALS cases are familial, with 20% accounted for by mutations within the superoxide dismutase (SOD1) gene. This represents 2% of total ALS attributed to SOD1 mutations. The discovery of SOD1 in familial ALS (FALS) has prompted investigations into the contribution of free radical damage to sporadic ALS. Other proposed pathological mechanisms include factors such as autoimmune phenomena, oxidative stress, excitotoxicity, viral infection, cytoskeletal abnormalities, and loss of trophic support (Louvel *et al.*, 1997). Based on these hypotheses, several clinical trials have examined the benefit of immunosuppressive agents, neurotrophic factors, and neuroprotective agents in treating ALS patients. Thus far, riluzole, which is a glutamic acid release inhibitor and Na$^+$-channel blocker, has proven to be the most promising (Louvel *et al.*, 1997).

10.3.4.3   *Spinal muscular atrophy (SMA)*

Spinal muscular atrophy (SMA) is a progressive childhood movement disorder with a wide range of severity. SMA has been subdivided into types I, II and III which indicate the most severe (type I) to milder forms of the disease (type III). Clinical features correspond to classical lower motor neuron diseases that include symptoms of muscle atrophy (wasting) and paralysis. Similar to ALS adults, children with SMA succumb to respiratory failure. The pathological finding in SMA is severe degeneration of α-motor neurons within the anterior horn of the spinal cord.

The molecular genetics of SMA are not straightforward. Linkage analysis has mapped SMA to chromosome 5q11.2–q13.3, but three candidate genes reside within this region. These genes include survival of motor neuron (SMN), neuronal apoptosis inhibitory protein (NAIP), and a subunit of the transcription factor p44. Current literature has strongly implicated SMN and NAIP in the molecular pathogenesis of SMA. In SMA, 95% of the patients have a deletion of exon 7 in the SMN gene. However, this finding has produced confusion and controversy because: (1) the deletion of exon 7 of SMN does not account for the different severities of SMA, and (2) asymptomatic siblings can have homozygous deletions of exon 7 of SMN. This lack of phenotype–genotype correlation has provided the impetus to investigate other genes that could serve to modify SMN. The most likely candidate is NAIP in which deletions of exons 5 and 6 have been identified. Since NAIP normally inhibits programmed cell death, this suggests that SMA may be a disease of aberrant apoptosis. One current hypothesis is that a deletion of SMN is required for SMA but that the status of the NAIP gene may regulate the severity of the phenotype (Crawford and Pardo, 1996). Transgenic animal models will be vital to testing this hypothesis.

10.4   Multifactorial neurogenetic diseases

There are a number of significant nervous system disorders that are multifactorial in nature, but have clear-cut genetic components. These disorders can be partially linked to a single gene, or have several genes involved in producing the disease. In some situations, there have been a variety of potential mechanisms advanced, but none has been proven to be a singular cause and the neurogenetic component is based on the observation that the disease appears to "run in families". Examples include schizophrenia, manic depression, Parkinson's disease, Alzheimer's disease, alcoholism, and Tourette's syndrome. For the sake of simplicity, this discussion will focus on two of these disorders for which there is a clear genetic correlation; namely, Alzheimer's disease, and Parkinson's disease.

10.4.1 Alzheimer's disease (AD)

Alzheimer's disease (AD) is a neurodegenerative disorder characterized by the slow, progressive loss of cognitive function, including memory loss, impaired judgement and declining reasoning skills. The disease generally occurs late in life (after the age of 60). Neuropathologically, AD is definitively diagnosed at autopsy by the presence of microscopic neuritic plaques (largely composed of a deposition of amyloid-β peptide) and neurofibrillary tangles (composed of paired helical filaments of the protein, tau). The molecular mechanisms of plaque and tangle formation and even their precise role in AD remain a subject of controversy. However, there are frank mutations in three different genes that lead to familial forms of AD (see Table 10.1); these genetic forms of AD are indistinguishable in their end-stage clinical and pathological presentation. Moreover, there is a fourth gene that appears to be involved in modulating the susceptibility to AD (Lannfelt, 1997; Hardy and Gwinn-Hardy, 1998; Price *et al.*, 1998).

10.4.1.1 Amyloid precursor protein mutations

The amyloid precursor protein (APP) is a large protein of unknown function that resides on chromosome 21 (21q21). While the cellular role of APP is unknown, it is clear that it serves as the precursor for the smaller aberrant amyloid-β peptide (40 to 43 amino acids) that is found in neuritic plaques considered to be the hallmark of AD (Goate, 1998). Interestingly, this same chromosome is duplicated in trisomy-21 patients (Down's syndrome), and these individuals invariably display signs of AD at an early age. Several different mutations have been identified in APP that result in fully penetrant, autosomal dominant genetic forms of AD. In each case, the mutations alter the normal processing of APP such that it produces high levels of amyloid peptide. These inherited mutations produce early onset disease (45 to 60 years of age) and very severe forms of the disorder. In spite of the illustrative nature of these mutations, they are very rare (approximate 20 documented families worldwide) and therefore account for a minuscule fraction of the individuals affected by AD.

10.4.1.2 Presenilin-1 and presenilin-2 mutations

The presenilin genes (PS-1 and PS-2) are located on chromosomes 14 (14q24) and 1 (1q31–42), respectively, and are the sites of a very large number of different mutations that lead to inherited, early-onset AD (Kovacs and Tanzi, 1998; Renbaum and Levy-Lahad, 1998). These proteins, like the previously discussed APP, have an unknown function, but are membrane-spanning proteins associated with the endoplasmic reticulum. There are approximately 50 different mutations (disproportionately associated with PS-1; only two reported PS-2 mutations) that all produce early-onset AD. While the gene products are distinct, there may be a functional link between APP and the PS genes in that mutations in the PS are associated with increased production of the amyloid-β peptide. While mutations in these proteins have a higher contribution to AD (estimated as high as 50% of the early-onset cases), they still represent the minority of all of the AD cases.

10.4.1.3 Apolipoprotein E alleles

Epidemiological and linkage studies have consistently identified the naturally occurring alleles of the apolipoprotein E (ApoE) gene, located at chromosomal position 19q13, as a major contributor to late-onset AD (Corder *et al.*, 1998). Specifically, this protein is a normal component of physiological lipid metabolism and occurs in the normal population in three forms – the ε2, ε3, and ε4 alleles. It is important to note that these are not mutations; rather, they are naturally occurring genetic variants. Any given individual inherits two copies of the gene and so has any combination of two of the alleles. From the standpoint of AD, the rare ε4 allele has been shown to be enriched in populations of AD patients and was subsequently demonstrated to increase the risk of manifesting the disease by as much as eightfold. There is also an apparent gene dosage effect in that two copies of the ε4 allele are worse than one copy. Conversely, the ε2 allele is associated with protection from AD. To date, the ApoE gene remains the strongest link to the previously untapped late-onset AD. Potential roles for the involvement of ApoE gene product in the pathogenesis of the disease include binding to the amyloid-β peptide, modulation of tau phosphorylation and therefore neurofibrillary tangle formation, and alteration of APP processing to produce the amyloid peptide.

10.4.2 Parkinson's disease

Parkinson's disease (PD) represents the second most prevalent neurodegenerative disease (following AD) and is characterized by generalized movement disorders: resting tremor, bradykinesia, rigidity, and loss of postural reflex. The underlying neuropathology is linked to the loss of dopaminergic and noradrenergic neurons of the substantia nigra-pars compacta and locus coeruleus, respectively. Microscopically, PD is also characterized by the deposition of intracellular Lewy bodies (composed predominantly of the α-synuclein protein). Like AD, PD is generally considered a sporadic disorder, although some genetic autosomal dominant inheritance has been reported (reviewed in Clayton and George, 1998; Mizuno *et al.*, 1998). The genetic contribution to PD was firmly established with the identification, within a large Italian kindred, of a genetic polymorphism responsible for an autosomal dominant, familial form of PD (Polymeropoulos *et al.*, 1997). This polymorphism resides on the long arm of chromosome 4 (4q21–q22) within the coding region of the gene for α-synuclein. This single point mutation, also found in three unrelated Greek kindred with familial PD, produces a single amino acid change (ala53 to val53). Like many of the previously discussed genetic variants of neurological disease, the mutant α-synuclein form of PD still cannot be explained from a functional/biochemical standpoint. Moreover, it is unlikely that α-synuclein mutations play a major role (in terms of numbers of cases) in the global incidence of PD. However, given the presence of α-synuclein in Lewy bodies and the full penetrance of this genetic abnormality, it is clear that these new findings will provide pivotal guidance in the characterization of the non-genetic forms of PD.

10.5 Conclusions

Advances in molecular genetics and recombinant DNA technology have produced a wealth of information on polymorphisms impacting on nervous system function. A casual examination of Table 10.1 illuminates 35 different genes that are implicated directly in neurological diseases. Several of these have many distinct mutation sites producing hundreds of different polymorphisms. The vast majority of these were identified and characterized within the 1990s – The Decade of the Brain. In nearly all cases, however, identification of the disease-causing polymorphism has yet to produce the underlying link between gene defect and disease. The tasks for the coming decade and beyond will involve: (1) continuing the identification of genetic polymorphisms involved in nervous system function; (2) providing mechanistic explanations for the progression from polymorphism to neurological disease; and (3) translating the knowledge of genetics and disease progression into effective therapeutic interventions.

References

Bidichandani, S.I., Ashizawa, T. and Patel, P.I., 1998, The GAA triplet-repeat expansion in Friedreich ataxia interferes with transcription and may be associated with an unusual DNA structure. *American Journal of Human Genet.*, **62**, 111–121.

Bone, L.J., Deschenes, S.M., Balice-Gordon, R.J., Fischbeck, K.H. and Scherer, S.S., 1997, Connexin32 and X-linked Charcot–Marie–Tooth disease. *Neurobiol. Disease*, **4**, 221–230.

Budarf, M.L. and Emanuel, B.S., 1997, Progress in the autosomal segmental aneusomy syndromes (SASs): single or multi-locus disorders? *Human Molec. Genet.*, **6**, 1657–1665.

Bulman, D.E., 1997, Phenotype variation and newcomers in ion channel disorders. *Human Molec. Genet.*, **6**, 1679–1685.

Burright, E.N., Orr, H.T. and Clark, H.B., 1997, Mouse models of human CAG repeat disorders. *Brain Pathol.*, **7**, 965–977.

Campuzano, V., Montermini, L., Lutz, Y., Cova, L., Hindelang, C., Jiralerspong, S., Trottier, Y., Kish, S.J., Faucheux, B., Trouillas, P., Authier, F.J., Durr, A., Mandel, J.L., Vescovi, A., Pandolfo, M. and Koenig, M., 1997, Frataxin is reduced in Friedreich ataxia patients and is associated with mitochondrial membranes. *Human Molec. Genet.*, **6**, 1771–1780.

Casari, G., De Fusco, M., Ciarmatori, S., Zeviani, M., Mora, M., Fernandez, P., De Michele, G., Filla, A., Cocozza, S., Marconi, R., Durr, A., Fontaine, E.B. and Ballabio, A., 1998, Spastic paraplegia and OXPHOS impairment caused by mutations in paraplegin, a nuclear-encoded mitochondrial metalloprotease. *Cell*, **93**, 973–983.

Clayton, D.F. and George, J.M., 1998, The synucleins: a family of proteins involved in synaptic function, plasticity, neurodegeneration and disease. *Trends Neurosci.*, **21**, 249–254.

Collinge, J. and Palmer, M.S., 1997, Human prion disease, in Collinge, J. and Palmer, M.S. (Eds.) *Prion Diseases*, pp. 18–56, Oxford University Press.

Comery, T.A., Harris, J.B., Willems, P.J., Oostra, B.A., Irwin, S.A., Weiler, I.J. and Greenough, W.T., 1997, Abnormal dendritic spines in fragile X knockout mice: maturation and pruning deficits. *Proc. Natl Acad. Sci., USA*, **94**, 5401–5404.

Corder, E.H., Lannfelt, L., Bogdanovic, N., Fratiglioni, L. and Mori, H., 1998, The role of APOE polymorphisms in late-onset dementias. *Cell. Molec. Life Sci.*, 54, 928–934.

Crawford, T.O. and Pardo, C.A., 1996, The neurobiology of childhood spinal muscular atrophy. *Neurobiol. Disease*, 3, 97–110.

des Portes, V., Pinard, J.M., Billuart, P., Vinet, M.C., Koulakoff, A., Carrie, A., Gelot, A., Dupuis, E., Motte, J., Berewald-Netter, Y., Catala, M., Kahn, A., Beldjord, C. and Chelly, J., 1998, A novel CNS gene required for neuronal migration and involved in X-linked subcortical laminar heterotopia and lissencephaly syndrome. *Cell*, 92, 51–61.

Doyle, J.L. and Stubbs, L., 1998, Ataxia, arrhythmia and ion-channel gene defects. *Trends Genet.*, 14, 92–98.

Fransen, E., Van Camp, G., Vits, L. and Willems, P.J., 1997, L1–associated disease: clinical geneticists divide, molecular geneticists unite. *Human Molec. Genet.*, 6, 1625–1632.

Gabizon, R. and Taraboulos, A., 1997, Of mice and (mad) cows – transgenic mice help to understand prions. *Trends Genet.*, 13, 264–269.

Gleeson, J.G., Allen, K.M., Fox, J.W., Lamperti, E.D., Berkovic, S., Scheffer, I., Cooper, E.C., Dobyns, W.B., Minnerath, S.R., Ross, M.E. and Walsh, C.A., 1998, Doublecortin, a brain-specific gene mutated in human X-linked lissencephaly and double cortex syndrome, encodes a putative signaling protein. *Cell*, 92, 63–72.

Goate, A.M., 1998, Monogenetic determinants of Alzheimer's disease: APP mutations. *Cell Mol. Life Sci.*, 54, 897–901.

Greenberg, D.A., 1997, Calcium channels in neurological disease. *Ann. Neurol.*, 42, 275–282.

Harding, A.E., 1995, Molecular genetics of peripheral neuropathies. *Bailliere's Clin. Neurol.*, 4, 383–400.

Hardy, J. and Gwinn-Hardy, K., 1998, Genetic classification of primary neurodegenerative disease. *Science*, 282, 1075–1079.

Hoogeveen, A.T. and Oostra, B.A., 1997, The fragile X syndrome. *J. Inherited Metab. Diseases*, 20, 139–151.

Kobayashi, H., Garcia, C.A., Alfonso, G., Marks, H.G. and Hoffman, E.P., 1996, Molecular genetics of familial spastic paraplegia: a multitude of responsible genes. *J. Neurol. Sci.*, 137, 131–138.

Koenig, M. and Mandel, J.L., 1997, Deciphering the cause of Friedreich ataxia. *Curr. Opin. Neurobiol.*, 7, 689–694.

Koshy, B.T. and Zoghbi, H.Y., 1997, The CAG/polyglutamine tract diseases: gene products and molecular pathogenesis. *Brain Pathol.*, 7, 927–942.

Kovacs, D.M. and Tanzi, R.E., 1998, Monogenic determinants of familial Alzheimer's disease: presenilin-1 mutations. *Cell. Molec. Life Sci.*, 54, 902–909.

Lannfelt, L., 1997, The genetics and pathophysiology of Alzheimer's disease. *J. Internal Med.*, 242, 281–284.

La Spada, A.R., 1997, Trinucleotide repeat instability: genetic features and molecular mechanisms. *Brain Pathol.*, 7, 943–963.

Li, S.H., Gutekunst, C.A., Hersch, S.M. and Li, X.J., 1998, Interaction of huntingtin-associated protein with dynactin p150Glued. *J. Neurosci.*, 18, 1261–1269.

Louvel, E., Hugon, J. and Doble, A., 1997, Therapeutic advances in amyotrophic lateral sclerosis. *Trends Pharmacol. Sci.*, 18, 196–203.

Meiner, Z., Gabizon, R. and Prusiner, S.B., 1997, Familial Creutzfeldt–Jakob disease. Codon 200 prion disease in Libyan Jews. *Medicine*, 76, 227–237.

Melki, J., 1997, Spinal muscular atrophy. *Curr. Opin. Neurol.*, 10, 381–385.

Miller, R.J., 1997, Calcium channels prove to be a real headache. *Trends Neurosci.*, 20, 189–192.

Mizuno, Y., Hattori, N. and Matsumine, H., 1998, Neurochemical and neurogenetic correlates of Parkinson's disease. *J. Neurochem.*, **71**, 893–902.

Muller, U., Graeber, M.B., Haberhausen, G. and Kohler, A., 1994, Molecular basis and diagnosis of neurogenetic disorders. *J. Neurol. Sci.*, **124**, 119–140.

Murakami, T., Garcia, C.A., Reiter, L.T. and Lupski, J.R., 1996, Charcot–Marie–Tooth disease and related inherited neuropathies. *Medicine*, **75**, 233–250.

Pareyson, D. and Taroni, F., 1996, Deletion of the PMP22 gene and hereditary neuropathy with liability to pressure palsies. *Curr. Opin. Neurol.*, **9**, 348–354.

Philips, A.V., Timchenko, L.T. and Cooper, T.A., 1998, Disruption of splicing regulated by a CUG-binding protein in myotonic dystrophy. *Science*, **280**, 737–741.

Polymeropoulos, M.H., Lavedan, C., Leroy, E., Ide, S.E., Dehejia, A., Dutra, A., Pike, B., Root, H., Rubenstein, J., Boyer, R., Stenroos, E.S., Chandrasekharappa, S., Athanassiadou, A., Papapetropoulos, T., Johnson, W.G., Lazzarini, A.M., Duvoisin, R.C., Di Iorio, G., Golbe, L.I. and Nussbaum, R.L., 1997, Mutation in the α-synuclein gene identified in families with Parkinson's disease. *Science*, **276**, 2045–2047.

Price, D.L., Sisodia, S.S. and Borchelt, D.R., 1998, Genetic neurodegenerative diseases: the human illness and transgenic models. *Science*, **282**, 1079–1083.

Prusiner, S.B. and Scott, M.R., 1997, Genetics of prions. *Ann. Rev. Genet.*, **31**, 139–175.

Reilly, M.M., 1998, Genetically determined neuropathies. *J. Neurol.*, **245**, 6–13.

Renbaum, P. and Levy-Lahad, E., 1998, Monogenic determinants of familial Alzheimer's disease: presenilin-2 mutations. *Cell. Molec. Life Sci.*, **54**, 910–919.

Richards, R.I. and Sutherland, G.R., 1992, Dynamic mutations: a new class of mutations causing human disease. *Cell*, **70**, 709–712.

Richards, R.I. and Sutherland, G.R., 1997, Dynamic mutation: possible mechanisms and significance in human disease. *Trends Biochem. Sci.*, **22**, 432–436.

Roberts, R., Timchenko, N.A., Miller, J.W., Reddy, S., Caskey, C.T., Swanson, M.S. and Timchenko, L.T., 1997, Altered phosphorylation and intracellular distribution of a (CUG)n triplet repeat RNA-binding protein in patients with myotonic dystrophy and in myotonin protein kinase knockout mice. *Proc. Natl Acad. Sci., USA*, **94**, 13221–13226.

Ross, C.A., Becher, M.W., Colomer, V., Engelender, S., Wood, J.D. and Sharp, A.H., 1997, Huntington's disease and dentatorubral-pallidoluysian atrophy: proteins, pathogenesis and pathology, *Brain Pathology*, **7**, 1003–1016.

Sanguinetti, M.C. and Spector, P.S., 1997, Potassium channelopathies. *Neuropharmacology*, **36**, 755–762.

Scherer S.S., 1997, The biology and pathobiology of Schwann cells. *Curr. Opin. Neurol.*, **10**, 386–397.

Siddique, T. and Hentati, A., 1996, Familial amyotrophic lateral sclerosis. *Clin. Neurosci.*, **3**, 338–347.

Singer, R.H., 1998, Triplet-repeat transcripts: a role for DNA in disease. *Science*, **280**, 696–697.

Strong, P.N. and Brewster, B.S., 1997, Myotonic dystrophy: molecular and cellular consequences of expanded DNA repeats are elusive. *J. Inherited Metab. Diseases*, **20**, 159–170.

Sudarsky, L. and Coutinho, P., 1995, Machado–Joseph disease. *Clin. Neurosci.*, **3**, 17–22.

Tarleton, J.C. and Saul, R.A., 1993, Molecular genetic advances in fragile X syndrome. *J. Pediat.*, **122**, 169–185.

Terwindt, G.M., Haan, J., Ophoff, R.A., Frants, R.R. and Ferrari, M.D., 1997, The quest for migraine genes. *Curr. Opin. Neurol.*, **10**, 221–225.

Timchenko, L.T. and Caskey, C.T., 1996, Trinucleotide repeat disorders in humans: discussions of mechanisms and medical issues. *FASEB Journal*, **10**, 1589–1597.

Welch, W.J. and Gambetti, P., 1998, Chaperoning brain diseases. *Nature*, **392**, 23–24.

Wellington, C.L. and Hayden, M.R., 1997, Of molecular interactions, mice and mechanisms: new insights into Huntington's disease. *Curr. Opin. Neurol.*, **10**, 291–298.

Wolf, U., 1997, Identical mutations and phenotypic variation. *Human Genet.*, **100**, 305–321.

Yagishita, S. and Inoue, M., 1997, Clinicopathology of spinocerebellar degeneration: its correlation to the unstable CAG repeat of the affected gene. *Pathol. Internatl*, **47**, 1–15.

INDEX